THE ART

OF THE

WINEMAKER

THE ART

OF THE

WINEMAKER

A GUIDE TO THE WORLD'S GREATEST VINEYARDS

Consultant Editor Serena Sutcliffe

COURAGE BOOKS

An Imprint of
Running Press
Philadelphia, Pennsylvania

A QED BOOK

First published 1981
© Copyright 1981 QED Publishing Limited

This book was designed and produced by
QED Publishing Limited
32 Kingly Court
London W1

Art director Alastair Campbell
Production director Edward Kinsey
Editorial director Jeremy Harwood
Editor Kathy Rooney
Editorial research Patricia Monahan
Designers Clive Hayball, Heather Jackson
Paste-up Dennis Thompson
Illustrators Marilyn Bruce, Chris Forsey, Edwina Keene, Elly King, Perry Taylor
Filmset in Great Britain by Oliver Burridge and Co Ltd
Colour origination in Hong Kong by Hong Kong Graphic Arts
Printed in Hong Kong by Leefung Asco Limited

Courage Books's Canadian representatives: General Publishing,
30 Lesmill Road, Don Mills, Ontario M3B 2T6. International representatives:
Kaiman & Polon, Inc., 2175 Lemoine Avenue, Fort Lee, New Jersey 07024.

Library of Congress Cataloging in Publication Data:

Great vineyards and winemakers.
  Art of the winemaker.

  Reprint. Originally published: Great vineyards and
winemakers. QED Pub., 1981.
  Includes index.
  1. Wines and wine making.   I. Sutcliffe, Serena.
II. Title.
TP548.G694 1985   641.2′22   84-27657
ISBN 0-89741-341-8
ISBN 0-89471-341-8 (cloth)

Published by Courage Books, an imprint of Running Press Book Publishers,
125 South Twenty-Second Street, Philadelphia, Pennsylvania 19103.
This book can be ordered by mail from the publisher. Please include $1.75 for
postage. But try your bookstore first.

Consultant editor
SERENA SUTCLIFFE M W

Other contributors
ROBERT BARTON-CLEGG
LEN EVANS
RALPH B HUTCHINSON
DAVID PEPPERCORN M W

Photography
*The photography for this book was specially
commissioned and carried out with the help of the
vineyard owners and the winemakers on the estates.*

The photographers
JON WYAND (France, Germany, Italy)
MICHAEL FREEMAN (United States)
COLIN MAHER (France)
JEAN-PAUL FERRERO (Australia)

*QED would like to thank the many people who have
helped in the preparation of this volume. Too numerous
to mention individually are the owners, winemakers
and others at the vineyards who gave so generously of
their time to both our authors and photographers.
Without them the book would not have been possible. We
would also like to thank Anders Ousbeck and Ian
Jamieson* M W *in London and Jonathan Lyddon in France.*

$\mathscr{C}$ONTENTS

*I*NTRODUCTION

Any choice of great wines will, by its very nature, be considered by some as both arbitrary and subjective. The great vineyards and winemakers in this book are therefore representative of the great wines of the world. They are merely good examples of fine wines of their genre and the list of those included in this volume should in no way be thought of as definitive.

The aim of this book is to show what wine can become when produced with infinite care, great knowledge, and a certain amount of altruistic devotion. A wine tastes as it does as a result of how it is grown and made. Apart from the hand of man, the vineyards should be in the right place, with sun, rain, and sometimes even irrigation playing their part.

What are the pre-requisites for great winemaking? The contributing elements change in order and emphasis with the region, and sometimes with different winemakers within the same region. The factors with which to juggle are—soil, situation, microclimate, type of vines, age of vines, and the people themselves. For it is the winemaker who, at the beginning, decides where to plant, what type of grape variety to choose, when to treat the vines, when to pick them, what machinery to use in the cellar or winery, how to vinify the wines, when to bottle them, and when to release them to the waiting public. Sometimes these decisions can be taken on the basis of the winemaker's knowledge and experience alone, sometimes financial factors enter into the question, and a piece of machinery or new barrels must be foregone because the capital is not there. Some wineries, for example, lack nothing for money, shining with new equipment and the latest aids to perfect winemaking, while others take a certain satisfaction in the undoubted challenge of producing great wine on a limited budget.

The human element is, thus, immense. But there are certain areas of the world where the immovable factors, such as the inestimable importance of the soil in Bordeaux, for instance, are far stronger than elsewhere. In California and Australia, on the other hand, soil is of far smaller impact on the final product, and man manipulates all. In climates with a tendency to extremes of hot or cold, the microclimate becomes the first consideration—thus, in Germany, a microclimate offering good exposure to the sun and protection from wind is essential.

Two great winemakers of Bordeaux, the late Henri Woltner at Château La Mission Haut Brion in Graves, and, among the new generation, Christian Moueix, who is responsible for some of the greatest wines of St Emilion and Pomerol, both put the soil at the top of their list of essentials for making distinguished wine. Moueix estimates that, in his region, the soil contributes up to 80 per cent of the quality of a wine. Then, in the remaining 20 per cent, he put the age of the vines in a place of primary importance. This would not happen if he was making white wines instead of red, as vine age is of minimal importance in

Important factors in wine-making are soil, situation, microclimate, type and age of vines. However, the skills and expertise of the winemakers are also extremely influential.

white wine production. However, Woltner put the weather in second place—in a maritime climate, rainy, cold summers, or glorious, sunny autumns, can make or break a vintage. Woltner's third item of importance was the people who make the wine, as they take the decisions which inevitably lead to the final taste. In a way, Christian Moueix endorsed this, as his third vital ingredient was attaining the right balance of the grape varieties, which is a human decision.

Choosing grape varieties immediately presupposes that there is a mixture or a blend of varieties, where in some great wines one grape variety alone is responsible for the ultimate product. Examples of a blend of grape varieties making great wine include the Médoc, Bordeaux, where the usual combination is Cabernet Sauvignon, Cabernet Franc and Merlot, Australia, where Cabernet Sauvignon and Shiraz (also called Hermitage or Syrah) frequently make great wine, and Chianti Classico, where at least one red variety and two white are added to the dominant Sangiovese. Examples of a single grape variety being responsible for a great wine would include Pinot Noir for red Burgundy, Chardonnay for white Burgundy, Syrah for Hermitage in the Rhône, and Riesling for a top estate on the Rheingau, while, in addition, all fine Alsatian wines are the product of one grape variety alone.

Grape varieties have to be selected for their suitability to the environment, and this, like everything in winemaking, is achieved through a judicious pairing of technical knowledge and experience. Some areas of the world are easier to work in than others in this respect, because their comparatively new industry, and accompanying shortage of tight laws, allows them more flexibility. Thus, if one grape variety is not found to do well on a hot valley floor as happened, for instance, with the Pinot Noir in the Napa Valley, California, the variety can be changed, either by uprooting or grafting another variety on to the existing plant. Germany, however, also has great flexibility of choice with regard to grape variety as a wide range is permitted there.

The grape varieties which make fine wine belong to the family of *vitis vinifera*, which originally came from the Middle East. The problem with *vitis vinifera* is that it is susceptible to a devastating aphid called phylloxera, and in nearly all cases (except in pure sand and some extremely schistous soils) the only way to avoid the vine succumbing to this pest is by grafting onto immune American root stocks. The matching of the root stock to the grape variety is a skill as great as matching the whole with the soil and climate. Many crossings of the original American root stocks have been evolved and are now produced in Europe.

Long experience has shown which grape varieties like which soils. The Cabernets like well-drained gravelly soil, Merlot ripens well where there is clay, Chardonnay and Sauvignon like basically chalky, limestone soil, while Riesling does well in slate and also in loess and loam

mixtures. This is not to say that all these grape varieties sometimes cannot flourish on variations of these combinations, and California has certainly proved that, there, climate is more of a deciding factor than soil.

The grape lies at the heart of all winemaking. Fine wine cannot be made with poor raw material. In some parts of the world, the adage 'Even a great winemaker cannot make great wine on bad soil, but a bad winemaker can make poor wines on good soil' is truer than in others, but there is no getting away from the prime importance of the condition and health of the actual grape when it comes into the winery from the vineyard. Nothing will elevate a poor grape variety, not even the most scientific vinification, but poor viticulture and vinification can ruin even the noblest of grapes.

Firstly, extreme care must be taken in the vineyard. This means planting to the right density for the soil and climate, deciding whether to allow grass to grow round the vines or not, and ensuring adequate drainage. Too much leaf growth and consequent bad air circulation increases the risk of rot, if the weather becomes humid. Spraying at the right moment against mildew, oidium and botrytis (which is only desirable on certain white grape varieties and then only when they are ripe) is essential, as is similar treatment against insects and other pests such as red spider or moths. Before planting a vineyard, the soil should nearly always be rested and should always be totally clean and disinfected. Fertilizer should be used sparingly if quality is sought in the resulting wine, and certain mineral deficiencies or excesses can be corrected in the soil.

There are some weather conditions which man can do little about, others which can be combatted, even if in a small way. Examples of the latter are frost, which often deals its heaviest blows in vineyards on valley floors or near woods, and which can be fought by systems of spraying water over the vines, which freezes and protects them, or by placing heating apparatus amongst the vines. Hail, also, which can fall very locally, touching one vineyard but not its neighbour, can be encouraged to fall as rain by light aeroplanes breaking up the storm clouds. However, there is no antidote to bad weather at flowering time, which can cause the blossoms or the embryonic berries to fall immediately afterwards. This is known as *coulure* in France and is responsible for vintages of small quantity. Another problem, *millerandage* is caused by unequal flowering and will result in a bunch of grapes having some berries that are normal, and others that are small and green and will never ripen. These unripe berries should not be made into wine and should remain unpicked (the best machines now do this) or be selected out by human hand. This also applies to all rotten berries.

Favourable microclimates can be caused by many different factors. Protection can come from mountains or hills or from a nearby water surface, which slightly raises the ambient temperature of the surroundings. South or semi-south facing slopes obviously attract the most sun. Winds can be a problem, and here trees or bushes can be used as windbreaks. But one can be less didactic about climate and the suitable grape varieties than about almost anything else in winemaking. Cabernet, for instance, flourishes in the maritime, temperate climate of Atlantic Bordeaux, but also in hot Napa and Australia. The Pinot Noir is thought to prefer the relatively cool climate of Burgundy and not to like high temperatures—in California it did not do well in the Napa Valley and prefers cooler Monterey, but a wonderful example of Pinot Noir has emerged from the Hunter Valley in Australia. The Riesling is meant to owe some of its steely greatness to the northern climes of Germany, but wonders of delicacy and breed have been wrought with this grape variety in the Napa Valley, while the Chardonnay adapts itself beautifully to a whole range of climates. The underrated Sémillon produces some of the finest dry white wines of the world in the Hunter Valley and in Graves, at the same time as making fabled sweet wines in Sauternes and Barsac—but no-one would suggest that the Hunter Valley and Bordeaux share the same climate.

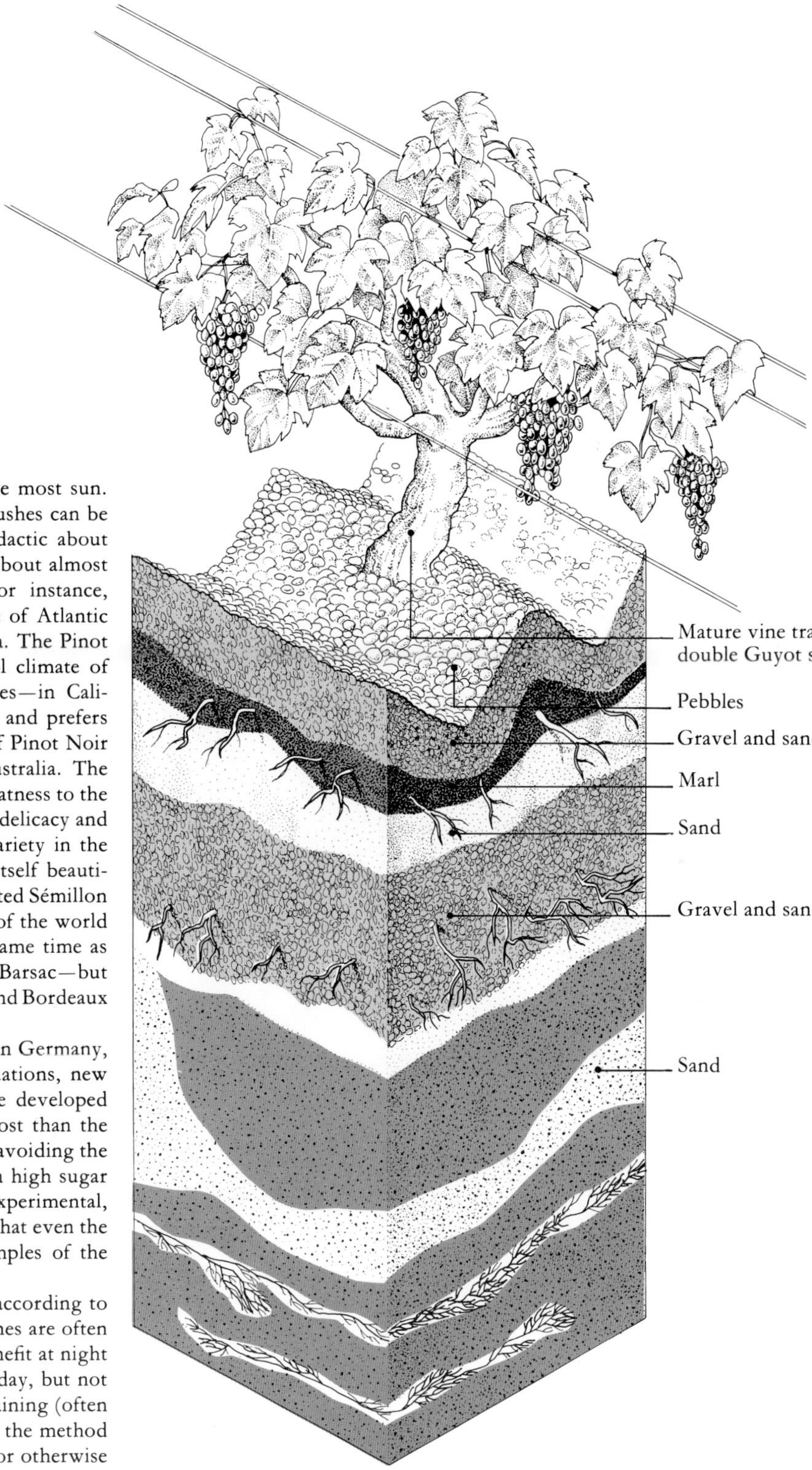

Where the climatic conditions are hard, as in Germany, and where there can be more vintage fluctuations, new grape varieties (crossings of *vitis vinifera*) are developed and tried out. Some are more resistant to frost than the traditional varieties, others ripen earlier, thus avoiding the onset of bad weather, while still others attain high sugar levels. Some of these new crossings are experimental, others have come to stay. But it is true to say that even the finest does not approach the very best examples of the traditional 'noble' varieties.

The system of training the vine is chosen according to terrain, soil and climate. As a general rule, vines are often trained nearer the ground in cool areas, to benefit at night from the heat amassed in the soil during the day, but not too low if there is danger from frost. High training (often on a pergola system) is used in hot areas. But the method must also be chosen according to the vigour or otherwise

Mature vine trained on double Guyot system

Pebbles

Gravel and sand topsoil

Marl

Sand

Gravel and sand

Sand

Soil is one of the main factors in growing vines for fine wines. Different grape varieties like different soils. For example, Chardonnay and Sauvignon Blanc like a chalky soil while Cabernet Sauvignon prefers well-drained, gravelly soil. The main grape grown in Germany, the Riesling, likes slatey soil, but also grows well in loess and loam. The diagram (ABOVE) shows in general terms the kind of soil mixture in which a mature Cabernet Sauvignon vine might grow well.

Loudenne/Cab Sauvignon

Mouton/Cab Sauvignon

Mirassou/Cab Sauvignon

Ridge/Cabernet Sauvignon

Phelps/Cab Sauvignon

Montelena/Zinfandel

Vicchiomaggio/Sangiovese

Renarde/Pinot Noir

Clair-Daü/Pinot Noir

Giscours/Cabernet

Masi/Corvina

Torgiano/Canaiolo

Loudenne/Merlot

Pétrus/Merlot

Hugel/Gewürztraminer

Fine wine depends not just on the skill of the winemaker or the location of the vineyard. One of the prime factors is the inherent qualities of the grape variety or varieties used. Of the several thousand strains of vitis vinifera about 30 are capable of producing top

quality wines. One of the greatest red wine grapes is the Cabernet Sauvignon which is the main grape planted in the Bordeaux area. It is also one of the most important varieties grown in California and in Australia. It is often used in combination with other grapes

—such as Merlot, Cabernet Franc, or, in Australia, Shiraz or Hermitage. Wines made with the Cabernet Sauvignon grape benefit from ageing in both wood and bottle. The main red wine grape of Burgundy is the Pinot Noir. It is not grown extensively

elsewhere in the world. Zinfandel is a variety native to California, now believed to be related to the Sicilian Primitivo grape. It grows well in dry, but not too hot areas. The main grape for Chianti is the Sangiovese. The renowned Brunello di Montalcino wine is

made with a variety of the Sangiovese. Merlot is the second main red wine variety of Bordeaux. It ripens earlier than the Cabernet Sauvignon and gives a softer wine. Other types of red wine grape include the Corvina and Canaiolo. Germany's top quality grape

Johannisberg/Riesling

Kloster Eberbach/Riesling

Von Buhl/Riesling

Bürklin-Wolf/Riesling

Egon Müller/Riesling

Doktor/Riesling

Bisch Weingüter/Riesling

J J Prüm/Riesling

Milawa/Riesling

Kurfürstenhof/Silvaner

Bürklin-Wolf/Scheurebe

Clair-Daü/Aligoté

Schlumberger/Gewürztram
*is the Riesling. About 17,000
acres of Riesling are grown
there, mainly on the Mosel and
in the Rheingau. When
affected by noble rot it can
produce some outstanding late
harvested wines. The other
main variety which can
become affected by noble rot is*

Bouchard Père/Chardonnay
*the Sémillon which is the most
important ingredient in the
great sweet white wines of
Sauternes and Barsac. Other
important types of white grape
are the Silvaner, which is
grown especially in the Rheinhessen region of Germany. The
Gewürztraminer is one of the*

Louis Latour/Chardonnay
*main grapes of Alsace. This
spicy, pungent grape ripens
late. The Scheurebe, named
after its creator Dr Scheu, is
a cross between the Riesling
and Silvaner varieties. Chardonnay is the great white
grape of Burgundy; it is also
popular in California. It*

Dom Leflaive/Chardonnay
*makes a white wine which
ages well and gives particularly good results when grown
on chalky soil. The Aligoté
variety is also grown in Burgundy; it usually gives a
greater yield than the Chardonnay. Another important
white French grape is the*

du Nozet/Sauvignon Blanc
*Sauvignon Blanc which is
grown in the Bordeaux area as
well as on the Loire. It also
grows well in California.*

15

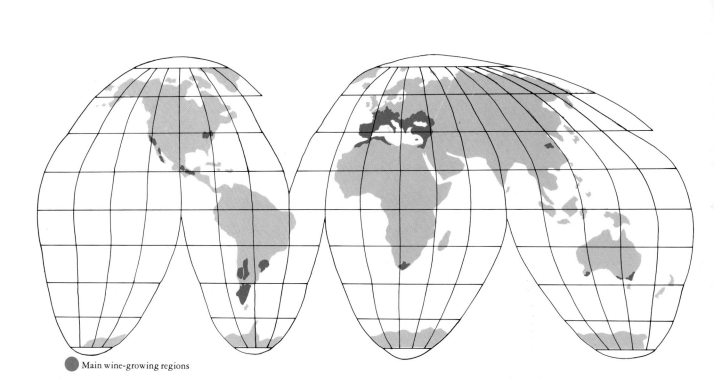

● Main wine-growing regions

The world's main wine-growing regions are shown on this map (ABOVE).

of the grape variety, or the steepness of the terrain, and this can be more important than climate. In Bordeaux and Burgundy, the vines are trained along wires, a system called in French *taille guyot*, and this has been adopted in many parts of the world, especially for the Cabernets, Merlot, and Pinot Noir and Chardonnay. In southern regions, or sometimes to combat wind, the vines are trained up one stake, or as a totally free-standing bush; this method is termed *à gobelet*. On the steep slopes of the Mosel, the vines are trained up one tall stake, with the fruit-bearing canes bent over in a circle.

How vines are pruned varies with the grape variety, as some are more productive than others, and need 'taming', while others need encouraging, with less rigorous pruning. Pruning directly affects the yield, and an experienced pruner will judge whether a vine is strong or not, and know the age of each vine, and will adapt pruning accordingly. An old vine, which produces little quantity anyway, needs to be nurtured, while a vigorous, young vine probably needs restraining. Too generous pruning, resulting in high yield, will 'dilute' the final quality of the wine, but too strict pruning can be disastrous for the quantity of the crop if it is followed by, say, frosts in spring.

The time necessary between flowering and harvesting is usually reckoned as 100 days, although recently in Bordeaux this has been averaging 110 days. Late-ripening varieties and grapes intended for making sweet wines will obviously need up to a month more, several months in the case of *Eiswein*. Machine picking has made huge strides in the last decade, with the machines doing less

and less damage to the vine as they pass and becoming more selective. They are used commonly in California, where the new vineyards could be trained to suit them, and their use in Bordeaux is visible in the lesser areas and likely in the future for most of the fine growths. In Burgundy, with the relatively low training and the small plots, they are only just making their appearance, and in steep areas such as the Mosel they are, of course, out of the question. But there is no doubt that, as labour problems increase, the machines will be made to adapt for most grape varieties and methods of cultivation, even if they will never be able to cope with steep gradients.

Enormous progress has been made in judging the best moment to pick the grapes in order that they enter the *cuvier* or winery in the best possible condition. The right balance between rising sugar content and falling acid content must be found for each region, each year and each grape variety, and the pH level has to be taken into account. Many winemakers used to be so concerned with the degree of ripeness (sugar) that they forgot that at the same time the acid was falling, resulting finally in a wine that tended to be 'flabby'. In a large vineyard, it is difficult to pick all the grapes at the optimum moment; this is easier to achieve in an estate with several grape varieties which mature at different times. Judging when to pick the late-harvested grapes is always a risky business, as the weather can break and ruin everything. Most picking is done once only per vineyard, but in the case of top Sauternes/Barsac, or *Beerenauslese* or *Trockenbeerenauslese* wines, pickers might go through the vineyard four or more times, gathering each bunch, or sometimes each berry, at its optimum point. Grape must—that is the grapes before fermentation—consists essentially of sugary juice. Thus, the composition of a must would be approximately the same with some grape variety and regional variations. Red grapes are usually lightly crushed and de-stemmed, but sometimes a proportion of the stems is conserved to add some tannin to the subsequent wine. Fermentation usually starts naturally, sometimes with either the winery or the must itself needing to be heated, or with selected, cultivated yeasts, which can facilitate absolute control over the fermentation process. In a natural fermentation, the wild yeasts, *kloeckera apiculata*, begin the process, and then the *saccharomyces ellipsoideus* take over. The basic formula for alcoholic fermentation is that 100 parts glucose or lèvulose ($C_6H_{12}O_6$) equals 51.1 parts ethyl alcohol ($2C_2H_5OH$) plus 48.9 parts carbon dioxide ($2CO_2$). There are other side-products of fermentation, such as glycerine, succinic and acetic acid, esters and acetaldehyde.

Fermentation can take place in open or closed receptacles, made of wood, concrete (lined with enamel, glass or unlined) or stainless steel. Many people say that the material is not nearly as important as cleanliness at all stages, to avoid bacteria. A small amount of sulphur dioxide, the great disinfectant in the winemaking process, is used to cleanse the must initially. The temperature should be controlled throughout the alcoholic fermentation process if quality wine is to be obtained. Normally, maximum temperatures for red wine are 28°C to 30°C, but sometimes this is lower, thereby prolonging fermentation. Most of the colour from the grape skins is usually extracted during the first days of fermentation, the 'tumultuous' stage when the sugar is transformed into alcohol and carbonic gas far quicker than at the final stages.

The temperature can be controlled by a number of devices, ranging from cold water running down the side of the vats to a system of pipes resembling a milk cooler. Naturally, the gas from the fermentation forces up the solid matter of the grapes, the skins and pips and other debris, to the top of the vat or other fermenting vessel, and this forms a cap, or *chapeau*. If the vats are open, this cap is dangerous if exposed undisturbed to the air, as it dries and

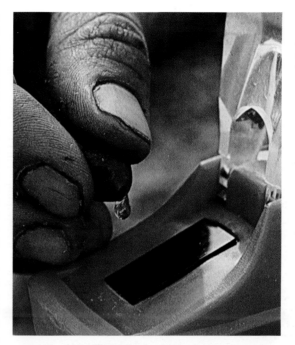

Sugar levels are checked at many stages in the wine's development. A refractometer can be used in the vineyard or in the cellar.(LEFT).

attracts bacteria. The skins should also remain in constant contact with the fermenting must to extract maximum colour. An undisturbed cap can also cause the middle of the vat to rise alarmingly in temperature. So, the cap is kept in contact with the fermenting juice by means of *remontages*. In this process, the must is drawn off from the bottom of the vat and pumped over the top or sprayed over the surface, or the pipe is even immersed in the liquid and pumped. Open vats can also have a system of a permanently submerged cap, using a sort of perforated lid, but *remontages* are still necessary.

Chaptalization—and deacidification if necessary—usually takes place during this stage of the fermentation, and can be effected with a *remontage*. Chaptalization means the addition of some extra sugar which is converted into alcohol, thus raising the alcohol level when this is thought to be necessary (and when allowed by law). This takes

1

The reason for pruning vines is to control the quantity of fruit which each vine produces, because this affects the quality of the grapes and thus of the wine. The vines are pruned annually. Pruning itself demands experience, as each vine should be assessed before

2

pruning takes place. If the pruning is too drastic, the vine will tend to produce wood growth rather than fruit, while too many fruit buds on vines growing in poor quality soil will give poor quality grapes. Pruning and training methods differ from area to

3

area and according to the type of grape. The double Guyot system (1) is widely used in Burgundy and Bordeaux, while the Gobelet system (2) offers good protection against wind and is therefore used in the Rhône. The échallas method (3), which requires

4

new vine growth to be tied to stakes, is also used on the Rhône. In the Lenz Moser system (4) the vines are trained fairly high off the ground. The method is used in Northern Italy. Californian and Australian estates tend to use various of the European

ways of pruning and training vines.

Reg[...] of the vineyard, its loca[...] the kind of grape vari[...] and type of wine, there are certain main stages in the winemaking process, with variations for red and white wines.

Care and attention is vital at all times, beginning with planting and tending the young vines (1). Pruning (2) takes place at certain times of the year throughout the vine's life, after the harvest and in the spring and early summer. Likewise, care of the vineyards is of prime importance at all times. This includes tilling the soil between the rows of vines (8). Spraying against pests and diseases such as mildew or oidium is also important.

The ti[...] the harvest depends on the weather and is determined by the ripeness of the grapes. While some picking is now mechanized, most top estates still pick by hand. As the pickers work their way along the rows of vines, the picked grapes are transferred from small hand bask[...] larger containers. In F[...] e traditional container is the hotte (7). The grapes are then put into a trailer (12) before being taken to the cellar or winery (13). Vinification of red and white grapes differs. White grapes are usually destemmed and slightly crushed before being pressed (18). The gentle pressure possible with a horizontal press is best for white grapes. The juice is then allowed to clarify before fermentation. This process is called débourbage. A centrifuge (19) is also sometimes used. The must is then usually put into a fermentation tank (14). Fermentation times vary according to the type of grape and wine[...] fter fermentation, the wine is often put into barrels for ageing. The best type of barrel varies. Racking (transferring wine off the lees from one container to another) (15) is performed at various stages. Topping up (20) is necessary especially

when the wine is in small barrels because of the amount of evaporation. After barrel ageing (21), the wine is blended, bottled (22) and, finally, shortly before the wine is sold, the labels are put on (23). Red grapes are put into a crusher-destemmer (9) before being placed in the fermentation tanks. These are traditionally made of wood, but stainless steel is becoming more popular. During fermentation the cap of skins is mixed with the fermenting must—often with poles (3). Pumping over is another way of breaking up the cap to improve the wine's colour and prevent the development of undesirable bacteria. Chaptalization (4) —the addition of sugar which is turned into alcohol—takes place during fermentation (10). Not all estates chaptalize every year, and the practice is illegal in some countries. After fermentation the free-run juice is removed (5), and the remaining matter is taken from the vat (6) and put into the press (11). Barrels are thoroughly cleaned often using sulphur (17) to ensure that they are free of bacteria before the new wine is put into them (16). Small oak barrels (15) suit fine red wines. As with white wines, work in barrel includes topping up (20). When wine is in small barrels, the bung hole may be to one side (21) for a time to help lessen evaporation. Blending between barrels takes place before bottling (22) and labelling (23).
In all winemaking, two important precepts are cleanliness and care. Wine is a living organism, and vigilance and attention are essential so that it can develop.

place more often in temperate climates than in hot climates, where the strength of the sun produces more than sufficient sugar in the grapes. In Italy, chaptalization with sugar is forbidden, and concentrated must is used to raise the alcohol level. For red wines, sugar from cane or from beetroots can be used, but the former must be used for white wines. Chaptalization normally cannot be envisaged by law unless the must in question already has the minimum alcoholic degree required by the ruling authorities for that particular wine. It is possible for a wine to have its alcohol increased by 1 per cent or even 2 per cent, but above that, the overall balance of the wine is disturbed. Chaptalization slightly lowers the fixed acidity of a wine, although the addition of concentrated must raises the fixed acidity level. The ratio of alcohol to extract is also higher in chaptalized wines, and, when tests are made for excessive chaptalization, this is taken into account. Making a sugar syrup with water is forbidden when chaptalizing, but the sugar is mixed with a small proportion of the fermenting must and added to the mass. Occasionally, experienced winemakers will chaptalize gradually, throughout the fermentation process, instead of once only, but the temperature has to be watched carefully in this case to see that it does not rise unduly.

The alcoholic fermentation should progress steadily, neither careering forward nor 'sticking', as this would attract bacteria. It is particularly important with white wines that the fermentation should be long and at a low temperature, as this greatly increases bouquet and flavour. The whole process with white wines is in a different order, with pressing before fermentation. The stalks are usually not removed except when automatic harvesters are used, and the grapes are normally pressed in cylindrical presses, either pneumatic or screw type. For fine wine, this pressing must be gentle. After pressing, the murky juice is then run off, and usually left to 'stand' and clarify with the solid matter falling to the bottom of the vat. In French this process is called *débourbage*. Centrifuges are sometimes used at this stage in modern wineries.

When the *débourbage* is completed, fermentation will thus start with a cleaner must and will probably result in fewer rackings. Racking means the transfer of wine off old lees in one cask to another, clean one, at a later stage. Fewer rackings are particularly desirable in delicate, dry white wines where minimum handling is advisable to guard against oxidation—the great enemy of white wine. White wine fermentation is usually between 15°C and 20°C when elegant wines with marked bouquet are desired, but in some parts of the world the temperature can go below this.

After fermentation, both red and white wine can be run off into barrels or kept in vat. Fine winemaking relies a good deal on oak barrels, for a greater or lesser period is spent in new or used oak. With red wine which has only undergone fermentation, but not pressing, the correct term is free-run wine. This *vin de goutte* can be mixed with a

proportion of wine obtained after pressing the *marc* or lees, which is called *vin de presse*. The amount will vary with the characteristics of the year in fine wine regions, and accepted practice with given grape varieties. The higher tannin, volatile acidity and mineral content in the *vin de presse* mean that it has to be carefully 'married' with the *vin de goutte* to give a balanced wine.

The *élevage* or bringing up of a wine is vital to its subsequent taste and quality. Wines for quick consumption are bottled early, perhaps within the first six months after the vintage, and there is no real maturing process. But many fine wines are matured, or aged, before bottling, and have to be tended all along the way. Top red wines, dry and sweet white wines (except in a dry white wine, such as a Sauvignon, which needs early bottling) are usually put into oak barrels after fermentation.

A second, or malolactic, fermentation is usually desirable, almost always in red wines and often in whites, except where a wine lacks acidity and the malic acidity needs to be conserved in order to give the wine freshness.

The size and the wood from which barrels are made influence the way the wine develops. In Germany, many barrels — often called ovals — are carved or painted. This example (RIGHT) depicts the Kurfürstenhof estate, nearby river and the town of Heidelberg.

In Germany they wish to conserve the malic acid as it enhances the particular fruity acidity of their wines. The complicated process of malolactic fermentation, not yet fully understood, has the effect of reducing acidity, as it involves the transformation of the rather green, appley malic acid in the wine into the milder lactic acid, the acid found in milk. In the process, bacteria attack the malic acid and turn it into lactic acid and carbon dioxide gas. This can happen concurrently with the alcoholic fermentation, or follow afterwards. It is difficult to have absolute control over this phenomenon, but warm conditions can be used to encourage the malolactic bacteria to begin their work, and in the most technical enterprises, malolactic bacteria can sometimes be cultivated and added to the must right at the beginning of the whole fermentation process. But it is not an exact science, and many anomalies can occur from cask to cask. Usually, once one cask has begun its malolactic fermentation, there is enough bacteria in the air to encourage the others to begin. Prevention, where this is desired, is usually effected by the use of sulphur

dioxide. If the malolactic fermentation occurs in bottle, the taste and appearance suffer. Such wine would have some bubbles as, during malolactic fermentation, 1 gram of malic acidity is transformed into 0.67 gram of lactic acid and 0.33 gram of carbonic acid.

So, maturing wine is left in cask or vat, usually the former for fine wine. This results in significant evaporation, and regular topping up is necessary to prevent oxidation. This should be done very carefully, with wine of the same quality, itself kept in perfect conditions, through the bung-hole at the top of the cask. At some stage, the cask is usually rolled three-quarters over, in French *bonde de côté*, and then topping up is not necessary.

If wines are stored in vat, they can be protected from the air by means of a covering of carbon dioxide or of nitrogen. Wines in cask are always subject to a small amount of oxidation through the wood, but this is desirable in the maturation process, particularly with grape varieties such as Cabernet, Merlot, Pinot Noir, Syrah, Chardonnay and Sémillon. With Riesling, there are those who seek this, and those who do not. If the casks are new oak (which can come from different origins, each imparting its own properties), flavour and tannin will certainly be added to the wine. When a wine is big and solid, this can add a new dimension, but a light wine can be overwhelmed by being put in new oak.

During a wine's life in cask, it will be periodically racked. It will also probably be clarified, or fined, removing solid matter, excess tannin and unstable elements in the wine. There are many products used to achieve this, including egg white (for very fine red wines), gelatin, fish glue or isinglass (for white wines), and casein. For a wine to be stable and protected from oxidation, it should contain both free and bound sulphur dioxide. On the whole, the free sulphur dioxide content is lower in red wines than in white. The total sulphur dioxide content (free and bound) has to be quite high in the case of sweet white wines, to prevent risks of re-fermentation.

Heating, refrigeration, sorbic acid, ascorbic acid, meta-tartaric acid, citric acid, and bentonite (against excess protein) are included in other stabilizing treatments, although their use in fine wines is very limited, where the aim is to produce less treated wines, handled with great care. When prices are high, more labour-intensive methods can be used than with cheap wine, and more expensive equipment to avoid problems like oxidation can be called into action. Refrigeration is used to cause precipitation of potassium bitartrates or calcium tartrates as, unfortunately, the public does not like crystals in a bottle of wine, although they are completely harmless and do not affect the wine—on the contrary, they are an indication that a wine has not been over-treated. A wine that has a long cask life will throw most of its deposit naturally.

Choosing the right bottling time for individual wines is of paramount importance. This can vary from a few months after the vintage for light, fresh white wines, to two years for top red Bordeaux, three years for Château Yquem, and four to five years for Brunello di Montalcino. Wines bottled too late for their particular composition and character can tend to lack vigour and dry out. Some grape varieties then need time in bottle to achieve their full potential—Cabernet, top Chardonnay and Sémillon are examples, and Riesling when there is the correct balance between ripeness and acidity. Balance is, in fact, the essence in fine wine, with fruit, acidity and tannin all necessary in harmonious proportions for red wine, and in white wines there should be fruit, acidity and 'fat' (or *gras*) to support ageing. Wines with no extract rarely age well. Tannin needs time to soften, but if there is no balancing fruit, it can dry out a wine. The esters and aldehydes, which essentially provide the great bouquet of fine wine, need time in bottle to develop.

Bottling itself should be carried out in conditions of great hygiene, and should be homogenous, with casks being 'assembled' before bottling, so as to avoid cask-to-cask differences. Pre-bottling filtering should be light in fine wines, so that no flavour or matter is removed.

Sometimes, this process is avoided altogether, but then the consumer must expect a deposit to form in a bottle—again, a sign that a wine has not been over-treated and therefore to be welcomed. In nearly all fine wines, bottling under cold, sterile conditions is desirable, rather than any form of pasteurization, flash-pasteurization or hot bottling, which ensures a totally sterile wine but is not intended for wines expected to age in bottle. Nevertheless flash-pasteurized wines are often indistinguishable on nose and palate from those that have not been through this process. It is obvious that winemaking is not all poetry. It is science, mixed with a degree of art. There are certainly some winemakers who have a 'feel' for wine that others lack. It also behoves a winemaker seeking quality to be somewhat self-critical, and to taste a good deal of wine outside his own domain. The emphasis on science naturally becomes even stronger when climatic conditions are not propitious for fine winemaking, when all man's ingenuity is called into action. All the technical qualifications possible can never replace experience, and experience in one area of the world does not equal experience in another. Regional differences can be more than quirky—

1

2

3

Most wines spend some time in barrel, the length of the period depending on the type of grape and of wine. The size of the barrel also varies— most fine French-style red wines age in small barrels, while Chianti ages in larger vessels. The sweet white wines of Bordeaux also age in small oak barrels, while Rieslings prefer larger ones. Today very few estate actually still make their own barrels. One of the few estates to continue the practice is Chateau Margaux in Bordeaux. The 25 new oak staves (1) are selected. Three

they can be unrecognizably different. Within each region, there are nuances, and people attempting different things. So, there will inevitably be differences of opinion, and indeed, contradictions, in this book. Conflicting views and theories will be expressed by individual winemakers, all experienced and all right in their way, since they have found what suits their particular conditions and aims. The proof is in the fine wine they are making, evidence that the theories they hold do produce the desired results.

What are the parameters of a great wine when such very different tastes and styles are represented in this book? Above all, the wine must be individual, have a unique character which expresses the very best that the respective region and grape variety or varieties can attain. Harmony and balance are essentials, with no jarring note in the overall composition of the wine. The very greatest wines should have a potential for ageing, for improving with maturity and developing more complex flavours and increased depth of taste. However, a few of the wines in this book do not possess this ultimate bonus, because by their very nature, certain grapes are not destined for old age and show at their most seductive when young. The

winemaking principles and ideals are expressed in the respective idiom of each country, as it is felt that this is the most accurate way of pinpointing the winemaking processes and more faithfully reflects what a winemaker is doing.

Perhaps the most exciting aspect of the book is that some of these great vineyards and winemakers would not have been operating a decade ago. While some properties have been making fine wine for 200 years or more, others did not have a vine planted 20 years ago. This is the most encouraging thing of all for those who have the great good fortune to taste these pinnacles of winemaking achievement. It shows that there is constant interest in fine winemaking and revitalization at the highest level.

Great winemaking will only continue and evolve if there is the audience to appreciate it. Perhaps, to appreciate fine wine fully, there is a need for some background knowledge and some idea of what is being looked for. It is to be hoped that this book will contribute to filling this need. May those of us who do not make great wine, but who like to think we can recognize it when we taste and smell it, always face the liquid in the glass with anticipation and humility.

different widths are used. The staves are then put upright into a hoop (2). The second hoop is put on using a combination of heat and humidity (3). In order to put on the third hoop, the staves have to be brought together (4). For this, a hand-winch is used (5), *then the third hoop can be hammered on (6). The barrel is then reheated (7). Barrel-making requires great skill and precision. Oak, and particularly French oak, is considered best for wine barrels; some estates in California and Australia* *even import French oak for this purpose.*

THE VINEYARDS

FRANCE: INTRODUCTION

rance is perhaps associated with the production of more fine wine than any other country in the world. This fact has enabled many French wines which fall short of this description to be sold more easily than they otherwise might have been, but in an annual production of about 70 million hectolitres, it is inevitable that there is a huge variety of wines made of all standards. There is virtually no type of wine that is not made in France from dry red and white, sweet whites, rosés of all shades and flavours to the best sparkling wine in the world and a wide range of fortified wines that are little known outside the country itself.

Many of the noblest grape varieties in the world, capable of making the finest wines, reach their apogee in France. This is not to say that these grape varieties cannot make excellent and fascinating wines in other continents, where the French version is often taken as the standard with local differences adding their own dimension to the final taste. There are few winemaking avenues that are not explored in France, although great sweet, late-harvested wines from the Riesling grape are not in the French wine-making tradition, with Alsace nearly always making Riesling wines dry. It is of no constructive use to waste valuable drinking time wondering whether French wines are really the best. It is impossible to tell if a great, aged northern Rhône wine made from the Syrah is really 'better' than a blend of Hermitage and Cabernet from the Hunter Valley of New South Wales, or if a classified growth from the Médoc is intrinsically a 'better' wine than a renowned Cabernet Sauvignon from the Napa Valley of California. What matters far more is that each wine should be good in its own way, faithfully reflecting the aims and intentions of the winemaker and contributing its own special taste.

The nexus that links all winemakers should be the search for quality, and in this France has played her full part. The *appellation contrôlée* system established in France was not the oldest organization to define vine-growing areas or draw up rules aimed at improving quality; that distinction probably goes to the Chianti Classico zone in Italy, the first area to be defined by governmental decree, in 1716. However, the French laws were extremely far-reaching in both intention and reality. Throughout the 1930s, the *appellation contrôlée* areas were worked out and set into law, and, although there have been additions, there have been few changes to the original list, with only a handful falling into disuse. The areas of production, and their quality scale, particularly with regard to such matters as *grands crus*, *premiers crus* and *commune* wines in Burgundy have stood the test of time. If no law can impel a wine-maker to make perfect wine, it can indicate the potential of a certain piece of land.

Thus, while the fact that a French wine is *appellation contrôlée* does not ensure that it will be delicious, it will help ensure that a wine comes from the place it says it does.

FRANCE

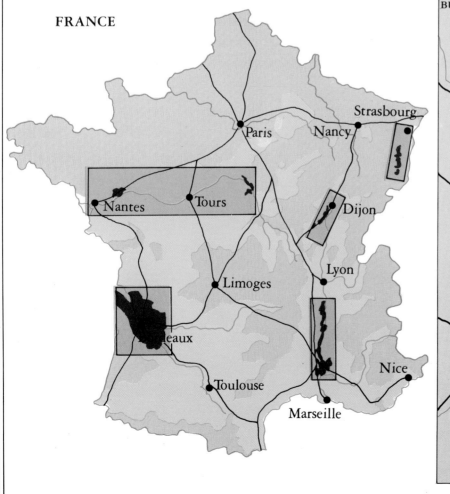

BURGUNDY

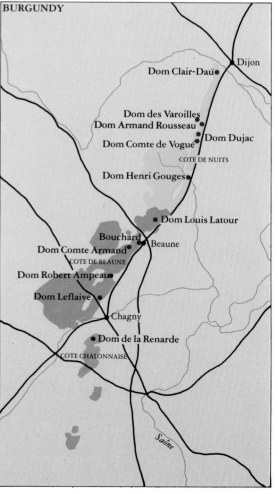

The *appellation contrôlée* laws lay down basic viticultural and vinification instructions that should lead to better quality. No law is of use unless it is enforced, but the law, coupled with an effective inspectorate, does make the likelihood of drinking fine wine rather stronger than if winemaking was left to the whims of all who attempted it. The *appellation contrôlée* rules cover the area of production, the grape varieties permitted (these are sometimes divided into principal and accessory grape varieties), the minimum alcohol level of the wine (which should be achieved before chaptalization, in the areas where this process is permitted under *appellation contrôlée* law), and the maximum yield allowed per hectare. The law also covers the methods of planting, pruning and treating the vines, the vinification of the wine and, in some cases, conditions of ageing. The level below *appellation contrôlée*, known as *vins délimités de qualité supérieure*, have the same laws, but they are wider than those governing *vins à appellation contrôlée*. Yields are therefore larger, there are less strict conditions to the zones of production, and sometimes grape varieties of more modest pedigree are permitted.

There was, however, an important flaw in most *appellation contrôlée* laws; there was no provision for tasting the wine before it went on sale. A decree in 1974 changed this, and provisions were laid down for all *appellation contrôlée* wines to pass through a tasting panel before they could emerge with the certificate of *appellation d'origine contrôlée*.

Tasting panels usually consist of a grower, a *négociant* and a broker, thus giving a balance of all aspects of the wine trade. It would be true to say that the panels can be somewhat lenient, and the few wines that are refused can always be presented again.

Yield is linked to the search for quality, on the assumption that a 'stretched' yield will give wines of little character or concentration. There is some flexibility within the law, which is only logical in a country like France where varying weather conditions present different conditions every year. There is a basic permitted yield for each *appellation*, then a modified yearly permitted yield, based on the conditions of the year, with an upper limit of production, which is the yearly permitted yield plus a percentage, usually 20 per cent, which represents an absolute maximum.

The *appellation contrôlée* laws have been honed to close loopholes. The improvements to the law passed in 1974 also ensured that the grower must opt for the classification he wishes for his crop in its entirety (within a specified area, of course), at the time when he declares what he has harvested. He cannot, therefore, declare a proportion of his crop as the *appellation* wine, up to its legal limit, and everything over that limit as a lesser classification. Here, the law recognizes that over-production affects the whole crop, not just the surplus yield. Naturally, as with every bureaucratic machine, there are anomalies and aspects of

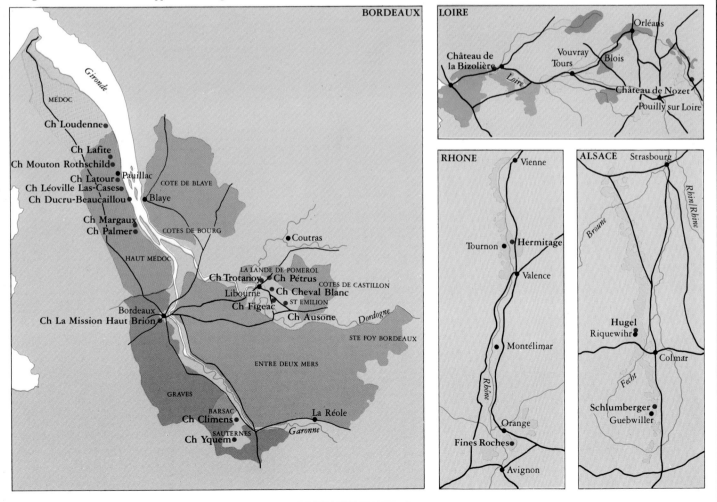

Vintage chart
Bordeaux

The Bordeaux area is large, with different weather conditions on each side of the Gironde, as well as varying grape composition in the wines. Apart from the great red wines, there are also distinguished dry white wines and luscious sweet white wines. Bordeaux probably matures with more majesty than any other red wine, attaining finesse and subtlety with time. Years with good balance between tannin, fruit and acidity mature the best and stay at the top for longest.

1945 Massive year of small yield and great concentration. Remarkable whites, both dry and sweet.

1946 Hardly seen, but fading.

1947 Great summer heat caused some vinification problems, but the best wines have great richness and opulence. Wonderful Sauternes.

1948 Less generous than 1947, needing time to soften, with some remarkable wines. Whites have really disappeared.

1949 Wonderfully harmonious wines, which matured to great elegance. Classic white wines.

1950 Large, somewhat variable crop, with the best only good, but now at the end of their days. Very good sweet whites.

1951 Forget it.

1952 A good vintage, rather tannic when young, but the best have got through this stage. St Emilion and Pomerol are very good. Some good white wines, both dry and sweet.

1953 A vintage of the utmost charm, delightful all its life. Beautiful wines which will not linger for ever. All whites very good.

1954 Only fair and mostly gone. Terrible sweet whites.

1955 Very good red wines, some hard at first but they have lasted well. Marvellous white wines.

1956 Reds mostly very sad, a few presentable whites at the time.

1957 A moderate year with big variation—it will not improve now. Some very pleasant sweet white wines.

1958 Fairly light wines, many of them attractive, but hardly built for lasting. Whites equally charming, but not stayers.

1959 The immense heat meant richness, but the low acidity has made some look tired. Plummy and full, rather than finesse. Equally rich, full-blown whites.

1960 Light red and white wines that should all be drunk by now.

1961 The greatest, grandest, most intoxicating vintage for red wines since the Second World War and maybe of the century. Massive concentration, stunningly balanced by fruit and bouquet. They will live to a great age, but that is not so say they are not delicious in middle age. The white wines do not have the same balance, good as some are.

1962 Very good wines, with relatively high acidity causing many to last well. Lovely wines of breed in Sauternes-Barsac.

1963 The year of rot which produced sickly, unhealthy wines.

1964 Rain at the vintage prevented this from being great. In the Médoc some wines are a bit dry, but there are some really fine wines in St Emilion, Pomerol and Graves. Dry whites quite pleasant, the sweet whites were mostly washed away.

1965 The grapes were simply not ripe—ignore.

1966 A very classic vintage, with deep colour and tannin, and therefore destined for long life. The whites are good, without being luscious.

1967 A vintage that was at its best younger, and now generally rather hard with a dry finish, best from St Emilion and Pomerol. Outstanding Sauternes-Barsac, combining lusciousness with style and breed.

1968 Rain and cold did their worst. A few wines were salvaged, which were light and fruity in youth. Sauternes was a write-off.

1969 Very mean wines, lacking fruit and charm. Light, not very consequential Sauternes.

1970 An exceptional year, combining abundance with quality, the wines need time to open up. Lesser growths often needed a decade, and the great wines much longer, to develop complexity. The tannin is in balance with the fruit and acidity. Sauternes are good, if lacking ultimate breed and finesse.

1971 St Emilion and Pomerol made the more stunning wines, opulent, fruity and of great interest. In the Médoc, the great ripeness has sometimes caused the wines to be a little low in acidity, causing browning and some drying out. Sauternes and Barsac of great finesse and distinction.

1972 A very 'green' and unripe year, making most wines raw and mean. More modest properties sometimes made better wine than the top names. Acid white wines, with many sweet wines not appearing at all.

1973 A large vintage, producing wines generally medium to light in body, the best with attractive fruit and charm. They show best when comparatively young, and not many will go on improving into the 1980s. Sauternes very nondescript.

1974 Really rather a mediocre year, with the wines with the odd exception lacking fruit, charm and individuality. Sauternes was very sad and rarely seen.

1975 A very good vintage indeed, concentrated, tannic, but with the marvellous fruit now beginning to emerge. Some classic wines for the future with consistent quality. Some Sauternes were too alcoholic for perfect balance.

1976 A very hot summer, followed by rain during picking, produced wines with as much tannin as 1975, although of a different composition, with low acidity. Probably not real stayers, but they have interest and attraction. The top wines have class and opulence, and there are lovely St Juliens and Pomerols. Exquisite white Graves and Sauternes-Barsac.

1977 September saved the vintage after a disastrous summer. The wines are modest, the best fruity and charming in youth. the worst thin and a bit acid. Frost caused the Merlot to suffer. Good selection can turn up some very pleasant wines for drinking in young to middle life. No great hopes for Sauternes.

1978 Wonderful early autumn created this magnificent vintage, which looks as if it has it all—fruit, 'fat', body and structure. Lovely, rich wines, very consistent at all levels. In Sauternes very ripe grapes were picked, but the lack of autumn humidity preclude the formation of *pourriture noble*.

1979 A huge, late vintage. The wines look superb, especially St Emilion and Pomerol, where the Merlot was very favoured. But lovely, glossy wines everywhere, even in extreme youth. The best châteaux in the Médoc were selective, and the great yield did not affect the *grands vins*, which have very good body. Very good signs that there will be classic, *pourriture noble* Sauternes.

1980 Far too early to predict. Some rain during vintage, but those that could wait, picked up some nice weather at the end. The colour of the musts was uniformly good.

Burgundy

Burgundy vintages do matter. And with the huge diversity of growers and producers in this region, it is worth remembering that a good producer can make a success of a moderate vintage, while a poor producer will not suddenly mend his ways when a superb vintage comes along. With regard to white Burgundy, it is a matter of taste whether you like that old Chardonnay flavour, but the greatest white Burgundies do not begin to show age for up to 20 years.

1945 Very good, small vintage. Extremely concentrated. They can still taste very robust.

1947 An unusually hot year, and with no temperature control the fermentations did sometimes go wrong. The best wines are gems, rich and powerful.

1948 Some great wines, very complete, lasting well.

1949 Some wines still have lovely fruit, balance and charm.

1952 Rather underrated wines, some of which are still in good order. Excellent, firm whites.

1953 Balanced and harmonious wines, the best of which have lasted with elegance. Very good whites.

1955 Wines with some warmth and style. Some have blossomed extraordinarily well, others are showing age. Excellent whites.

1957 There are wines with good body and their flavour and character have emerged with honours.

1959 The hot summer caused these wines to be rich, and low in acidity. So, although they may not be set for the longest of lives, they have great generosity and fullness. The lack of acidity makes most whites too old.

1961 Small crop of excellent wines, intense and concentrated. Taste marvellous at beginning of 1980s and will go on. Some whites still show superbly.

1962 Wines of enormous class—real aristocrats. Balance and length. Because of this, the vintage will produce some great stayers. Wonderful whites.

1963 A write-off really, with nothing of note except a few whites from top sites.

1964 Some very good wines, with backbone and character —both reds and whites look good at the beginning of the 1980s. Some wonderful, opulent whites.

1965 Forget it.

1966 A really well-rounded vintage, with fruit and body, with wines that still have a future before them. Elegant, stylish whites.

1967 Mixed vintage, with the best possessing elegance. The large crop, however, sometimes resulted in wines that were over-chaptalized and have 'fallen apart' and browned. Whites can have a good deal of breed.

1968 The reds are not to be considered, a few whites were passable.

1969 A simply superb classic Burgundy vintage, unfortunately only a small one. The reds have breed and *race*, with the fruit held by good acidity. Will produce the aristocratic wines of the future. The whites are equally classy.

1970 A large vintage of relatively soft, fruity, charming wines. Tempting and easy to drink, but, on the whole, neither reds nor whites will age well.

1971 A small vintage producing extremely ripe wines of concentration and richness. Sometimes they were too ripe and alcoholic for ultimate balance. Great natural sugar.

Looked lovely when young, but the low acidity could cause them not to stay the course. Some hail in both Côtes has browned some wines and advanced their maturity considerably. The whites have great flavour and power, but might not last indefinitely.

1972 A vintage that looked very 'green' and unripe when young, but the best reds have developed quite superbly in bottle, and have great character and individuality. Will last extremely well. The whites have softened in acidity, but lack the gift of fruit; some, however, have character and depth.

1973 Very large vintage, with resultant light wines, often fruity with charm. Drink rather than keep, and they will give much pleasure. The whites had flowery appeal and some finesse, but with low natural acidity, so will not last long.

1974 Good colour reds, often firm but lacking real individuality. The best, however, are good buys. The whites can be more interesting. Altogether a dependable but not exciting vintage.

1975 Rot ruined this vintage, and it is rare to find a red wine that does not have a 'tainted' taste. Some whites were more pleasant, but all should be drunk quickly. Chablis was, however, very good.

1976 A Burgundy vintage really unlike any other. The very hot and dry summer produced wines of exceptional tannin, and some variations on normal vinification procedure were often advisable to take account of this. The enormous tannin envelops the fruit, and it is hoped that these wines will eventually attain a balance between the two. The whites can be a bit overblown, rewarding if not kept too long. Chablis, however, produced another fine vintage.

1977 A miraculous vintage,

Bordeaux produces some of the world's finest red wines with great capacity for ageing. Many estates keep some very old wines. These come from pre-phylloxera vines.

saved by the fine September weather. The reds are mostly light, a few are too acid, but many have projected fruit and are good for relatively young drinking. The whites can have real finesse and are very good buys.

1978 A superb vintage, better balanced than either 1971 or 1976, and giving wines which combine structure, fruit and 'fat'. Perfect for laying down. The whites are as good, concentrated and with great interest, in both the Côte de Beaune and Chablis. Both reds and whites should be kept for their true complexity to emerge.

1979 A large vintage, although localized hailstorms on the Côte de Nuits made for limited, but serious, damage. The red and white wines have charm and fruit, and the best are very nice indeed.

1980 Variable, with only the vineyard owners who treated against rot succeeding in making good wines. But there will be good red and white wines, probably suitable for medium-term drinking.

Southern Rhône— Châteauneuf-du-Pape

Very good years in Châteauneuf-du-Pape are: **1945**, **1947**, **1949**, **1955**, **1957**, **1959**, **1961** (superb), **1962**, **1964**, **1966**, **1967** (tannic and full), **1969**, **1970** and **1971**. **1972**s need careful selection, the **1973**s were rather weak, the **1974**s somewhat dull, and the **1975**s should be missed, due to rot. **1976** is mixed, with the best good; **1977** a bit thin but some pleasant wines for medium-term drinking. **1978** is a superb vintage, absolutely classic for the area, with the power and fruit to last beautifully. The **1979**s are immensely attractive, not quite as big as the **1978**s.

Northern Rhône— Hermitage

Very good years in Hermitage are: **1929**, **1945**, **1947**, **1949**, **1952**, **1953**, **1955**, **1957**, **1959**, **1961** (superb), **1962**, **1964**,

1966, **1967**, **1969**, **1970** (particularly good) and **1971**. **1972** produced some excellent wines at Hermitage. **1973** was pleasant, **1974** only moderate, and there was rot in **1975**. **1976** has great attraction, but may not be for the longest of lives, and **1977** was sometimes not quite ripe, but there are amazing exceptions on the best sites. **1978** is a magnificent vintage, with exceptional extract, concentration and fruit. **1979** combined abundance with great quality, sometimes even rivalling the 1978s.

Loire—Savennières

Years of which myths are made are **1921** and **1947**, with the **1947**s still tasting magnificently. Other very good vintages were **1949**, **1953**, **1955**, **1961**, **1970**, **1973**, **1975**, **1978** and **1979**; excellent years include **1959**, **1964**, **1969**, **1971** and **1976**.

Loire—Pouilly Blanc Fumé

These wines should always be drunk young. **1980**—small crop, nice quality. **1979** and **1978** very good. **1977** a little acid. Other good years were **1976**, **1975**, **1974**, **1973** and **1971**.

Alsace

In Alsace, the great years keep beautifully, the lesser years should be drunk young, and certain years favour certain grape varieties. **1945**, **1947**, **1953**, **1959** and **1961** were all great years, as were **1964**, **1966**, **1967**, **1969**. **1970** was good and **1971** great, probably the best since **1961**. **1972** was rather unripe, **1973** pleasant and for young drinking, **1974** a fill-in year, **1975** elegant and balanced. **1976** gave extraordinarily high Oechslé degrees, and the wines have real noble rot character, a majestic year. **1977** light and for young drinking, **1978** small quantity but some most attractive wines, **1979** very good quantity and quality with some top-class wines, and **1980** good, but *coulure* virtually wiped out the Muscat, with the Gewürztraminer faring little better.

the law which seem irrelevant to winemaking needs, the result of either politics, or the distance of a Champs Elysées office from the vineyard or cellar-floor. But the good intentions and basic structure are there, and wine-making is a vital part of French economy and agriculture.

Wine is made in France by all sorts and all manner of people. There are aristocratic owners of imposing châteaux or domains, some personally involved in the making of their wine and others content to leave it to capable managers. There is a mass of small growers, some growing wine full-time, and others combining it with other work. There are co-operatives, some huge and some more modest, taking in the produce of growers, usually in the form of grapes. There are shippers, or *négociants*, some of whom own vineyards themselves, but whose main work is buying in wine, preparing it for sale, and marketing it. A *négociant-éleveur* places particular stress on this aspect of 'bringing up' the wine he buys in, often buying in grapes so that he can supervise the whole winemaking process. There are many family businesses, and many properties owning vines which have not changed hands for generations, but the laws of French inheritance often break up estates and cause domains to split.

Continuity can be a good thing, or it can be stultifying, depending on the attitude of the owner. There is no doubt that winemaking in France used to be steeped in tradition, with a somewhat suspicious view of innovation. But the last 20 years have seen a growing awareness of the advantages of improved technology, and its increased influence is visible everywhere. The oenology faculties of Universities such as Bordeaux, Montpellier and Dijon have contributed a major part of current winemaking knowledge, and this oenological wisdom has often been exported along with French cultivated yeasts and French oak barrels.

Generally speaking, the western side of France has a temperate, maritime climate, without great extremes of temperature. This includes the viticultural areas of Bordeaux, the largest fine wine-producing area in the world, with an annual production of over 3 million hectolitres of *appellation contrôlée* wine, and the south-west, and a good part of the Loire Valley wine areas, particularly Muscadet near the coast, inland to Anjou and Touraine. The vineyard areas of eastern France have a more continental climate, colder winters and more marked changes of temperature. These areas include Burgundy, Alsace and Champagne. South of Burgundy and Lyon is the Rhône Valley, which starts in the north with a climate somewhat similar to southern Burgundy, but finishes, round the Rhône estuary, with a distinctly Mediterranean climate. The long, sunny summers here have a great influence on the wines.

These main viticultural zones remain faithful to families of grapes, which over the centuries have been found to suit the soil and climate. The west of France is the land of the Cabernet, one of the world's noblest grape varieties for

Fermentation is an important stage in a wine's development.

making red wine. The main characteristics of the Cabernet are its breed and longevity, producing wines which have the potential to become some of the most fascinating and complex ever put into bottle. There are two main types of Cabernet grape, the Cabernet Sauvignon and the Cabernet Franc. The Cabernet Sauvignon is the more tannic in youth and the one which ages with the greatest splendour, while the Cabernet Franc is fruity and perfumed. The Cabernet Sauvignon predominates in the make-up of the greatest red wines of the Médoc, those from the districts of Margaux, St Julien, Pauillac and St Estèphe, and is an important part of the great red wines of the Graves. The Cabernet Franc is a valuable part of the grape variety mixture in these wines, and even more important in St Emilion and Pomerol, where the Cabernet Sauvignon is generally hardly used and does not ripen well. The Cabernet Franc is also much used in the red wines of the Loire, particularly in Chinon and Bourgueil and the best rosé wines. Cabernet does particularly well on well-drained gravel or pebbly soil.

The other great red grape variety of south-western France is the Merlot. The Merlot is an important component part of the red wines of the left bank of the Gironde, but the most important element of nearly all the greatest St Emilion and Pomerols, particularly the latter, where the clay in the soil needs the ripening efficacity of the Merlot. The Merlot produces wines of luscious, rich fruit which are most seductive. The Merlot and Cabernet stretch into all areas of the south-west, including the areas of Bergerac in the Dordogne, Cahors and down into the Pyrenees, sometimes with one or other disallowed, and sometimes mixed with rather stranger, traditional local varieties. Some Cabernet has also crept across the south of France to Provence, but this is a comparatively modern innovation as the variety tends to be used only by those seeking to make rather more serious red wines than was usual for this area.

The dominant white grape variety of the west is the Sauvignon, which generally produces crisp, dry white wines destined for young drinking—typical examples are the dry white Entre-Deux-Mers wines of modern times, some white Graves, and the Sauvignon wines of Touraine and Haut-Poitou. The Sauvignon even creeps inland along the Loire to the centre of France, and is uniquely responsible for Sancerre and Pouilly Blanc Fumé. However, the Sauvignon is often mixed with the Sémillon grape variety, either to produce sweet wines such as Sauternes or Barsac, or Monbazillac from the Dordogne (here mixed with the less noble variety, the Muscadelle), or to produce some of the greatest dry white Graves. Whether vinified dry or attacked by the noble rot of Sauternes and Barsac, to produce wonderful luscious, sweet white wines with their highly individual bouquet and taste, the Sémillon is a grape of the highest class.

The other widely planted white grape variety of the west is centred round Nantes where the Loire river meets the sea. This is the Muscadet, which makes the light, refreshing dry white wine of the same name. This is the only place in France, or the world, where the Muscadet is grown, although it originally came from Burgundy, where it did not make very special wine. Touraine and Anjou, the historic heartland of the Loire river, are famous for their dry, demi-sec and sweet white wines made from the Chenin grape, which can be affected by noble rot.

Eastern France largely bases her fame as one of the world's best viticultural areas on the Pinot Noir for making red wines and the Chardonnay for making dry white wines. The great red Burgundies of the Côte d'Or and the Côte Chalonnaise are made entirely from the Pinot Noir, and the rich, elegant whites depend entirely on the Chardonnay for their quality and character, whether the area be the Côte de Beaune, the Côte Chalonnaise, the Mâconnais, or Chablis, the somewhat isolated small area mid-way between Dijon and Paris. Champagne also depends upon the Pinot Noir and the Chardonnay, together with another member of the Pinot family, the red Pinot Meunier.

The Pinot Noir, often planted on marly soil, makes red wines of immense flavour, the best of them ageing to a glorious, rich, almost earthy splendour. The Chardonnay, which likes calcareous soil, and which has subtle clonal differences between the Mâconnais and Chablis, makes white wine which—with the exception of basic Mâcon Blanc—is well-suited to ageing in bottle, developing a wonderful, heady scent and great texture with maturity. There is also the more modest Aligoté, unique to Burgundy which makes relatively simple white wines. The other red grape variety of Burgundy, largely concentrated in Beaujolais to the south of the area, is the Gamay. The Gamay makes scented, fruity red wines that are amongst the easiest for the wine-taster to appreciate. Intended for young drinking, they are charming and tempting, and win friends the world over.

Alsace is almost a law unto itself, a corridor between countries, with a mixed French and German heritage affecting the choice of grape varieties grown. In a viticultural zone almost entirely devoted to white winemaking, the Riesling, the Gewürztraminer, and the Pinot Gris (Tokay d'Alsace) dominate, with some Sylvaner and Muscat. The latter is, unusually, vinified dry as are all Alsatian white wines, except in late-harvest years. The only true concessions to French winemaking traditions are a small amount of Pinot Blanc, the Chasselas, which is responsible for the simplest Alsatian wines, and the Pinot Noir, for making the rare reds and rosés.

The red wines of the northern Rhône owe their existence to the powerful, long-lasting Syrah, which clings to the granite slopes. The dry white wines are made from the Marsanne and the Roussanne, and occasionally from the perfumed Viognier which is found nowhere else in the world. The southern Rhône is dominated by the Grenache, but there are a group of other red grape varieties that, between them, are responsible for the whole range of southern French wines, from the great, long-lasting Châteauneuf-du-Pape to the holiday wine of Côtes de Provence and the inexpensive, everyday wines of the Languedoc-Roussillon. These include the Cinsault, the Mourvèdre and the Carignan. Thus the great vista of French wine drinking is unfolded, and whole panoramas of tasting experience await the drinker bent on new discoveries, or merely renewing acquaintance with some of the finest wines produced today, for France remains perhaps the most important—and certainly the most renowned—winemaking country in the world.

CHÂTEAU MARGAUX

Château Margaux is often the ultimate in first growth finesse and breed. It emphasizes all that great scent in a wine can mean. It is a wine that ages with grace and distinction, typifying the essence of <u>race</u>— **or breed—in a wine.**

The modern history of Château Margaux begins in 1925 when a syndicate which included Fernand Ginestet, purchased the property from the Duke de Tremoille. Ginestet eventually began acquiring the shares of the other members, and in 1949 his son, Pierre, completed the acquisition by selling off the family's St Emilion property, Clos Fourtet.

However, the enormous cost of running the estate, huge death-duties which had to be paid in the early 1970s and the great wine slump of 1974 proved too much for the Ginestets and the property came up for sale. The American company, National Distillers, offered a price of 82 million French francs, but reluctance to see another first growth pass into foreign hands caused the French government to veto the offer. It later approved a lesser offer of 72 million French francs from a French supermarket group called Felix Potin, which was owned by a Greek millionaire, the late André Mentzelopoulos.

The estate occupies approximately 250 hectares in the commune of Margaux and—as with many large wine-making properties—not all of it is contiguous. Some of the vineyards, such as the one which makes the white wine, are located away from the main property.

Château Margaux has a total of 80 producing hectares, of which more than 10 are planted with Sauvignon Blanc for their white wine which is called Le Pavillon Blanc. This is probably the only classed growth in the Médoc— apart from a small amount at Château Talbot—that makes a white wine, unlike their distant neighbours in the Graves where this practice is quite common. The Margaux plot planted with Sauvignon is not suited to either the Cabernet or Merlot as it consists of somewhat chalky soil, unlike the *graves* or stony soil mixture generally found on the ridge. Furthermore, no Sémillon is added to the wine, which is wholly Sauvignon.

The remaining 70 hectares of Château Margaux are devoted to the grapes of the Château wine and are planted with approximately 75 per cent Cabernet Sauvignon, 20 per cent Merlot and the remainder with Petit Verdot. However, this is not necessarily the composition of the wine in the bottle, because what is added to the final blend depends largely on factors of maturity and taste.

Quality vineyard land depends on three factors—the soil, configuration of the slopes and the sub-soil. These factors should never be considered separately as they are each necessary in combination for the proper growth and inherent quality of the wine.

The soil for the vine must be penetrable and permeable to water and air so that the winter rain can drain off and not soak the soil in springtime. If the soil is too water-logged, the rootlets will not form, the vine will suffocate and the growing cycle will be impeded from the start.

Château Margaux owes the exceptional quality of its wine to an excellent combination of these factors. The soil is relatively poor,being gravel mixed with sand, but it has excellent drainage. Except for those on the top of the ridge, the vineyards are on varying degrees of slope with good exposure to the sun. Further down the soil becomes much sandier, then a thin layer of silt is followed by a clay and calcareous mixture known as the *calcaire de plessac*.

The soil mixture of the vineyards lower down the slopes, at the level of the château itself, contains more clay and calcium like the subsoil of the vineyards higher up. According to Philippe Barré, the *régisseur*, this land is special to Margaux and produces a Merlot grape of very high quality and substance.

Much of the Margaux soil is weak, so that in poor or mediocre years the wine lacks body in comparison to, for example, a Château Latour in Pauillac, where the soil is richer. However, the wine still shows the exceptional finesse characteristic of Margaux. In good years, Margaux becomes a really great wine because all the elements and qualities in the soil are in harmony.

Selection of the grapes begins in the vineyards where the pickers pick each vine twice for the ripe grapes. An additional selection which removes under-ripe or rotten grapes, leaves and so on is made as soon as the grapes pass into the destalking-crushing machine, from where they are pumped into one of the 20 oak vats for the fermentation-maceration process which lasts on average 20 days.

Only about 8 or 10 per cent of the press wine is added to the free-run wine for malolactic fermentation in the vat. The rest is used in the Château's second wine and the table wine which the employees are permitted to buy at cost.

Margaux are convinced that all their wine needs new oak, so new barrels are always used. This is as essential to their first growth status as the beautiful oak vats that line the walls of the *cuvier*. Managers and winemakers feel that anything less is not appropriate to a great growth and that such a procedure is necessary for wine made to be laid down for a long time.

The malolactic fermentation is strictly controlled at 18°C to avoid variations in temperature; this favours greater finesse. If the temperature is allowed to fall to 14°C or 15°C, this secondary fermentation would not be completed. It is absolutely vital that this process is finished and the wine put into barrels by the end of December. For both the alcoholic and malolactic fermentation, the must is cooled, if necessary by passing it through pipes cooled by a refrigerant. According to the cellar-master, simply pumping over the must is not sufficient to lower the temperature to a reasonable level.

Both Philippe Barré, the *régisseur*, and Jean Grangerou, the cellar-master, feel very strongly that the vineyard should be

The Château Margaux label depicts the château itself. It specifies that the wine has been bottled at the château. Margaux was given premier grand cru *status in the famous classification of 1855.*

Gironde

Margaux

CHATEAU MARGAUX

Louens

Cantenac

Château Margaux is one of the largest properties in the Médoc. The estate includes woods and land for animals and crops as well as the vineyards. The château and chais *are very close to the village church. As in all estates, the vineyards are divided into small plots of land called* parcelles *for ease of working. This is a convenient way to treat and pick the vines.*

La Bégorée

N

La Côte

CHATEAU MARGAUX

Virefougasse

Soussans

Margaux

● Vineyards for red grapes

● Vineyards for white grapes

In the Commune de Soussans (ABOVE) *the vineyards are planted to white grapes. The soil here differs from that in the remainder of the vineyard area. It is rare to see white grapes grown in the Médoc.*
Some sections of the vineyard *are rested. This takes place when the vines are uprooted and before replanting. Most estates in the Médoc are made up of plots of vineyards on different sites, only rarely are long stretches of vineyards all in one piece. In the Médoc, vineyard plantings take* *advantage of the area's best gravelly soil.*

33

Philippe Barré (BELOW) *is director of the Margaux estate. He began to work for Margaux under the late André Mentzelopoulos.*

Jean Grangerou (BELOW) *is the* maitre de chais *at Margaux. He followed his father in the post. This type of family tradition is quite common at top châteaux in Bordeaux.*

The château at Margaux and its gardens (BELOW) *have been extensively restored since 1974 when the estate was bought by André Mentzelopoulos.*

allowed to rest for a long time after old vines are torn up. Depending on the piece of land, it can be left for up to seven or eight years. When land has been planted for several hundred years, it is vital to let it rest or even cultivate something else for a few years. Philippe Barré maintains that the soil must be allowed to reconstitute itself because it is obviously exhausted. Vines do not need soil that is rich, but they do need soil that is not empty. Further, if planting takes place too soon, there is the risk that any virus left in the soil, even after it is disinfected will attach itself to the new vines. There is also a type of mould called *pourridié* on old roots that will be transferred to the new vine roots if the old ones are not torn up or allowed to decompose. Château Margaux is fortunate in having enough land to allow them to give the soil a long

and beneficial rest.

A classed growth is not allowed to use the grapes from young vines within four years of planting. At Margaux young grapes are only considered for use around five or six years after planting. It may seem early compared to some châteaux, who claim not to use their new vines for 10 years, but the winemakers believe factors other than age are involved. In certain pieces of land, some vines will mature faster if, for example, the soil contains more gravel than clay. Also, the Merlot will arrive at maturity quicker than the Cabernet Sauvignon thus making it available at an earlier age. The climate also has a great influence and so there are some years when the grapes attain maturity earlier than others.

Margaux will soon market a second wine called Le Pavillon Rouge which will be released only when it is ready to drink. Under Ginestet, there was a Château Margaux non-vintage, but perhaps because of consumer uncertainty or because no-one knew when to drink it, it was not a great success and was soon taken off the market. There are varying opinions about a château marketing a second wine, but here it is felt that wine of insufficient quality for the *grand vin* is still too good to be sold off as a simple Bordeaux red. The new ownership has brought both obvious and subtle improvements to the winemaking. The most visible improvements are the four new 155 hectolitre oak vats which mean that the new wine has to be handled less. Making good wine is always very much a matter of selection. The new proprietors—and with them new money—have permitted, and indeed demanded, a greater stringency in the selection process which has given outstanding results and improved the quality of the wine.

André Mentzelopoulos (LEFT) *did much to revive the quality of the wine at Margaux and restore the château's reputation before his tragic death in December 1980.*

This very old vine (LEFT) *is growing on the pebbly soil characteristic of the area. When an old vine is no longer productive, it will be uprooted and the soil allowed to rest for up to eight years.*

The process of décuvage, *taking the wine off the lees,* (TOP) *takes place after fermentation in the large oak casks has been completed. In the background is the press which is used to press the lees and produce the* vin de presse *or press wine, which*

may be added to the free-run wine. This large chais (CENTRE) *provides excellent air circulation. It is above ground, a typical location for large Médoc cellars. The height of the cellar means that casks do not have to be stacked. The estate keeps a*

large collection of old wines (BOTTOM).

CHATEAU PALMER

Château Palmer makes almost lush, opulent Margaux wines with enormous fruit and gloss. They combine seductive charm with the ability to age. These qualities were shown to perfection in the famous 1961 but have shown with remarkable consistency throughout the 1960s and 1970s.

The modern development of Château Palmer dates from 1938, when the property was purchased by the present owners: the families of Mahler-Besse (Dutch), Allan Sichel (English) and Miailhe (French). Ginestet, who sold their share to Mahler-Besse after the Second World War, were also part of the original consortium. They are all famous names in the wine trade, and their wise and unanimous decisions have brought the wine the excellent reputation it enjoys today.

As with many of the Bordeaux estates, little is known about the origins of the land as vineyard, but the first mention of this particular property is as Château de Gascq, owned by members of the Gascq family in the early part of the eighteenth century. General Palmer, the son of a wealthy English brewer from Bath, entered the picture in 1814 when he met an attractive widow on a long journey to Paris. The lady was anxious to sell the property because her husband's inheritance had to be divided among his family. By the time they reached Paris, Palmer had purchased the Château de Gascq for one quarter of its value, a bargain due much less to the General's business acumen than to the urgency of the sale.

Palmer renamed the property Château Palmer and began to expand it. He purchased large plots of land, including one vineyard called 'Boston' and another named 'Dubignon'. By 1831, he owned approximately 162 hectares, of which half were under vines. Unfortunately, General Palmer made many mistakes, including appointing a dishonest manager. So, as the General's fortunes declined, so did the size of the property. What remained was mortgaged to the bank in 1843, two years before he died.

The very wealthy Pereire family bought the property in 1853 and, between that year and 1856, began and completed the building of the elegant and charming château which, incidentally, has not been lived in since the mid 1930s. The better business sense of this family did much to raise the quality and reputation of the wine, but the First World War, the series of poor vintages in the 1930s, and the growing difficulty of making decisions because of the ever-increasing size of the family again caused the vineyard to decline. In addition, more bits and pieces of land were sold off, reducing the total area to approximately 36 hectares, the size of the property at the time of purchase by the present consortium.

It is very important to note that in all the sales of different plots of land, the land and vineyards retained were those best for grapes. The present cellar-master and owners are convinced that it was sheer luck that the château was constructed where it was, and that the best vineyards were not sold off, because they surrounded the building or perhaps because they were situated on the highest ground. It is now broadly speaking acknowledged in viticulture that

the location and soil structure make a most important contribution to the quality of the wine. If Palmer still retained the whole 160 or more hectares, it could not make the quality wine it does, unless much of it went into a second wine. No second wine is made at this property.

The 40 hectares now under vine are situated on one of the highest parts of the gravel ridge that runs along the Gironde estuary from the best part of Graves south of the city of Bordeaux to St Estèphe where it begins to disappear. This bank is composed of gravelly, well-drained and aerated soil with a subsoil of limestone, the water-table of which provides the vine with moisture when needed during periods of growth or drought.

The ridge is at this point less than 1.5 kilometres wide. The vineyards extend 800 metres on the inland side and 500 on the river side. The château itself is next to the road from Margaux to Cantenac, but actually two kilometres from the Gironde which is not quite visible even from the top of the château's turrets.

The balance between the grapes in the vineyards has been changing over the years, but has now settled at approximately 55 per cent Cabernet Sauvignon, 40 per cent Merlot, with the remainder Cabernet Franc and Petit Verdot. Owners and winemakers believe they have now found the right mix and will stay with it.

Average production is only 25 hectolitres per hectare, and it is characteristic of Palmer that they produce very little in comparison to most other châteaux. Many of the vines are between 30 and 35 years old, and maintaining a very old vineyard is again typical of Château Palmer. Some plots are as much as 60 years old, but produce only 15 hectolitres per hectare. Beyond this stage, they are no longer profitable and are uprooted. The plot rests for a year before being replanted, and there are generally two hectares resting while 40 are growing.

Such a short rest for the soil typifies thinking in the Médoc. Everyone admits that it is economically impracticable to leave a vineyard barren for very long. However, it is also felt that modern scientific methods of treating and turning the soil mean it is no longer necessary. Furthermore, as Claude Chardon, the *régisseur*, suggests, soil that is allowed to remain fallow for too long becomes too rich, thus causing too high a yield, which is detrimental to the overall quality of the finished wine.

Palmer waits as long as possible before harvesting in order to achieve maximum maturity. If the season has been relatively poor, the pickers are told to leave the unripe grapes on the vines. Palmer is the only Bordeaux classed growth that still does the destalking by hand. The winemakers feel this brings less astringency to the wine and that the four extra workers needed for the task are well worth the cost. After picking, the grapes are crushed very lightly to

The distinctive label of Château Palmer depicts the château building. The hot summer of 1976 produced wines which had a high level of tannin.

prepare them for being pumped into the vats. In practice, this does little more than break the skins so as to induce an easier but more gradual start to the alcoholic fermentation.

The walls of the *cuvier* are lined with 14 large red oak fermenting vats. This is one of the traditional quality factors the Château intends to keep. There have never been any major problems with this type of vat at the Château, so they see no reason to change. Once a vat is worn out, it is replaced. The newest is only a few years old.

A relatively long fermentation-maceration period of 25 days on average ensures colour, tannin, extract and aroma. The press wine is added to the free-run wine for malolactic fermentation which is also carried out in the vats. The wine is then pumped into Bordeaux *barriques* where it matures for almost two years before bottling.

According to the *régisseur*, too many new barrels for the new wine give it too much tannin from the wood. All winemakers have their own views on this, but Château Palmer strikes a balance fairly common in the Médoc by using 50 per cent new barrels each year. 'Any more', says Chardon, 'and the wine could take on an oaky taste not looked for in the wines of this region.'

Another quality factor is the decision to keep the barrels' bung-holes on top until July or August. This practice enables more air to reach the wine and it is felt that this oxygen is necessary to the development of the young wine. This is an expensive practice, because evaporation means the barrels must be topped up every two days. Putting the stoppered hole to the side stops evaporation and thus does not necessitate the addition of new wine. The wine is not filtered and it is only fined with egg white. This is best for red wine because it adds suppleness while maintaining that important element, finesse.

Château Palmer is one of the few properties to enjoy the long experience of several generations of *régisseurs* and cellar-masters from the same family. Claude Chardon recently took over the title from his now retired father Pierre. His grandfather began working at Palmer no less than 103 years ago, and Claude's son is being trained to take over his father's duties eventually. This will make four generations from the same family. Claude's brother Yves also works as a winemaker.

Claude Chardon enjoys talking about winemaking. He feels that 'The way to make good wine is with love. The oenology of a great wine is not brought about by an oenologist . . . the real quality is brought to the wine by the soil and by nature.' What do the owners and cellar-master look for in a wine? Claude Chardon replies 'Always the very best. Beautiful colour and suppleness —a complete wine. Each year we hope it will be the very best.'

*Yves Chardon (*LEFT*) is part of the winemaking family who have been associated with Château Palmer for several generations.*

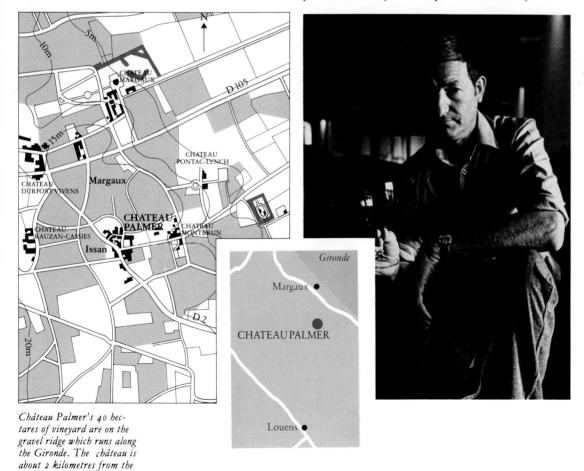

Château Palmer's 40 hectares of vineyard are on the gravel ridge which runs along the Gironde. The château is about 2 kilometres from the river.

These secateurs (BELOW), used for cutting the grapes, are ready to be cleaned before the harvest. At Château Palmer picking takes place as late as possible so that the grapes are at maximum maturity.

These containers called hottes (BELOW) are carried on the backs of pickers who walk up and down the rows of vines collecting the grapes gathered into baskets by other pickers. The contents of the hottes are then put into a trailer or cart and taken to the chais.

The flags flying from Château Palmer (BELOW) reflect the estate's multi-national owner-ship. The estate has been under joint French, Dutch and British ownership since 1938.

This sample tap on the fermentation vat (BELOW) is
used for taking samples for
analysis and tasting during
fermentation.

At Château Palmer, large
red painted oak fermentation
vats (BELOW) are used. There
are 14 vats in all. This traditional practice is felt to aid
the quality of the wine. Many
châteaux today use stainless
steel vessels for temperature
control and cleanliness.

At Palmer the grapes are destalked by hand. The grapes
are rolled gently on a wooden
grille (LEFT), the grapes fall
through with their skins
slightly broken, while the
stalks remain on top. This
manual destalking has almost
died out, while this type of
traditional wooden press (FAR
LEFT) has now been largely
superseded by more modern
presses.

There is a separate fermentation graph (LEFT) for each
fermentation vat to show the
density (sugar content) of the
wine. The downward direction
of the graph shows how the
sugar is turned into alcohol.
The graph is monitored daily,
and, finally, the date of devatting the wine is recorded.

When taking a sample from a
cask which is on ¾-bung
(ABOVE), a wooden pin is
removed and the flow of the
wine from the hole is caused
by exerting pressure on the
cask head using the hammer as
a lever.

CHATEAU DUCRU-BEAUCAILLOU

**Château Ducru-Beaucaillou is St Julien wine par excellence.
Always noble and refined in taste, it has enormous fruit and harmony
with the balance—imparted by both the estate's beautiful situation and
the skilful winemaking techniques—to age majestically.**

The domain was originally called simply Beaucaillou which means 'beautiful pebble', but it was purchased by someone called Ducru in the early nineteenth century and, as was the fashion of the time, his name was added to the label. In 1866 the Johnston family, who were of Irish origin and had made their fortune from Bordeaux wines earlier in the century, purchased the property and added the two square towers at both ends of the single storey château built by the Ducru family, which is pictured on the label.

The Johnstons sold out to another Bordeaux *négociant* called Desbarta de Burke in 1928. In 1942 Francis Borie, grandfather of the present owner, Jean-Eugène Borie, purchased the property. Jean-Eugène Borie owns several other vineyards as well.

He is directly responsible for the winemaking at all his properties and he can take the credit for the fact that Ducru-Beaucaillou is considered one of the top Saint Juliens today. The cellar-master is André Prévot. The famous French oenologist, Professor Emile Peynaud, acts as consultant.

The château itself is the most easterly in the area and is only a short distance from the River Gironde. The 43 hectares of vineyards which surround the château to the north, west and south are at the eastern edge of the undulating gravel plateau which forms the plantable vineyard land of this *appellation*. This position adds greatly to the soil drainage. This is an important factor in the quality of the Médoc growths because there is heavy rainfall in this region. It is vital for good vineyard land in this area that the rainwater should filter through the topsoil, which is usually several metres thick, to the subsoil which is usually made of marn-limestone. The level of the watertable varies according to the time of year from the lower level of the gravel bed to the upper level of the subsoil. Drainage from the Beaucaillou vineyards is exceptionally good because they are at an altitude of about 14 metres, overlooking the low-lying area next to the river.

The presence of gravel in the soil is very obvious in the Ducru-Beaucaillou vineyards. The stones are the size of hens' eggs and permeate the soil to a depth of several metres. The pebbles are made of quartz, silex, limestone, chaille and alterite. Further down there is a layer of clay and below that one of limestone.

A detailed cross-section shows the typical soil structure of the Saint Julien area in greater detail. A thin layer of light grey coloured sand and gravel on the surface is followed by a narrow band of marn. At a depth of between 30 centimetres and 60 centimetres, there is a hardened layer of finer gravel which gives way to a dark grey zone of coarse gravel down to a depth of 120 centimetres. Beneath this there is another thin, hard layer of sand and, further down still, more sand which contains patches of blue, yellow and rust. This layer rests on top of another layer which consists mainly of clay.

In the Ducru-Beaucaillou vineyards about 65 per cent of the vines are Cabernet Sauvignon, 25 per cent Merlot and the remainder Cabernet Franc with a small amount of Petit Verdot. The best quality vines are the older ones and for this reason they are kept for as long as possible. However, they have to be uprooted when they are 50 years old and sometimes sooner. At the other end of the scale, the new vines are usually at least 12 years old before the grapes are incorporated in the *grand vin*. When the vines are younger than this, the wine is made separately and sold simply as Saint Julien.

Jean-Eugène Borie believes that the density of the vines in the vineyard helps to determine the wine's quality. He believes that for a red wine to have a full flavour and strong character, the grapes should be relatively small. If there are too few vines over an area, each one will be too vigorous and the grapes will be too large. This makes the juice less concentrated and characteristic. There is a clear connection between the quantity of skin and the juice of the grape.

Borie also believes that the high quality of the tannin has a major bearing on the quality of the classed growths. He tries to vinify his wines to include a generous amount of tannin because he believes this ensures the richness of the wine and adds to its longevity.

The cellars at Château Ducru-Beaucaillou contain 22 concrete vats with an average capacity of 180 hectolitres. They are lined with epoxy and a layer of calcium tartrate. To Borie, the material from which the vats are made is not important, but it is essential that the vats are clean. Concrete vats have the advantage that they are air-tight and can be cleaned easily with steam and hot water which are applied under high pressure.

Concrete is not a good heat conductor. This is beneficial for the malolactic fermentation which needs warmth. However, during the preceding stage of the alcoholic fermentation, there is the risk of overheating. At Ducru-Beaucaillou, the winemakers regularly hose down the vats and, if the temperature reaches between 28°C and 30°C, the wine is passed through a cooling tube into another vat.

The stalks are removed completely from the grapes, as is the practice in almost the whole region. This reduces the volume of the harvest by about 30 per cent, eliminates the possibility of a green or stalky taste, increases the alcohol content by as much as half a degree and deepens the colour, as the stems absorb colouring matter. These advantages far outweigh the disadvantages which are that it is more difficult to press the *marc*, fermentation is slower and the acidity level is higher.

The distinctive label of Château Ducru-Beaucaillou shows a picture of the château building. The two square towers were added in the late nineteenth century.

Jean-Eugène Borie, the owner of Château Ducru-Beau-caillou (BELOW), and his family live at the château and are one of the few owners of classified châteaux in Bordeaux to live on their estate. This means that the winemaking is very personal.

The vatting period—including alcoholic fermentation and maceration—varies from year to year. However, at Ducru-Beaucaillou it usually lasts longer than the average of three weeks prevalent in the rest of Médoc. This happens particularly in lighter years when more tannin is needed to give the wine structure. The winemakers want to produce wine which is to be drunk at least seven or eight years after the harvest, while some are laid down for up to 20 years.

Another variable factor in the quality of the vintage is the amount of *vin de presse* which is added. This wine adds tannin, body and backbone. Less is needed in a year when the juice is more concentrated than when the harvest produces lighter wines. The *vin de presse* is blended in after the malolactic fermentation and before the wine is put in the barrel.

At Ducru-Beaucaillou between 30 and 50 per cent of the oak casks are new. Borie feels that the wine cannot take more. The wood imparts flavours which complement those natural to the wine, but it is felt that too much new wood would give the wine a woody taste.

The wine is stored in cellars beneath the château where the family lives for most of the year. This happens very rarely in the Bordeaux region because the storage cellars are usually in or near the winery and therefore separate from the château.

Racking takes place when tasting indicates that it is necessary and at intervals of between six weeks and three months. The barrels are rolled to one side—*bonde de côté*—during May of the first year. This allows the wine to develop with far less exposure to oxygen. For fining, five egg whites per barrel are used. These are added during the second winter and racked off one month later. Bottling takes place the following June and July.

Production varies tremendously from year to year. In 1977 it was as little as 75 *tonneaux* and in 1970 it was as much as 220 *tonneaux*. The latter is luckily probably the best vintage in that decade. Ducru-Beaucaillou has no second wine, and any wine below the *grand vin* standard is sold as generic Saint Julien.

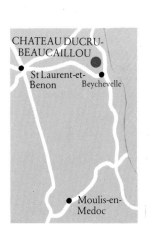

CHATEAU DUCRU-BEAUCAILLOU

St Laurent-et-Benon

Beychevelle

Moulis-en-Medoc

The château is surrounded by its 43 hectares of vineyard. The estate is close to the river and thus benefits from the best of the gravel on the Médoc plateau.

These traditional Bordeaux casks (FAR LEFT) have a capacity of 225 litres. The first year maturing cellars are directly under the main part of the château at ground level with the second year cellars below ground level under both wings of the château. Some of the vines at Ducru-Beaucaillou are up to 50 years old (LEFT). Vines of this age do not yield much quantity but give good, concentrated quality.

The château and gardens at Ducru-Beaucaillou (LEFT) are among the most carefully maintained in the whole of the Bordeaux vineyard area. The long, low architectural style is typical of the Gironde.

CHATEAU LÉOVILLE LAS-CASES

Léoville Las-Cases is a second growth in the front rank of Bordeaux's top wines. The wines marry the immense fruit of St Julien with weight and backbone, giving them the ability to open out gradually. They are wines of immense class and great depth of bouquet and flavour.

Both land and buildings of Léoville Las-Cases were owned by the Abbadie-Léoville family until 1900, when a majority shareholding was sold to a group of Bordeaux businessmen that included Théophile Stawinski, the grandfather of the present owner. Between that year and 1930, Stawinski purchased the other members' shares and formed the *Société Civile de Léoville Las-Cases*. Today Paul Delon owns a 63 per cent share and descendants of the Marquis de Las-Cases 37 per cent of the property.

The estate is now managed by Michel Delon who took over in 1976 when his father became ill. He takes an active part in the winemaking. His assistants are Jacques Depoizier and Michel Rolland, the cellar-master.

The vineyards of Las-Cases, which comprise about half of the original Léoville property, are in several parts, but the most important lies in a stone-walled enclosure between the road from Bordeaux to Pauillac and the Gironde about a kilometre away. This 45 hectare area—part of the famous Médoc gravel bank—is separated from Château Latour and the commune of Pauillac by a small stream known as the Juillac.

Most of the remaining 35 hectares are located across the road in two groups between Château Pichon-Longueville-Lalande and Château Léoville Poyferré. The soil is sandier and less stony than in the enclosed vineyards. A third plot, called Le Petit Clos, lies to the south of the commune and next to Léoville Barton.

Total potential vineyard area is 80 hectares, but of the 75 actually planted, only 70 are producing as five are not yet four years old. This is a legal requirement for French *appellation contrôlée* wines. Grapes are not used for the Château wine until their quality—assessed by tasting the finished wine—is high enough. Vines less than seven or eight years old cannot normally produce grapes to the required standard, so their grapes are used to make the second wine, Clos du Marquis.

Approximate composition of the vineyards is 65 per cent Cabernet Sauvignon, 15 per cent Cabernet Franc, 15 per cent Merlot and less than five per cent Petit Verdot. Proper care of the vineyards, essential for good wine, involves ensuring good drainage, disinfecting where and when necessary, and providing adequate nourishment. The soil must be properly treated after uprooting, otherwise the new young vines can be killed by the disinfectant. This can happen if the soil is not worked enough or if there is too little rain. Analysis of the soil allows appropriate steps to be taken to prevent this. A helpful sign that the disinfectant has disappeared is grass growing on the plot. Scientific viticultural techniques today permit planting in the spring following an uprooting in the previous autumn, but Las-Cases takes the double precaution of waiting a year and a half before replanting, as well as using the latest soil-treatment techniques.

The rootstock is the widely-used riparia-glory and the '101-14' rootstock from the same family. These have given best results in these vineyards and satisfy the winemakers' desire for stock that does not push production and gives the wines good tannin, colour and finesse.

Old vines make good wine, but only if they are healthy. Each plot is carefully monitored for quality and quantity. If production falls to 15 hectolitres, and the vines are in a poor state, the plot is torn up. However, if they are healthy, they are left, because the winemakers prefer a small quantity of high quality wine to a large quantity of lesser distinction from younger vines.

The Château's second wine, Clos du Marquis, is produced from the young vines and wine judged of insufficient quality for the Las-Cases. However, in some years the quality of the two wines is similar because, in order to produce a marketable quantity, Delon is forced to sacrifice one or two vats of wine destined for the Las-Cases label. In an average year Clos du Marquis amounts to five or six per cent of the year's harvest.

When the grapes are brought in by Spanish pickers, a rigorous selection is carried out by four people stationed around the bin into which the harvest is emptied for towing to the winery. One of the winemakers stands at the entrance of the bin leading to the Amos destemmer-crusher keeping an eye on the quality of the grapes and looking for leaves and other debris. A large stainless steel screw conveys the grapes to the destemmer. The speeds of both the machines can be adjusted. Delon believes that this control is necessary and that each grape variety requires a different speed.

Yeasts are added to the first few vats to promote fermentation. They are of the genus *saccharomyces*, species *cerevisiae* or *byanus* and can resist the numbing effect of alcohol and can transform the sugar into alcohol up to a volume of approximately 13 per cent. As soon as fermentation begins, pumping over or *remontage* is carried out to aerate the must and provide the oxygen which the yeasts need in order to reproduce. Chaptalization is recommended in moderation. This procedure is generally practised, especially for vintages which are low in natural sugar.

The main winery contains 11 wooden vats ranging from 160 to 260 hectolitres in capacity, of which four are less than 10 years old. According to Delon, the price paid for using wood should be measured in terms of convenience rather than wine quality. Many of the classed growth proprietors feel the same way, but some, such as Lafite, Margaux and Mouton, continue to use the wooden vats because the public expect them, and they are prestigious. Las-Cases has several concrete vats which are also

Léoville Las-Cases was no exception to the high quality of the 1978 vintage throughout Bordeaux. The wine develops slowly and takes 10 years or more before reaching its full potential.

The consistently high quality of Château Léoville Las-Cases is the result of close teamwork. Two of those responsible for the winemaking are Jacques Depoizier, the owner's assistant, and Michel Rolland, the maître de chaîs (BELOW).

used every year and the winemakers claim they have never noticed any difference between wines from the two sorts of vat. The most important factor is cleanliness of all material and equipment.

The harvest should be completed in a maximum of two weeks and the vatting period in between two and three weeks. The duration of the latter depends on fermentation and maceration, which in turn is determined by the nature of the vintage and even the grape variety. Delon tends to run off the wine from the older Cabernet Sauvignon vines before the younger and allows the latter to macerate a day or two longer. This is done according to the vintage. For example, in some years the must from older vines can absorb too much tannin, whereas in lighter years it might be desirable to extract as much tannin as possible.

Wooden shovels are used to transfer the *marc* from the vats to the press so as not to damage the pips and release bitter tasting elements into the wine. The *marc* is pressed twice, about 60 per cent being extracted from the first press and about 30 per cent from the second. These wines are aged separately in barrels and blended with the free-run wine during the second year. The proportion of press wine added to the free-run ranges from 12 to 15 per cent, depending on the quality of the vintage.

Following malolactic fermentation in a second vat, a preliminary blending takes place before the wine is put into barrels. The winemakers believe that the optimum number of new barrels depends on soil constitution more than vintage. The present proportion is 35 per cent. They are now experimenting with 40 per cent to see what effect this has on the wine, but Delon knows that much more than this would give the wine a 'woody' taste.

The barrel's bung-hole is rolled to the side in May. Rackings are performed four times in the first year and three in the second. The last is at bottling during June and July of the second year.

Delon has very clear-cut ideas about what makes a successful growth. He is convinced that it is possible to produce the maximum allowed yield (40 to 45 hectolitres per hectare) while maintaining top quality. He is, however, well aware that this quantity cannot be obtained without extremely well-kept vineyards, minute attention to detail and cleanliness at all levels. He turns to advantage the best that contemporary technology and business methods have to offer. In this way he epitomizes what a modern grower and winemaker can achieve. His aim is to make the wine of Léoville Las-Cases known through its quality.

For Michel Delon and his assistants, the wine of Léoville Las-Cases evolves very slowly, taking 10 years or more to reach its full potential. They attempt to make wine that is supple, round, well-balanced and seductive. Michel Delon is very sensitive to the intensity and finesse of the bouquet, and, judging from tastings of many different vintages, a pleasant spiciness seems to be a recurring characteristic of the wine.

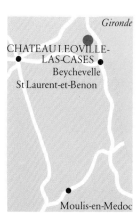

Gironde

CHATEAU LEOVILLE-LAS-CASES
Beychevelle
St Laurent-et-Benon

Moulis-en-Medoc

This section of the Léoville Las-Cases vineyard (LEFT) shows its proximity to the river Gironde. Most of the best gravel ridges in the Médoc are close to the river with the amount of gravel lessening further inland.

The estate buildings at Léoville Las-Cases (FAR LEFT) are relatively modest. The main chais is on the other side of the road. The chef de culture, M Nemetz (LEFT), is holding some of the large pebbles characteristic of the soil, particularly in the enclosed section of the vineyard.

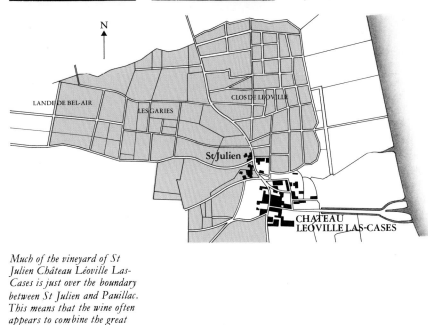

N

LANDE DE BEL-AIR   LES GARIES   CLOS DE LEOVILLE

St Julien

CHATEAU
LEOVILLE LAS-CASES

Much of the vineyard of St Julien Château Léoville Las-Cases is just over the boundary between St Julien and Pauillac. This means that the wine often appears to combine the great fruit and charm of St Julien with the body and structure of Pauillac.

CHATEAU LAFITE

Château Lafite is a wine that magically combines intensity of flavour, the ability to age with enormous distinction, and a delicacy and breed that makes it the epitome of a first growth. Mature Lafite always has a complex, penetrating bouquet and great length on the palate.

Baron James de Rothschild purchased the Château Lafite property in 1868 and it has remained in the Rothschild family ever since. The Baron died in Paris without ever having visited the château, but his widow restored and extended the existing building after moving there from Paris. The oldest part of the château was built in the sixteenth century.

At present, Château Lafite is owned by four members of the Rothschild family, Baron Elie de Rothschild, his brother Baron Alain, and two cousins the Barons Guy and Edmond. Baron Elie is the president of the group, but a few years ago delegated his responsibility for administering the Château to Baron Eric, the son of Baron Alain. Eric spends much of his time in Paris, and is assisted by Yves Le Canu who has responsibility for marketing, public relations and publicity.

The resident manager is Jean Crété, who supervises the technical operation from the plant to the bottle. His cellar-master is Robert Revelle. There are 70 employees at Château Lafite, including a blacksmith, painter, mason and electrician. In all, the estate has about 14,000 square metres of covered floor space which include the château, *chais*, storage, and workers' quarters. About half the employees actually live on the property.

Lafite is not far from the Gironde river, and Jean Crété believes that the wines of properties close to the river have more finesse than those further away. The moisture in the air as well as the soil and the proximity to the Gironde help create the best conditions for making great wine.

Lafite has 88 hectares of producing vineyards. Even though the soil is quite gravelly, there are variations in soil composition in different parts of the vineyard which produce certain particular characteristics in the wine from that area. The grape varieties are chosen according to the particular plot, but the balance is about 65 per cent Cabernet Sauvignon, 20 per cent Cabernet Franc and 15 per cent Merlot. The rootstock or *porte-greffe* is generally the widely-used *riparia* except for the extra dry or calcareous plots where it does not grow well.

Density of the vines is quite important, and the wine-makers believe this is necessary to maintain development and vegetation of the plant at a low level and to ensure high quality. They plant between 8,000 and 8,500 vines to the hectare. Density must be maintained at a relatively high level, or the vine may grow too vigorously and too long, whereas the grapes on a vine which 'suffers' a little will tend to ripen sooner and thus benefit more from the sun in the months of September and October.

Selection is also important at Lafite. Good, healthy, ripe grapes are necessary, and pickers are made to throw the others away. This is one of the reasons why only families of the employees and people from the region are used as pickers. Harvesting takes two to three weeks, depending on the size of the harvest, the weather, and the efficiency of the team. It is usually finished by 10 October and often sooner.

In the view of the *régisseur*, Jean Crété, another important point is the order in which the grapes are picked. The Merlot, which takes about two days to pick, is ready to be harvested about eight days before the others. With the Cabernets, there are always some plots of land which are in a more advanced state than others because the grapes on a vine will ripen in reverse proportion to the quantity of the fruit. This means that the more grapes there are on the vine, the longer the ripening process takes. It is also very important to harvest a plot of vines at the best moment, and, to this end, an acidity-sugar analysis begins two weeks before maturity to follow the development of the grapes.

The winemakers believe that it is important not only to keep each variety separate during fermentation, but also to pick according to the age of the vine and the nature of the terrain. It thus becomes necessary to move around the groups of harvesters as each different *cuve* is being filled in order to keep the must as homogeneous as possible, and also to see that any change in the quality of the grapes coincides with a change of vat. This is often rather difficult.

At the end of the vinification process there are as many different types of wine as there are vats—27 in all. This procedure allows the winemakers to make the best selection. The quality of the wine in each vat is monitored very closely, although the winemakers know through experience which vineyards produce the best. The wines not chosen for the *grand vin* are used for the second wine which is called Moulin des Carruades.

One particularly successful practice adopted at Lafite is to employ a small team to gather the ripe Merlot two days before the main harvest begins. Two vats are thus filled, allowing seeding with yeasts to begin the fermentation. This is common practice in the Médoc, as the first few vats are always difficult. This practice also allows the winemakers to try out the equipment, so that when the full group of pickers arrives, fermentation and equipment are already working with a steady rhythm.

In spite of scientific methods which are able to reconstitute the soil in one or two years, Lafite prefers to allow the vineyard a more natural regeneration using manure, some fertilizer, grass and up to four years of rest. Watching how well the grass grows is often a way of testing the quality of the soil.

It is six years before the vines are old enough to yield grapes for the *grand vin*. Like most classed growths, they try to retain the old vines, and, according to the *régisseur*, they are normally kept for up to 60 years. There is even a plot with vines 100 years old.

The wines of Château Lafite have elegance, bouquet and depth. The excellent autumn weather helped create a particularly good vintage in 1978.

The winemakers are very careful to avoid excessive tannin, and the fermentation-maceration vatting period is adjusted according to the quantity of tannin in the vintage. A light year's vintage will remain in the oak vats for between 20 and 22 days.

When the wine is drawn off the skins there is a preliminary blending, but only of wine of similar quality and grape variety. This selection is made by tasting and helped by the vineyard's consultant, Dr Emile Peynaud, in order to determine the qualitative category of each vat. The wine is then put into clean wood vats to undergo malolactic fermentation. This secondary fermentation takes about a month, and if there is any difficulty in starting the process, a small quantity of bacteria is added.

After the wine is made, a sample is taken from each vat. Eric de Rothschild, Dr Emile Peynaud, Jean Crété and Robert Revelle sit around a table, analyze each one for colour, nose and tannin, and determine which will be Lafite or Moulin de Carruades. In some years, the selection leaves comparatively little for the first growth because it is essential to maintain a certain level of quality for the Lafite.

Lafite is aged in new barrels. Château Lafite and Château Margaux are the only châteaux to have their own cooper-age. This enables the winemakers of Château Lafite to select the wood they prefer from the French forests of Allier and Tronçais. There are different qualities of wood, and the winemakers believe these two types possess an aroma of tannin that marries best with the wine of Lafite. Enough wood is ordered a year in advance to make almost 1,000 barrels per year. A proportion of the wine destined for Carruades is also conserved in new barrels, but the wine of Duhart-Milon—another Rothschild property—is not.

According to Jean Crété, Lafite normally does very well in the new wood but he is beginning to think that the wine of a light vintage would be better in a smaller proportion of new barrels. After bottling, the casks are kept for Carruades or Duhart-Milon, after which they are sold to small growers in the region.

The glass bung is left on the barrels until the first racking, which takes place after three months. It is then replaced by a wooden plug. The glass is used to allow the carbon dioxide gas in the new wine to escape until winemakers feel enough has left the wine to roll the barrels to one side. This greatly lessens evaporation.

The storage *chais* still have a hard earth floor because it is felt that this breathes well and gives a better aroma than

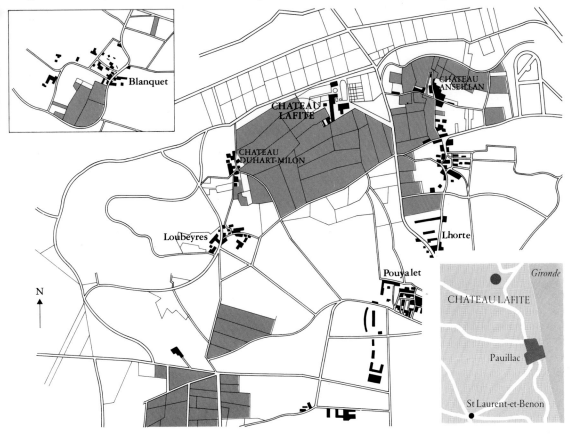

Château Lafite has one of the largest areas under vine in the Médoc. The part near the chais has, for the Médoc, a relatively steep slope. The size of the vineyards means the estate needs to employ large teams of pickers.

*Jean Crété (BELOW) was for
some years régisseur at
Léoville Las-Cases. He has
worked at Château Lafite
since late 1975, and 1976 was
the first vintage he supervised
from the start. The vineyards
and chais can be seen in the
background.*

concrete. The compacted earth ensures more regular humidity for ageing in the second year.

Rackings are performed every three months the first year and every four months during the second year, when the wine is in the second *chais*. Fining with egg-white takes place in January or February of the second year and is left to act on the tannin for 45 days. This process gives the wine a degree of suppleness. After racking from the albumen, the wine is generally clear enough to need no further racking before being bottled in July.

At Lafite the average yield per hectare is rarely more than 25 hectolitres—this is because the further a vineyard rises in the hierarchy of classed growths, the more sacrifices are necessary to maintain this exalted position. What makes Lafite a great wine however, is not any trick by the cellar-master but a combination of conditions, procedures and knowledge that allows the winemakers at Château Lafite to extract the best from the climate and soil.

For Jean Crété, the qualities of Lafite are its elegance, bouquet and aroma. Other châteaux can obtain more powerful wine, but what is difficult to find elsewhere is the combination of elegance, distinction and longevity characteristic of Lafite.

*Lafite has one of the largest
collections of old wines in
France (BELOW). The bottles
are carefully recorked at
intervals. These examples
come from pre-phylloxera
years. The wines may still be
in fair condition as they have
not been moved.*

Robert Revelle (BELOW) *is*
maitre de chais *at Lafite.*
Behind him, the immense size
of the cellars can be seen. The
casks await the new wine in
January. At Lafite, all new
wine is put into new casks.

At Lafite, the château (LEFT)
itself is small and compact. It
is dominated by the huge
working chais.

Drainage channels (FAR
LEFT), *both natural streams*
called jalles *and man-made*
canals, help drain the vine-
yards. This is a typical pro-
cedure in the Médoc. In the
steepest part of the Lafite
vineyards (LEFT), *the large*
pebbles characteristic of the
soil in the Médoc can be seen.

Lafite is one of only two
châteaux to make their own
barrels, the other is Margaux.
This practice used to be more
common, but expense has
made it impractical for most
châteaux. Each year Lafite
makes enough barrels for the
new wine (LEFT).

CHATEAU LATOUR

The wines of Château Latour have a power, depth and capacity for ageing that indicate a first growth worthy of the classification, epitomizing wines from the commune of Pauillac and those founded on the Cabernet Sauvignon grape. Always of dark dense colour when young, the wines of Château Latour mature to complexity with majestic proportions.

Château Latour is managed by Jean-Paul Gardère, but is owned by the *Société Civile du Château Latour*. This has belonged to English interests since 1962 when the majority shareholding of 52 per cent passed to the London Pearson group. Harveys of Bristol retain 25 per cent and the rest is dispersed among French descendants of the Marquis de Ségur, owner of the château until the French Revolution when it was confiscated by the state. The family re-acquired the estate in 1840 and it was soon established as a company under the family's ownership and remained so until the sale to a group representing the English Viscount Cowdray.

A wine's quality derives mainly from the soil but, according to Jean-Paul Gardère, it is the great mystery of Latour that the wine is better than its neighbours: 'A wine is explained through its soil, grape variety, climate and the winemakers; the rest is literature. But the *grand vin* (first growth Latour) has the privilege of nobility of the soil that one can neither add nor subtract.' There is a large diversity in the soil, but its most important component is clay in addition to the layers of gravel.

Latour makes at least two and often three or four classes of wine. The *grand vin* is produced strictly from old vines, with a minimum of eight years of age. The second wine— Les Forts de Latour—is made from grapes of vines between four and eight years, and others that are not up to the quality of Latour. Even so, the winemakers feel that Les Forts should be accorded second classed growth quality. This necessitates making in many years a third wine labelled simply Pauillac Appellation Contrôlée. The fourth wine, utilizing whatever is left, is never sold, but drunk by the personnel as table wine.

Les Forts de Latour is made from grapes from two vineyards apart from those surrounding the château. The first, of less than a hectare, is located between the vineyards of Pichon-Longueville-Baron and Léoville Poyferré (St Julien). The second, containing about 4,500 vines, is located between Batailley, Haut-Batailley and Pichon-Longueville-Lalande. The managers of Latour consider these plots to be on a par with the second growth land of Latour, but not with the first growth.

Selection of the different wines is never decided on before the end of the summer following the harvest because it is felt that for a wine to reveal its ultimate quality it must first pass through the test of winter and summer. Even then, because Latour never sells *en primeur*, there are several years in which to follow the wine's evolution and 'downgrade' if this is felt to be necessary.

Vines are identified by a yellow tag until they are eight years old; any dead vines are replaced individually. A special team harvests the younger vines before the older ones so that the regular harvesters—local

people and families of the workers who live on the estate— will make no mistake. Château Latour is the only classed growth to harvest in this fashion.

The method of replanting vines individually is known as *jardinage*, and is practised quite extensively at Latour. The aim is to conserve the old vines as long as possible in the belief that even a low volume will yield high quality extract of exceptional richness. As the policy is one of quality rather than yield, when a group of vines is uprooted it is because the old vines are moribund. If 30 to 40 per cent still produce some grapes the plot is not uprooted. In the 18 years Jean-Paul Gardère has managed the domain, only about two hectares have been torn up.

When a plot has exhausted its potential, no account is taken of the young vines that have been planted in *jardinage* and every vine is uprooted to let the ground rest and be given proper treatment for two or three years. This policy was inherited, because when Gardère and Henri Martin (now proprietor of Château Gloria) arrived in 1962, they found themselves with 60 hectares of vines of all ages. They knew it would take several generations to rationalize the vineyard, but the only other solution was complete replanting of the whole of the vineyard.

For this reason it is impossible to state the exact composition of the vineyard. With 60 hectares and therefore over half a million vines, it would have been a monumental task to identify each variety at the start, for the fact remains that even though different plots were of Cabernet Sauvignon, Merlot and so on, in practising *jardinage*, the *chef de culture* often planted the vine or vines he had on hand. It is estimated, however, that 75 to 80 per cent is Cabernet Sauvignon, 10 per cent Cabernet Franc, and five per cent Merlot, with some Petit Verdot and Malbec. Production depends, of course, on the vintage, and varies between 200 and 250 *tonneaux* for the entire vineyard. In general, between half and two-thirds becomes Château Latour, with the rest divided among the other wines which are produced.

Latour was the first château in the Médoc to install stainless steel vats. A total of 14 were purchased in 1964 and an additional five in 1967. For the winemakers it is the perfect material, allowing an easier control of fermentation temperature and—even more importantly—allowing the vats to be kept clean and always ready for use. As Gardère so correctly points out, it is not the type of material that is of primary importance, but its cleanliness. In order to make good wine, good, healthy grapes and clean, efficient equipment are needed.

For the conservation and maturing process, however, Latour needs new casks. Jean-Paul Gardère feels that the stainless steel vats represent progress and the oak casks tradition. 'The tradition of today is the progress of yesterday and the progress

The rather austere label of Château Latour contrasts with the wine which is deep ruby red in colour and has power, unusual richness, breed and class. It ages superbly.

The Château Latour estate is in the Pauillac area near the Gironde.

of today is the tradition of tomorrow,' he says. It would be perfect if new wooden vats could be used each year because the wine is definitely influenced by the wood at this stage. However, this is economically impossible, and instead all the barrels are new each year.

The wood from the barrels gives the wine of Latour a nobility and richness it might otherwise lack. The proportion of new barrels used for Les Forts de Latour depends on the texture of the wine chosen, but ranges from 30 to 50 per cent.

Methods of vinification are classic. The grapes are brought into the winery in easily cleaned plastic tubs and then completely destalked and crushed before being pumped into the vats. The active phase of fermentation lasts from six to 12 days with a further maceration of 10 to 12 days. Pumping over of the must depends on the rapidity of fermentation, but averages four times in this period.

The whole operation takes three weeks, running off one vat each day. The *marc* is pressed by a pneumatic cylindrical press to yield the first *vin de presse*. This amounts to four or five barrels, but the quantity of press wine is about 20 per cent or nearly 12 barrels per vat. Only the first press is used in the *grand vin*; the rest is reserved for Les Forts de Latour, revinified the following spring with the wine for Les Forts, or used in one of the lesser categories, according to its quality. Press wine is rather like pepper in the kitchen—some is necessary, but not too much. It adds tannin, richness and structure to the wine.

The malolactic fermentation generally happens of its own accord *sous marc*, which means while the wine is still in the vats, immediately following the alcoholic fermentation. However, whether it is finished or not, the wine is put directly into barrels when drawn off. The winemakers do nothing to bring it about and consider it a natural phenomenon that begins and ends when it is ready. In 1979, this secondary fermentation of two-thirds of the wine was completed in the vats.

Blending of the different lots depends on the wine's evolution, but is generally performed before March of the year following the harvest. If there is an unresolved question about the press wine, it can be delayed, but is always added before June. The cask's bung-hole is rolled to the side in June or July before the summer heat sets in. Racking is carried out every three months and follows the general Bordeaux pattern.

Fining, performed in December and January of the following year, is also done in the classic manner. The wine, which has remained closed up until then, suddenly opens up and takes on colour. Wine will sometimes become 'younger' following the fining. Its development after this operation determines when bottling takes place, but September is the usual month for this. For the generally younger wines of Les Forts de Latour, bottling is brought forward to June or July to conserve the fruit and body.

For the winemakers at Latour, there is no mystery or

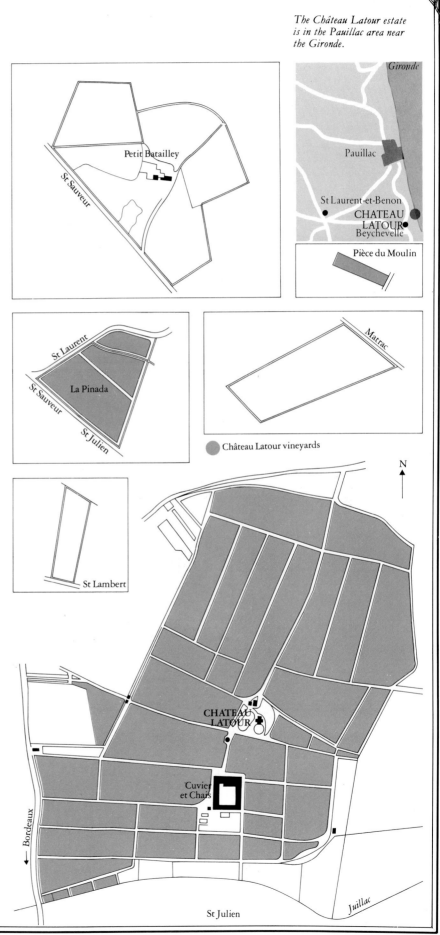

Pièce du Moulin

Château Latour vineyards

49

This plot of Cabernet Sauvignon vines (BELOW) shows the small château at Latour in the background. Latour has a large proportion of Cabernet Sauvignon vines. This contributes to the firm backbone and power which is characteristic of the wine.

Fermentation takes place in large modern stainless steel vats. Latour was the first Médoc château to purchase such vats and now has 19 in all. Stainless steel is easy to keep clean and fermentation temperatures can be exactly controlled (RIGHT).

secret formula behind the wine's quality. The explanation lies in the exceptional grapes supplied by the soil and nature—a discovery three centuries old. All the winemakers do is vinify these extremely high quality grapes with care and expertise. The wine does not make itself, but its quality is inherent.

What are the qualities of Château Latour? In the view of Jean-Paul Gardère, the wine has deep ruby colour, power, unusual richness, generosity and longevity. It is a wine that has superior breed, class, distinction and nobility. Château Latour was classified as a *premier grand cru* in 1855. The wines are among the best produced in the Pauillac area from the Cabernet Sauvignon grape. Château Latour is one of the world's great red wines.

Latour's oenologist, Jean-Louis Mandrau (RIGHT) tests for acidity at all stages of a wine's life. The level of acidity must be particularly carefully monitored during fermentation, although the must is also tested for sugar content. Wine with too low an acidity can be rather fragile. It is also increasingly being recognized that acidity, as well as sugar, is an important factor in deciding the best time to pick the grapes.

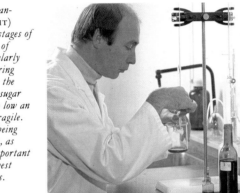

Château Latour produces a second wine from the property called Les Forts de Latour (RIGHT). This example shows a heavy deposit in the bottom of the bottle. This happens more often in years with high colour and extract, and is a completely natural phenomenon. However, the wine should be carefully decanted before being drunk.

Sulphuring the barrel (CENTRE LEFT) is always necessary to ensure that the barrel is totally disinfected before the wine is put into it. The régisseur *at Latour is Jean-Paul Gardère (ABOVE LEFT). The chef de culture, who supervises the care and*

treatment of the vineyards, is Guy Fauve (CENTRE RIGHT). He is carrying out a preparatory pruning with double secateurs. Jean Malbec (ABOVE RIGHT) is the maître de chais.

CHÂTEAU MOUTON ROTHSCHILD

Mouton Rothschild is a regal example of a first growth. Its high proportion of Cabernet Sauvignon grapes gives it an inimitable, intense, almost blackcurrant-flavoured nose. With its structure, deep, dark colour, and enveloping fruitiness, this wine ages magnificently.

In 1953, Baron Philippe de Rothschild celebrated the hundredth anniversary of the Rothschild acquisition of Château Mouton by featuring a medallion picture of Baron Nathaniel (1812-1870) on his labels. The Baron was a member of the English branch of the Rothschilds and was married to the sister of James de Rothschild who purchased Lafite in 1868. The Mouton estate was bought by Nathaniel from a Paris banker named Thuret who in 1830 had purchased it from Baron Hector de Branne.

Baron Philippe has been proprietor since the death of his father Henri in 1947. He has managed the property since 1922, when he began his fight to have Mouton accepted as a first growth Médoc. The Baron still presides over the domain, but it is now effectively managed by a three-member directorate composed of Philippe Cottin, head of La Bergerie the commercial arm of the organization, Lucien Sionneau, technical director, and Raphael Heras, administrative and financial director.

There are 150 producing hectares but not all are Château Mouton Rothschild. The domain also includes the adjacent properties of Château Mouton Baronne Pauline (also known as Mouton Baron(ne) Philippe, alias Mouton d'Armailhacq) which has 50 hectares, and Château Clerc-Milon with 28 hectares.

Winemaking is under the overall control of the directorate but falls mainly on Lucien Sionneau, assisted by Christian Prudhomme and Gilbert Faure the *chef de culture*. Raoul Blondin is *maître de chais*.

At the moment of optimum ripeness Mouton attempts to harvest the 150 hectares as quickly as possible. To accomplish this they increase their regular staff of 140 by 600 hired workers. Some 350 do the picking, but it takes the same number again to perform associated tasks such as the preparation of 750 meals three times a day. None of the harvesters is given accommodation, but buses are provided between Bordeaux and Pauillac. This procedure allows the harvest to be adjusted if the Merlot, which accounts for 12 per cent, is ripe appreciably before the Cabernet Sauvignon and Franc (87 to 88 per cent) and Petit Verdot (0.5 per cent). In 1979 the Merlot was ready eight days before the others, but it was possible for it to be picked by the numerous staff in a period of only two days.

The soil is composed of gravel, coarse sand, some clay and a base of alios. This is a red or blackish impervious layer under the looser sand, composed of sand caked together to form a slab. It often contains deposits of iron. This impermeable mass makes it necessary to use a rootstock which spreads out horizontally when it finally reaches this layer.

Mouton works or lightly ploughs the soil between the vinerows in order to preserve rainwater if the weather is dry. It is estimated that 40 cubic metres of moisture per hectare per day can be lost under these conditions, and breaking up the surface helps to conserve moisture as well as to improve absorption of any rainfall.

Each year samples of the soil are taken to determine which elements are lacking. The local agricultural laboratory analyzes the samples and issues a prescription of natural or chemical fertilizers to be added. It is important to replace no more nourishment than the vine has extracted in the course of its growth because this might cause the vine to overproduce and thus impair its quality.

The purpose of rotation is to allow an uprooted piece of land to rest for five years. During this time the soil is worked, treated and a green fertilizer, composed of oat, corn and other plant stalks, is mixed and buried so that it decomposes. An average of five hectares is rested at once, which means that Château Mouton Rothschild has 77 hectares planted at any one time.

In the Médoc, the grapes are not allowed to be used for *appellation contrôlée* wine until the vine is four years old. Even so, as with many châteaux, Mouton will wait a few extra years before even considering the wine for the *grand vin*. In practice, the grape variety and seasonal quality determines whether the juice will be used. Some vines have been planted for 15 years before their grapes have been deemed suitable.

In contrast to California where the generous heat, sunshine and dryness encourage planting in an east-west direction, the winemakers at Mouton believe they gain half a per cent of alcohol by arranging their rows in a north-south direction. The Bordeaux climate requires that the grapes be exposed to all the available sun.

The château and the other buildings devoted to making the wine are quite a showplace, with a well-appointed reception room, remarkable museum, spic and span new wine *chais* and 27 varnished oak vats. It is clearly designed to appeal to the 50,000 tourists who pass through it each year, but this traffic seems to have no adverse effect on the quality of the wine.

The winemakers admit the extra expense and trouble of maintaining wooden vats is costly, but Baron Philippe likes them and everyone agrees that it is part of the tradition of Mouton. Lucien Sionneau believes that, because of the growing difficulty of obtaining wood which has been seasoned for five to seven years, oak vats will disappear from the Médoc in 25 or 30 years. However, along with Lafite and Margaux, Mouton will probably be one of the last to change to an alternative material for their vats.

The grapes are transported to the cellars as rapidly as possible. This avoids oxidation of juice from grapes whose skins have been accidentally broken, and premature fermentation. The large number of harvesters

This label celebrates the 'year of victory' 1945. It was a great year which produced a small yield but wines with enormous concentration.

helps to prevent either of these processes from beginning. An entire vat can be filled in a single morning, even allowing for the destemming process and a last check to remove leaves, twigs, stones, or unhealthy grapes.

Curiously, it is never the first filled vat that starts to ferment, but the second. The winemakers rely on indigenous yeasts and if the must is too cold for them to begin working, it is stirred with an electric cane to warm it to 18°C or 20°C.

Chaptalization, or the adding of sugar to the must at the start of fermentation to increase the potential alcohol degree, is widely practised in France. It is particularly recommended in years when the grapes are deficient in sugar because of poor climatic conditions. It should be regarded as a complement to what nature has not given and supplied following an analysis of the must, rather than automatically. Monsieur Sionneau very correctly maintains that adding more sugar than necessary unbalances the wine, and those winemakers who add a few degrees in spite of the quality of the wine give the practice a bad name.

Fermentation and maceration for each vat takes an average of three weeks, but the fermentation rate in the later vats is increased because of the greater quantity of yeasts in the atmosphere. The must is pumped over between one and three times a day and cooled if the temperature reaches 29°C or 30°C. Maceration depends on the quality and maturity of the grapes and must be carefully monitored if their condition is unsound so that unwanted tastes can be avoided.

Following the *cuvaison*, the wine is run off the skins and seeds and pumped into clean vats to undergo malolactic fermentation. This often has to be induced by heating the cellars. The *marc* is pressed, but the *vin de presse* is kept separately until after the fining in the second year. Tasting samples containing different proportions then helps to determine how much should be added. This varies from a small percentage to all of it. Any remainder is fined with egg white once again.

A preliminary partial blending of similar styles and quality is carried out before the wine is put into cask towards the end of December or January. Mouton uses entirely new casks every year. Each cask now costs 1,000 francs, and the estate uses over 1,000 of these to age its wine. This is clearly an enormous investment that only the top vineyards can afford.

The winemakers believe that racking should be performed depending on the wine's evolution in the barrel and not in a predetermined fashion. There should not be too many rackings as this can harm the wine, and the best moment for bottling should be chosen partly in the light of the racking history. Each wine should undergo an equal number of rackings.

The time for bottling the *grand vin* varies widely from one year to the next. It is often bottled in July of the second year, as was the case in 1978. However, the powerful and hard 1975 stayed in cask for two and a half years. Bottling is performed as quickly as possible to avoid any difference in taste between different bottles of the same vintage.

Average production per hectare is between three and four *tonneaux*, which averages out at 27 to 36 hectolitres per hectare. However, this is not the amount bottled under the Château label, because wine of lesser quality does not go into the *grand vin*.

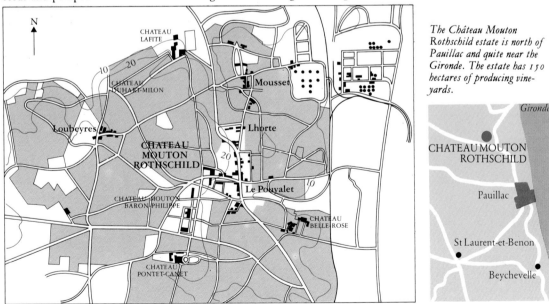

The Château Mouton Rothschild estate is north of Pauillac and quite near the Gironde. The estate has 150 hectares of producing vineyards.

1973, en hommage à Picasso (1881-1973)

Dessin inédit — *Andy Warhol*

PICASSO BACCHANALE, MUSÉE DE MOUTON

SPECIMEN

1975   à produit
cette récolte
241.000   bouteilles bordelaises et demies
9.245   magnums, jéroboams, impériales.

Philippe de Rothschild

Château
Mouton Rothschild

LE BARON PHILIPPE PROPRIETAIRE
APPELLATION PAUILLAC CONTROLEE

PRODUCE OF FRANCE

75cl   TE LA RECOLTE MISE EN BOUTEILLES AU CHATEAU

Philippe de Rothschild

Château
ton Rothschild
1973

R CRU CLASSÉ EN 1973

COND JE FUS
MOUTON NE CHANGE

IPPE   PROPRIETAIRE
UILLAC   CONTRÔLÉE
OF FRANCE
N BOUTEILLES AU CHATEAU

For their distinctive labels
(ABOVE) *Mouton Rothschild*
usually commission work by a
major artist. The 1973 label,
which is a homage to Picasso
who died that year, also commemorates the granting of
premier cru *classification to*
Mouton Rothschild. Baron
Philippe de Rothschild, whose
signature appears on the
labels, had fought for many
years to gain this distinction.
The 1975 label gives details of
the vintage, one which is
excellent for long ageing.

Good drainage is important
for vines. This drainage canal
(ABOVE) *is near some of the*
Cabernet Sauvignon vines at
Mouton Rothschild. This
variety makes up a large proportion of the planting on the
estate and helps to account for
the almost blackcurrant-like

nose of the wine. The vines are
planted in long rows on gently
sloping terrain.

The long-standing maitre de chais *at Mouton Rothschild is Raoul Blondin. After taking a sample, he checks the colour of the wine during its first year* (INSET). *The colour of the wine may fluctuate appreciably during its time in cask, particularly after an* addition of sulphur. *The colour of Mouton Rothschild is always deep because of the high proportion of Cabernet Sauvignon, which gives deep, red wines. Monsieur Blondin fills in the fermentation graph on an oak vat* (BELOW). *This is done daily.*

CHÂTEAU LOUDENNE

Loudenne soars above its <u>cru bourgeois</u> status and exemplifies what can be achieved with impeccable winemaking technique and a fine site near the Gironde river. The wines always have excellent colour, body, strong flavour and character, and age with great style.

In 1875 the British company Gilbey purchased Château Loudenne more as a point of departure for its dealings in Bordeaux wines than as a wine-producing property. In those days it was difficult to transport heavy casks over land to the port of Bordeaux for shipment to Britain. The vineyard's location next to the Gironde permitted the wines to be transported in bulk by a small boat to a larger vessel in mid-stream or to the quayside in Bordeaux. Even so, at that time, Loudenne possessed 80 to 90 hectares of vines out of a total area of 120 hectares, making it one of the major wine-producing properties in the Médoc. At that time even the low-lying land on the estate produced vines, but today this less suitable land grows grass for livestock.

The vineyard is now reduced to 40 hectares, mainly on the ridges and slopes of the property. Four more hectares of white wine grapes have been planted, bringing the proportion to 11 hectares of white wine production compared to 33 hectares of red.

As president of International Distillers and Vintners in France, which owns Gilbey de Loudenne SA, Martin Bamford is the driving force behind operations at Château Loudenne. He is also responsible for renovating and redecorating the charming pinkish-coloured château. Alain Bouilleau is the *régisseur*, while the head winemaker and oenologist is Jean-Louis Camp, a Burgundian by birth.

Loudenne is a bourgeois growth and located at the northern limit of the famous Graves-Médoc gravel bank. The thickest proportion of gravel is at the top of the slope, but further down there is more of a clay-limestone soil mixture. According to the winemaker, the structure is such that, because of good drainage thanks to the gravel, the vines do not suffer even after heavy rainfall. However, despite the pebbles, the soil is still difficult to work in wet weather. On the other hand, the deeper clay-limestone layer retains moisture which helps the vines in dry years such as 1976. However, in some sections of the vineyard, this layer is so hard that vine roots cannot penetrate it. This makes it impossible to plant.

Loudenne is one of the closest wine-producing châteaux to the Gironde. This gives protection from the frosts which occur further away from the river. The river is 6 kilometres wide at this point, which makes the climate as much as 7°C or 8°C warmer in winter. The Atlantic Ocean is a distance of about 40 kilometres away to the west.

The ratio of vines is estimated at 60 per cent Cabernet Sauvignon, 25 per cent Merlot and 15 per cent Cabernet Franc. The proportions have changed over the years in favour of Merlot because it gives the wine a certain amiability and suppleness as well as making it drinkable earlier. Production averages 150 *tonneaux* of red wine per year which amounts to a level of about 40 hectolitres per hectare overall. Château Loudenne still continues to fulfil its role as buyer and shipper of Bordeaux wines and so the operational part is much larger than that of most châteaux. However, for making the Château wine, the cellars have 12 resin-epoxy lined, 250 hectolitre concrete vats, which date from between 1923 and 1926. There are also several new stainless steel vats containing 150 hectolitres which are mainly used for making the white wine.

To the winemakers, the material from which the vats are made is not the most important factor in the vinification. What does count is the capacity, which should be between 100 and 200 hectolitres. The absolute maximum is 250 hectolitres. The size and shape should allow the juice to be pumped over relatively easily and the *marc* to be submerged daily so as to homogenize the must, extract colour from the skins and give the appropriate amount of oxygen to the working yeasts.

The second important factor in the vinification is complete control of the fermentation temperature. If the density is high—which means that there is still a high proportion of sugar to be fermented—it is important to use cooling pipes or running water when the temperature reaches 26°C. If the density is low and the fermentation almost complete, cooling is not necessary even at a temperature of 27°C. In practice, however, the temperature never exceeds 28°C or 29°C during fermentation.

Only ripe, healthy grapes are picked and the skins are kept unbroken for as long as possible to avoid oxidation of the juice on the way to the vat. The stems are removed completely and the grapes are crushed very lightly, so that between 70 and 80 per cent of the grapes are pumped into the vat whole. A weak sulphur dioxide solution is added immediately and mixed in well to prevent oxidation.

The first pumping over takes place the day after the vat is filled. The level of sulphur dioxide is checked and selected yeasts added. The winemakers feel this procedure is important to keep the fermentation under control so it does not proceed too quickly, which might lead to the risk of overheating.

Fermentation lasts between six and 10 days and, during this time, the must is pumped over the *marc* twice a day to extract colour and tannin gradually. Maceration, or the

further extraction of colour and tannin from the skins of the grapes, then follows in the same vat. This happens quickly enough so that only seven or eight days are required to extract what is needed for the desired balance between alcohol, acidity and fruit. This balance—and therefore the characteristics of the vintage—depends on the nature of the skins.

The total vatting period lasts approximately 20 days. As a rule, malolactic fermentation begins during maceration.

Château Loudenne produces wines which outstrip its cru bourgeois *status. The estate is owned by Gilbey, a British company.*

However, it is important for the wine to be run off the *marc* at the end of the period.

The first pressing of the *marc* is immediately incorporated with the new free-run wine. Any subsequent press wine is kept separately to be used later on depending on its quality and if it is needed. According to the winemakers, there should be little difference between the first press and free-run wines apart from a darker colour if the vinification has been done well. Furthermore, if the skins and pips which form the *marc* are submerged often enough and kept wet, bacteria will not develop and thus there will be no more volatile acidity in the press wine than there is in the free-run wine.

The different lots are partly blended and the wine put in another vat for two or three weeks so that some of the solid particles settle. If the wine is fairly clear, it is put directly into the barrel, but, if not, it is filtered. The winemakers feel sure that, at this stage, filtration has little if any effect on the wine.

Loudenne ages 25 per cent of its red wine in new oak barrels and the rest in slightly older wood. They believe that great wine needs to be aged in wood; but, after being used for four years, the wood no longer adds tannin and vanilla to the wine's character. Some wines are powerful enough to take completely new wood almost every year, but Loudenne wine would take on an unpleasant woody taste. At Loudenne each variety ages separately in the cask in order to develop individually until the final blending.

It can take up to two months for the carbon dioxide gas to leave the wine and barrel by escaping through the bung-hole when it is opened for the weekly topping up. As a result, the wine becomes more stable and the barrel is rolled to one side which virtually seals it. The winemakers then allow the wine to settle for three months before the first racking. Subsequent rackings follow the customary Bordeaux pattern, taking place every three months the first year and every four the second.

About 18 months after the harvest, trial samples of the

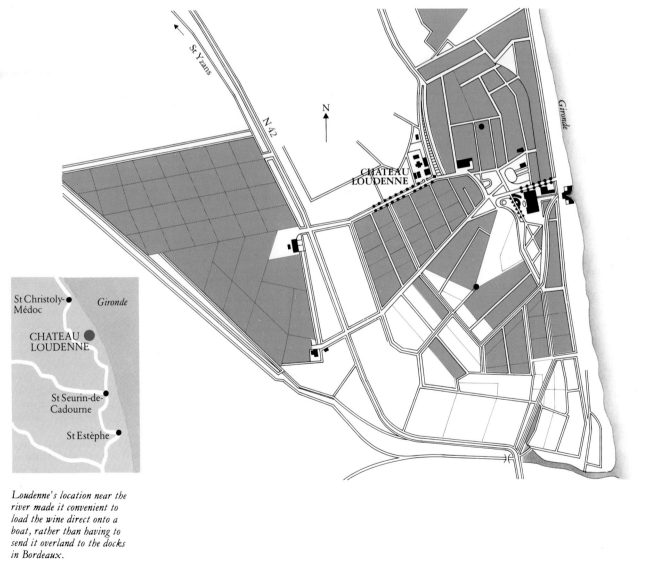

Loudenne's location near the river made it convenient to load the wine direct onto a boat, rather than having to send it overland to the docks in Bordeaux.

Roses planted at the end of the rows of vines (BELOW) give early warning of disease. These Cabernet Sauvignon vines (CENTRE LEFT) show the flat terrain on which much of the Loudenne property is situated.

The length of the rows of Merlot vines (CENTRE RIGHT) are typical of vineyard planting in the Médoc and contrasts with the smaller plots found in Burgundy. Francis Fouquet (BELOW) is the technical director of Château Loudenne.

different varieties of wine are prepared to determine which produce the best blends or whether it is necessary to discard some lots. Tasting determines which mixture is best and the final blending is carried out accordingly. Only when this is completed does fining with egg white take place. One month after this operation, the wine is drawn off and bottled.

Bottling sometimes takes place before the summer, but it depends on the texture of the wine. With a lightly constituted wine, it is best to transfer it from barrel to bottle earlier than a more powerful vintage. Furthermore, a light vintage is ready sooner and therefore needs to develop less in the cask.

The Château's 11 hectares of white Sauvignon and Sémillon grapes are only used for the white wine called Loudenne Blanc—AC Bordeaux. This is young, fruity and well-balanced with plenty of body and plenitude. Loudenne Blanc is best drunk within a few years.

When making white wine it is extremely important to avoid oxidation, so the grapes are picked, pressed immediately, the juice pumped into the stainless steel vats and the sulphur dioxide solution added immediately. Cold-fermentation at 18°C is controlled by running water on the sides, and a further precaution against oxidation is taken by first filling the vats with a neutral gas to replace the oxygen. The must is treated with bentonite to remove the proteins, then the solid particles are allowed to settle for 24 hours before yeasts are added. Alcoholic fermentation lasts for between 12 and 15 days, and malolactic fermentation is prevented so that the wines retain the lightness and aroma due to their higher acidity.

The white wine is not put into barrels but is vinified and treated entirely in the vat. It is fined, refrigerated and allowed to precipitate more impurities before being drawn off and bottled during February. The wines are very well made from a technical point of view and, because of the gravel in the soil, show some of the character found in the white wines of Graves further to the south.

The Château Loudenne red wine is fruity and perfumed It is never very darkly coloured, and always reveals a certain elegance.

The pink château (RIGHT) faces the Gironde. A decade ago vines were planted right up to the garden.

This picture (ABOVE) shows the spirit of France with a bottle of Loudenne wine. Château Loudenne kept the black label until the early 1960s.

During picking (RIGHT), the grapes are gathered by hand and put into baskets, which are then tipped into the hotte in which the grapes are taken to the waiting trailer or lorry. Here white grapes are being picked for the Loudenne Blanc wine.

CHATEAU LA MISSION HAUT BRION

La Mission Haut Brion is a majestic wine of great style and class. It combines some of the most intriguing aspects of a top Graves wine with a body and staying power which is almost Médocain in character. Powerful and strong in youth, the wines mature to give a deep, intense bouquet of many dimensions and long, full flavour.

Until the seventeenth century, the property that is now La Mission Haut Brion was part of the estate of Haut Brion, but in 1630, when the owner, Dame Olive de Lestonnac died, she willed a portion to the clergy of Bordeaux. It was later transferred to the religious order of Lazarites, who in 1698 built the small chapel now attached to the château, and called it Haut Brion La Mission. The order lost its ownership in the French Revolution and the domain then had a succession of owners.

Frederick Woltner purchased the property in 1919. Woltner had two sons, Fernand and Henri, and a daughter, Madeleine. It was Henri who, after graduating in oenology from the University of Bordeaux, managed the vineyard for 50 years until his death in 1974. After the death of Fernand in 1977, ownership passed to his son and daughter, to his sister Madeleine, and to Henri's daughter. Today, the domain is under the direction of Fernand's daughter, Françoise, and her Belgian born husband, Francis Dewavrin. The *régisseur* and winemaker is Henri Lagardère, who joined Henri Woltner in 1954; he is now assisted by his son, Michel, who is a trained oenologist.

La Mission, along with its neighbour, Haut Brion, enjoys a unique location surrounded by the city of Bordeaux, the suburbs of Pessac and Talence and the University of Bordeaux. This situation is responsible for some of the château's idiosyncrasies. The microclimate is partially man-made. The proximity to population and commerce has a moderating effect on the temperature, even at night. In addition, the centre of Bordeaux to the north-east plays a definite role in protecting the vineyard from an icy north wind. Nature's influence on the microclimate comes principally from the Garonne-Gironde.

The 'Domaine Woltner' has 26 hectares of planted vines and makes three Château wines—La Mission, La Tour and Laville. All have the 'Haut Brion' suffix. Laville is dry white Graves. Five hectares are given over to its Sauvignon and Sémillon grapes, but plans for a new road threaten to reduce this to just over three.

The soil of the red grape part of the vineyards is real *graves*, meaning it is composed of coarse gravel and sand and reaches to a depth of between 10 and 12 metres before meeting a limestone layer. As for topography, there are three *croupes*, or gently sloping hillocks, with the best plots located on or near the summit facing the sun.

This type of soil offered excellent natural drainage, even before the French National Railway intervened in the last century. Since they cut a trench through the vineyard for their one-track rail line to Arcachon and Spain, widening it over the years from three to 12 metres, the positive effect on the wine has been unmistakable. The soil in this area produces exceptionally thick-skinned grapes with much tannin and concentrated juice. Most of the vintages take a long time to evolve and have a characteristic Graves 'dusty' flavour often with a 'smoky' nose and taste. Both the Haut Brion and La Mission vines ripen eight to 15 days before other vineyards of the Bordeaux region.

Henri Woltner made La Mission Haut Brion what it is. He was not only a brilliant winemaker and innovator, but was known as one of the very best wine-tasters in the Bordeaux region. He was one of the first to use steel fermenting tanks and to understand the necessity of controlling fermentation temperature in order to keep the fruit and aroma in the wine. He was also one of the first to understand the benefits of malolactic fermentation and to practise it under more or less controlled conditions.

The winemakers at La Mission believe in the necessity of a second wine, and La Tour Haut Brion is now accorded that role. Some portions of the vineyard are not considered good enough ever to produce La Mission, but in general the ratio of La Mission to La Tour depends on the quality of the vintage and the age of the vines. The process of selection begins when the grapes are picked and sorted in the *chais*, before they go into the crusher-destemmer. This selection means that the wine is partially chosen before it is made. In 1979, an excellent year in the region, 40 per cent of the red wine became La Tour and 60 per cent La Mission.

At one time, Château Latour in Pauillac started proceedings against the domain for its use of the 'Latour' name. It was discovered that La Tour Haut Brion had a claim to the name going back to the thirteenth century, making it older than Latour in Pauillac. The suit was quietly dropped and no more said.

Having the large University of Bordeaux nearby means the château can call on an enormous reserve of potential pickers. This allows harvesting in eight to 10 days, when the grapes are at their optimum maturity.

Before the Second World War, Henri Woltner introduced steel vats like large thermos flasks with one-piece glass-lined interiors. These became too expensive and newer vats were made entirely of stainless steel. But some of the 'thermos flasks' still exist and continue to give excellent results when they are used.

Composition of the 21 hectares producing red wine is estimated to be 70 per cent Cabernet Sauvignon, 25 per cent Merlot and about five per cent Cabernet Franc. Of course, the relative proportions of the vines on the surface are not necessarily reflected in the composition of the wine.

Fermentation and maceration take the normal three or four weeks. Then, in December, when the malolactic is finished, a first *assemblage* or blending is carried out. Each grape variety is fermented separately, and it is determined before the *assemblage*

CHÂTEAU
LA MISSION HAUT BRION
GRAVES
APPELLATION GRAVES CONTRÔLÉE
Cru classé

1978

SOCIETE CIVILE DES DOMAINES WOLTNER
PROPRIETAIRE A TALENCE (GIRONDE) FRANCE
PRODUCE OF FRANCE
MIS EN BOUTEILLES AU CHATEAU 750 ML

The cross at the top of the label indicates the estate's lengthy ecclesiastical ownership. The 1978 vintage was magnificent in Bordeaux.

which vats will be reserved for La Mission and which for La Tour. Trial blendings are made in small amounts to determine how much press wine to use and, following this, the wine is put into barrels for the La Mission wine. Between 35 and 40 per cent of the barrels are always new.

Laville Haut Brion is indubitably the best white Graves and deserves to be regarded as one of the best dry white wines of France. It has only existed since 1928, when Henri Woltner observed that the red wine from a five hectare section of the vineyards normally reserved for La Tour Haut Brion was not of exceptional quality and decided it might make a better white wine. His reasoning evolved from the fact that the soil in this plot was of a dominant clay-limestone mixture, rather than gravel and sand. It is fine, typical of that found in many white wine vineyards all over France.

Laville is the only white wine in the Graves to have a predominance of Sémillon grapes over Sauvignon. According to Michel Lagardère, this adds to the wine's longevity. The grapes are brought to the winery and pressed in, at most, 45 minutes after being picked. Pressing with the stems helps to drain off the juice. The hydraulic press is the one used on the Cabernet and Merlot *marc* for the press wine and, while it is not the best type for white grapes, the production does not merit a separate press.

The juice obtained is treated with between three and six grams of sulphur dioxide per hectolitre, depending on the condition of the grapes. It is then pumped into a stainless steel vat and left for about 24 hours to allow the larger particles to settle. The vat is heated to induce fermentation and, if necessary, the wine is chaptalized. Bentonite, a type of natural clay, is added to fine out the proteins and then removed at the first racking.

Chaptalization is a support more or less necessary according to the quality of the vintage. A certain minimum alcoholic per cent is needed for Bordeaux wines to be conserved for any length of time. In France, two additional degrees is the maximum which can be legally induced by chaptalization, but this often interferes with the taste, imparting a certain dryness making it disagreeably heavy. The great winemakers of course realize this, and are very cautious with the practice.

For both La Mission and Laville Haut Brion, experience has shown that no more than one per cent should be added. Laville's ideal is between 12 and 12.5 per cent alcohol and, in many years this is achieved naturally.

Fermentation of Laville is normally started by indigenous yeasts. When it begins, the wine is put into barrels until they are three-quarters full, and then put into the special cold room designed for the white wine. The air conditioning maintains the air temperature around 12°C in order to keep the fermentation temperature under 20°C.

Fermentation takes three to four weeks and ends when less than two grams per litre of residual sugar remain. A quick *coup de froid*—a sudden lowering of the air tem-

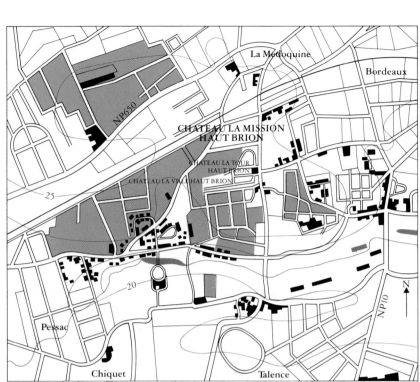

perature—serves to precipitate further impurities before a first racking is performed. The wine is left in the cold room for two months at a temperature of −5°C in order to reduce the need for filtering and fining.

Racking is done every two months, in contrast to the three for the red wines; and fining with fish glue takes place in April. After five weeks on the fining substance, the wine is racked into a vat and filtered; a cellulose and asbestos filter is used. According to the winemakers, there is no danger of asbestos fibres detaching themselves into the wine if the filtering is handled properly.

Bottling of Laville takes place in the May following the harvest, which means that the wine spends only seven or eight months in cask. Annual production of Laville averages 40 to 45 hectolitres.

Laville Haut Brion should not be drunk before it is five or six years of age and, unlike most Bordeaux dry white wines, it reaches its optimum when eight to 10 years old. It is a big wine—rich and concentrated, with balance and great breeding. The bouquet is intense, elegant and reminiscent of the great sweet wines of Bordeaux. The 1975 is the best vintage of the 1970s. With its tremendous finesse, it is a typically great Laville. To the winemakers, the 1976 is less typical, even though it is a massively rich wine with great Sémillion style and flavour. The qualities of La Mission Haut Brion include staying power and body which are unusual for a wine from Graves. It is a wine which has many dimensions, great style and class.

The estate is very close to Bordeaux. Much of the soil is gravel.

This colourful ceramic figure (RIGHT) is found in the chais where tastings take place.

Francis Dewavrin (BELOW) runs the estate administration. In addition to the vineyards, the estate includes woodlands and forest.

At the near end of the pretty château (RIGHT) can be seen the chapel built by the Lazarite order. The order lost control of the estate during the French Revolution.

This old vine (BELOW LEFT) is growing in the intensely gravelly soil from which the name Graves derives. The proximity of the town to the vineyards (BELOW RIGHT) is beneficial as it raises the temperature slightly and thus gives protection from frost.

This traditional wooden vertical press (CENTRE LEFT) is still used. These barrels (CENTRE RIGHT) have been placed bonde de côté *(with the bung on one side), which means they do not have to be topped up, as evaporation is lessened.*

These barrels (LEFT) are being soaked prior to the harvest in order to eliminate any bacteria and to see that they are watertight.

These wooden presses (LEFT) are in use, their lids are not fully down. This type of press is best suited to red grapes which are more robust than white because the tannin and colour which this type of press extracts are more necessary in red wine. A horizontal press which allows greater control over the pressing is better for white grapes, which need more delicate treatment.

The estate's ecclesiastical origins can be seen in the ageing chais *(ABOVE CENTRE). The château has an extensive collection of old wines, kept in numbered bins (ABOVE). This 1877 vintage (ABOVE RIGHT), is one of the last pre-phylloxera years.*

CHATEAU D'YQUEM

The wines of Château d'Yquem have the grandest stature of all Sauternes wines. Their enormous body and extreme lusciousness are married in the greatest years to an overriding <u>pourriture noble</u> character. The highly individual flavour that this imparts gives Yquem its massive complexity of tastes and allows it to age majestically.

The commune of Sauternes, like the other wine-producing regions of Bordeaux, has rolling hills and gentle slopes. The 173 hectare domain of Château d'Yquem occupies the most important area in Sauternes. The château itself, the cellars and other out-buildings are situated almost at the highest point in the whole of the surrounding area.

Yquem came into the family Lur-Saluces through the marriage of Louis-Amedée, Comte de Lur-Saluces and Josephine Sauvage d'Yquem in 1785. The Sauvage family had owned the vineyard since 1592. Today, Count Alexandre de Lur-Saluces, nephew of the former owner, Bertrand Marquis de Lur-Saluces, is the sixth generation of his family at Yquem. He has been in charge of the domain since the death of Bertrand in 1968. Another property, the Château de Fargues, a Sauternes bourgeois growth, has nine and a half hectares of vines and has been in the family since 1472.

The vineyards cover 100 hectares but, because of rotation and young vines, only an average of 85 hectares are available for the *grand vin*. The vineyards surround the château and face almost all directions—south, west, north and north-east.

According to Pierre Meslier, the *régisseur*, Yquem is at a geological crossroads. The southern portion consists of gravel over alios or brown freestone, which is the end part of a very hard layer originating in the region of the Landes to the south. The topsoil is composed of gravel, but, depending on the exact position, it is mixed with different proportions of sand, silt and clay.

All around the hill at the level of the water-table is an undulating layer of hard clay. In places where it is close to the surface, there are small springs. These can be located by the patches of poplar trees and grass instead of vines. One of the consequences of this hard layer is that it is difficult for rainwater to run off. To combat this problem, drains have been installed at various times over at least the last century.

There are over 100 kilometres of these drains in the vineyards which constantly need repair or replacement. Their condition is always checked in plots of land that are under rotation, but, if a drain becomes blocked in the producing vineyards, it has to be replaced immediately.

Up until 50 years ago the vineyards were planted with one Sauvignon vine after every four Sémillon vines in each row. This system was changed to four rows of Sémillon followed by one of Sauvignon until 25 years ago when the vineyards were again changed to have entire plots of each variety. The proportion of the two grapes remains about four to one, but Yquem now plants a little more Sauvignon to ensure that a proportion of 20 per cent of the juice comes from those grapes even in less productive years. The winemakers believe this is important for maintaining the style of the wine. When there is too much Sauvignon, the domain produces a dry white wine called simply 'Y'.

Each year between two and a half and three hectares are torn up and about the same area replanted so that the average age of the vines does not vary. The land is rested for two or three years depending on how well grass grows on the plot—this is a good indication of the condition of the soil. Depending on their state of health, the vines are uprooted when they are between 35 and 45 years old. As they are harvested between five and an average of 40 years old, the average age of the vines used at Château d'Yquem is around 22 or 23 years.

At harvest time—normally from mid October to November—120 pickers are divided into three equal groups. Each morning the group leader receives instructions as to which parcels of vines to pick. The grapes are transported immediately to the cellars and checked for quality and potential alcohol level. The group is then told whether they are picking grapes with too little sugar, just the quantity desired or—as sometimes happens—only grapes of very high potential. As the aim is a potential alcohol level of between 19 and 21 per cent, more highly concentrated grapes need to be balanced out by some that are less concentrated.

As each grape should be picked at just the right moment, each vine has to be picked several times. In an easy year when botrytis attacks most of the grapes at the same time, as few as four or five pickings are needed. However, when the opposite happens, on some vines each individual bunch may be picked as many as 11 different times.

Such exacting work requires particular experience, which most of the harvesters have. Yquem does not, as a rule, take on inexperienced pickers unless they are almost certain to return in the future. There are two people for each vine; one works on the right and the other on the left. An inexperienced picker works opposite one who is very experienced, so that the inexperienced picker can learn to recognize which grapes to pick. The selected berries should be of the right concentration, those that have been attacked by noble rot rather than grey rot, and those that have not been pecked by birds. In this case, a hole in the skin creates acetic acid in the juice and the only way to tell this without tasting is a nuance in the colour.

Botrytis cinerea is a type of fungus that attacks grapes and feeds on the juice by developing roots which penetrate the skin, but do not break it. It thus concentrates the juice and at the same time reduces the tartaric acid. It develops best under warm humid conditions, but the grape must be ripe for the final product to be of good quality. Development of the botrytis and

*Both 1967 and 1976 were excellent vintages
for the sweet white wines of Sauternes.*

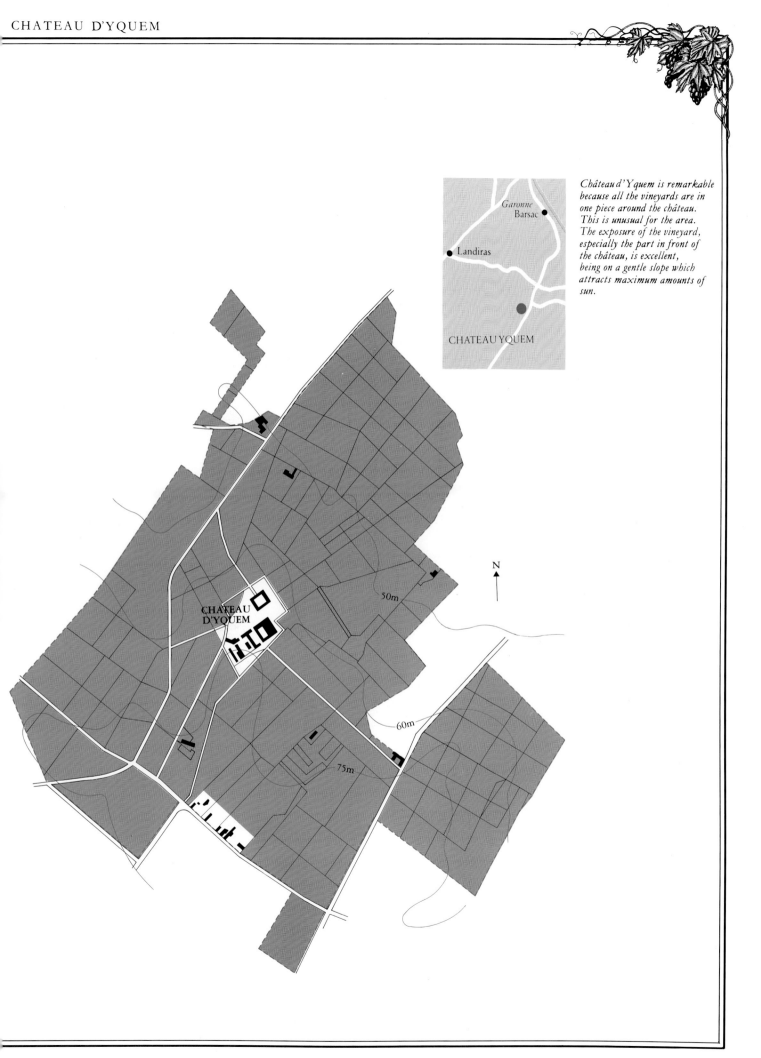

Château d'Yquem is remarkable because all the vineyards are in one piece around the château. This is unusual for the area. The exposure of the vineyard, especially the part in front of the château, is excellent, being on a gentle slope which attracts maximum amounts of sun.

Garonne
Barsac

Landiras

CHATEAU YQUEM

CHATEAU D'YQUEM

N

50m

60m

75m

other fungi on unripe grapes results in grey rot. Similarly, if weather conditions are not right and there is too much rain, the grape's skin may break, also resulting in grey rot or *pourriture vulgaire* as it is often called.

A good year in the Médoc or the neighbouring Graves area does not necessarily mean a good year in Sauternes. The harvest takes place after that in the other regions and in this part of France the weather can change dramatically from one week to the next. For example, 1967 was excellent for Yquem, whereas in 1978 and 1979 the harvests were only 15 and 40 per cent respectively of full potential.

Yquem has three hydraulic vertical presses which are similar to, but smaller than, the type used in Champagne. At present the winemakers believe these give the best quality, but they are experimenting with a pneumatic press because the small vertical presses date from 1912 and replacement parts are more and more difficult to find.

The grapes are pressed three times, but, unlike the red or dry white wine districts, the juice obtained at the third pressing is the best. This is explained by the fact that the most shrivelled or concentrated grapes are the last to release their juice. For example, if the juice from the first pressing has a potential alcohol content of 19 per cent the winemakers can expect 22 to 23 per cent from the third pressing. In addition, the juice of the last pressing is as clear as the first. This is one advantage of the vertical type of press.

Also contrary to procedure for dry white wines, where the juice is pressed then clarified before the fermentation in vat or cask, the must from the three pressings is mixed and pumped directly into new oak barrels. Because it is necessary to obtain juice that is clear so as to avoid handling and thus the possibility of impairing quality, the choice of press is important for this type of wine.

The régisseur *at Château d'*
Yquem is Pierre Meslier
(BELOW). *He is tasting a*
glass of 1976 Yquem, an
excellent vintage. Meslier
seeks to maintain the great
traditions of Yquem at the
same time as conducting
research into sweet white
wines and noble rot. At
Yquem the wine is kept in
wood for three years, the only
château which maintains this
practice.

Fermentation takes place in 225 litre casks, but, because of the lateness of the harvest, the juice is often quite cold. As it is difficult for fermentation to begin below 17 or 18°C, Yquem has a heater installed in the *chais* which maintains the temperature at around 20°C. An additional problem with the fermentation is that the casks are small in volume. This means that each is an individual unit and does not benefit from the massive volume of a vat. It is also necessary to watch the temperature of each barrel to prevent it exceeding 32°C.

It normally takes between two and four or even five weeks for the alcoholic fermentation to run its course. It has even taken as long as two months. A light sulphuring is necessary which helps eliminate the yeasts least suitable for fermentation and thus prevent bacteria from developing.

Fermentation stops of its own accord when the alcohol has reached about 13.5 per cent. Some species of yeasts can continue to work to a higher level, but those indigenous to the region are killed by the alcohol and a substance manufactured by the *botrytis cinerea* called botrycine. Sugar still remains because the non-fermented juice contains between 300 and 350 grams per litre. When fermentation stops, it means that 80 to 120 grams of sugar per litre have not been changed into alcohol.

Immediately following alcoholic fermentation—there is no malolactic fermentation with sweet wines—the wine is racked to separate the lees. The wine stays in the barrel for three and a half years and is racked every three months, or a total of 15 times. During this entire barrel-ageing period, the bung-hole remains uppermost, which makes it necessary to top up twice a week. It is estimated that 22 per cent of the total volume is lost over this period.

As sweet wines are more dense than red or dry white wines, much of the sediment is of the same density and does not settle on the bottom. The wine is fined, but there is no one product that works best all the time, so trials are made with all the fining agents to see which is the most suitable for the particular vintage. In general, the wine is fined twice in the course of its life in the cask.

As the harvest is so late and weather conditions never perfect, the winemakers always have to make a selection to establish what is good enough to carry the Château d'Yquem label. As it is impossible to harvest under completely ideal conditions, a good year is one in which 80 per cent of the harvest is of a high enough quality for Yquem. In more difficult years the percentage eliminated is greater, and in such years as 1964, 1972 and 1974, the entire harvest was declassified. The wine that is not good enough for the *grand vin* is sold as a simple Appellation Sauternes.

The characteristics of each vintage are, of course, different, but Yquem's reputation rests on the fact that a very high quality is maintained. This means that if the vintage does not have this quality, there is no Sauternes Château d'Yquem.

The medieval château (LEFT)
has been kept in good con-
dition. Concerts are held in the
picturesque central courtyard.

Yquem's maitre de chais is
Guy Latrille (FAR LEFT).
He has taken some wine from
the cask with a pipette and is
spitting it out after tasting.
The chef de culture, Yves
Laporte (LEFT) *is holding*
some soil—a mixture of
quartz, sand, alluvium silt
and clay. These Sémillion
grapes (FAR LEFT BOTTOM)
have noble rot. The character-
istic discolouring and shrivel-
ling can be seen, as well as the
fungus-like growth. The wine
is pressed gently so that it does
not discolour. Yquem retains
some old vines (LEFT),
although these are less im-
portant for white wine than
red. Old vines give concen-
tration and a smaller yield.

CHATEAU CLIMENS

The great, luscious flavour of the wines of Climens are always married to a flowery bouquet that lifts the wine above the merely sweet category. The dimensions of flavour and great length on the palate are intriguingly combined with all the complexities derived from <u>pourriture noble</u>.

Lucien Lurton, owner of Brane Cantenac in Margaux, Clos Fourtet in Saint Emilion and other châteaux in the Bordeaux region, purchased Château Climens in Barsac from Monsieur Gounouilhou, director of the daily newspaper *Sud Ouest* in 1971. Maurice Garros, the manager, supervises winemaking at all the Lurton properties, while septuagenarian André Janin has been the principal winemaker at Climens for more than 30 years.

The low-lying house opens onto a modest courtyard. It can hardly be called a château and the entire complex, including the cellars, is opposite the main part of the vineyards across a small road.

The domain has 30 hectares of vines, but because some are in rotation and others are too young for the Château label, only 27 are in production at any one time. The winemakers believe that a rest of three years is necessary between uprooting and replanting the vines.

Climens is located a few kilometres south-west of the town of Barsac in the section known locally as Haut Barsac. The soil in this part of the commune is a mixture of clay and limestone with more of the latter. At a depth of about 30 centimetres, there is a rocky layer which the vine roots have difficulty in penetrating. For this reason they are not as vigorous as vines grown in deeper, richer soil.

The vineyards are planted almost exclusively with Sémillon. According to André Janin, Sauvignon is not grown because it does not keep its aroma after three years in the barrel. Neither does it produce the degree of alcohol that Sémillon does. As rootstock, the well-known *riparia* does not grow well in this type of limestone soil, so Climens uses a stock known as the 420.

Harvest begins around 15 October. Several different pickings have to be carried out, so it usually lasts more than a month. It is becoming more and more difficult to hire enough workers from the local area, so Climens takes a number of Spaniards every year. They work well, but it is still necessary to check the pickers closely to see that grapes of the right concentration, with noble rot and unbroken skins are chosen.

It is much more difficult to have a good year in Sauternes-Barsac than in a red wine district for several reasons. The first is that the grapes have to be naturally well-ripened because the quality of the sweet wine ultimately depends on this. Secondly, it is essential that weather conditions in the autumn are conducive to the development of noble rot. *Botrytis cinerea*, the fungus that causes the phenomenon, flourishes best with morning fog followed by warm sun during the day. If the weather is too dry, the fungus will not develop, and, if there is too much moisture or rain, the grapeskin will develop grey or vulgar rot, from the same *botrytis cinerea*. Mist in this region rises from

the small river Ciron, which marks the eastern boundary of Barsac and the western limit of Sauternes. It flows into the Garonne, an additional source of humidity.

Another difficulty is the absence of a clear-cut visible difference between noble and vulgar rot. Many intermediate states can exist and it takes a practised eye to recognize the difference.

Further problems are caused by the fact that the botrytis does not attack all the grapes in a bunch at the same time. It is therefore necessary to pick each vine several times depending on how particular the winemakers are about the grapes' degree of concentration. The intervals between picking, which at Climens can vary from eight to 15 days, are also determined by this factor.

Botrytis cinerea causes a profound modification of the grape it attacks. Most obviously, its volume is greatly diminished: a vineyard that would yield 40 hectolitres per hectare from mature grapes is reduced to between nine and 12. Furthermore, while concentrating the juice and shrivelling the skin, the botrytis consumes up to 50 per cent of the sugar content. The acid is not concentrated, as it is a peculiarity of this organism that it decomposes tartaric acid, the most important acid in grapes. Finally, botrytis forms glycerol in the juice. This gives the wine thickness or viscosity.

Climens uses a horizontal press with two plates which squeeze the grapes toward the centre. Because rapid pressing gives too many solid particles in the must, this procedure is carried out very slowly. After the first pressing, which takes about an hour, the cage is opened, the mass of grapes is broken up, and the procedure repeated twice more. The entire operation takes four or five hours.

The must is then put into the usual 225 litre barrels for fermentation which lasts between three and five weeks. The winemakers say they can tell by tasting whether the quality is exceptional as soon as the fermentation is finished.

More than half of the barrels used for the new wine are made of new wood. As white wine is fermented without the skins, it lacks tannin and other substances which contribute to longevity and body. Fermenting this wine in new oak can, to some extent, substitute for the absence of skins and promote these qualities. It is for this reason that certain sweet wines, particularly the Sauternes and Barsacs, will normally age for a long time.

It is not necessary to heat the *chais* to start the fermentation, but the first barrels do begin slowly. When the wine has reached 14.5 per cent of alcohol, the winemakers decide if fermentation should be stopped. Small quantities of sulphur dioxide are added immediately to stop the activity of the yeasts on the wine.

According to the winemakers, it is not always necessary to fine the wine because it often clears itself. On the other hand, there

1971 produced wines of finesse and distinction in Sauternes-Barsac; for Château Climens it was probably the best year of the decade.

are years when the wine is cloudy and filtering is necessary.

The wine is racked to eliminate the lees every three months. The longer the wine stays on its lees, which settle slowly over a relatively long period of time, the more sulphur dioxide needs to be added to stabilize the wine. It is therefore necessary to get rid of deposits gradually as they form. Many of the particles have the same, or almost the same, density as the wine, so while barometric pressure often ensures that the lees stay at the bottom of the cask, a change in the weather can cause the lighter particles to rise back into the wine. It is partly for this reason that bottling takes place in winter after the wine has been in cask for three years.

Climens makes about 60,000 bottles per year. In the decade between 1970 and 1980, their best years were 1971, 1975 and 1976, with 1971 perhaps their greatest success.

André Janin (LEFT) *the chief winemaker at Château Climens is tasting young wine which is still cloudy. This cloudiness may clear naturally as the solid particles fall to the bottom of the cask, but filtering is sometimes needed. The wine is taken from the cask with a pipette.*
Christian Broustaut (LEFT BELOW) *is the future maitre de chais.*

Château Climens lies to the south of Barsac. It has some 27 hectares of producing vineyards.

These vines in early November (BELOW) retain a few grapes with noble rot. The vines are picked several times because the grapes ripen and are affected by noble rot irregularly. The vintage can take up to one month. Around the château the land is flat (CENTRE), characteristic of Barsac. The château and chais are similar to many others in the Bordeaux area. Both new and old oak barrels (BOTTOM) are used in the chais. The wood of those which have already been used is darker. More than half the new wine is put into new wood.

During misty autumnal mornings and evenings, the shutters on the chais are kept shut. This is to maintain an even temperature especially for the fermentation. Such weather conditions—misty mornings and evenings and warm afternoons—help to promote noble rot which needs warmth and slight humidity in order to develop (BELOW).

CHATEAU AUSONE

With an amazing individuality and sheer diversity of flavour, Château Ausone has the potential to be perhaps the most distinguished wine of Bordeaux. Nowhere is this shown more clearly than in the remarkable 1978 vintage. The great intensity of fruit, vast array of taste, and incredible length on the palate make Ausone a drinking experience that lasts.

Château Ausone has been owned by the same family since the property was purchased by the Cantenacs just after 1789. Today, two branches of descendants — the Dubois-Challons and the family of Marcel Vauthier—hold equal shares.

Ausone has sometimes been accused of producing wines unworthy of the top position in the St Emilion classification that it shares with Cheval Blanc. Like many of the Bordeaux estates, it is possible that Ausone went through a difficult period, but an eminent reputation is easily criticized. Moreover, since Madame Dubois-Challon took over on the death of her husband in July, 1974, and appointed a new *régisseur*, Pascal Delbeck, there has been substantial investment.

Ausone resembles a well-carved miniature. It has only seven hectares under vines and very little land left over for the château and park found on some Médocain estates. Such a small vineyard does not need large-scale equipment. Consequently, the cellars, the storage *chais*, and the fermenting vats are tiny compared to those found on the 60 to 80 hectare domains across the Gironde. The small-scale operation also allows for traditional methods like hand bottling from cask to bottle rather than the more mechanized operations found elsewhere. However, this may soon be replaced by a more modern system.

Annual production is also minute. 1970 was a prolific year but the recent yearly average is about 25 hectolitres per hectare—less than three *tonneaux*. 1977 produced no more than 19 hectolitres per hectare: 1978 produced 22; and the relatively abundant 1979 yielded less than 32 hectolitres per hectare. An entire year's output never needs more than a hundred barrels.

The vineyards are planted on slopes some of which are too steep for a tractor to negotiate, so the Château keeps a horse to work the difficult plots. Their situation, facing south-east, is quite exceptional. On a long summer's day, with no higher hills to block it, the sun warms the grapes from 5 am to 9 pm from front to back, producing a maturity that few others enjoy. It is an additional advantage that the altitude of the vineyards on the *côte* spares them from the damaging frosts that sometimes wreak havoc with those on the plain.

Even though the soil varies from one plot to another, the main constituents are clay, sand and limestone. This is known geologically as *molasse*. There is a dominance of clay and limestone and, according to Pascal Delbeck, the *régisseur*, Ausone is the only domain among the *premier grand cru* wines to have this soil. Indeed, some patches of earth in the vineyards have the consistency of child's modelling clay and others are more sandy and less cohesive. Thanks to the calcareous content, which promotes water retention, there is no problem during dry weather.

The vines are between 40 and 45 years old. Between 1950 and 1976 there was no replanting, and, in 1956, the vines escaped any permanent frost damage. From 1967, vines have been replaced as they died, a common practice in the Bordeaux area, but it was not until 1976 that entire plots were uprooted to be replaced after a year's rest and soil treatment. A very old vineyard is more likely to produce high quality wine, but the correlation of high quality and small quantity of production is by no means automatic. Unless a policy of rotation is maintained, production will fall below a minimum level of profitability, and the vineyard will be abandoned or need replanting, with a long interval before high quality production can resume.

The vineyards are divided almost equally between Merlot and Cabernet Franc (Bouchet), with a slight dominance of the latter. The Merlot is known for its earlier-ripening properties, but it is a particular characteristic of Ausone that both types reach maturity with equivalent acidity and sugar levels within a very short time of each other. Harvesting is performed by local townspeople, or gypsies who come each year. Because the owners possess a total of 38 hectares, including Château Bel-Air next door, they are able to choose the best moment for Ausone. Picking at Château Ausone takes on average three or four days.

In keeping with the small scale, there are nine wooden fermenting vats with a capacity of approximately 55 hectolitres each. It is a sensible tradition in winemaking to keep the lots as separate as possible for the period of vatting, and at least until the first blending or *assemblage*. This allows the development of each lot to be observed and the *grand vin* to be selected. However, the size of the vat is also determined by the quantity of grapes brought in each day. Filling vats sufficiently in one day helps to initiate the fermentation process.

The winemaking process itself is along traditional lines. At the time of crushing, between five and 20 per cent of the stems are left with the must, depending on maturity, acidity and tannin. Fermentation takes eight to 10 days and maceration a little longer, making a total *cuvaison* of 19 to 21 days. Fermenting temperature is controlled at levels of between 28°C and 30°C.

The *marc* is pressed three times for the press wine, but only the first is added to the free-run wine at the first blending. The press wine undergoes its malolactic fermentation in barrel and the free-run wine in the vats. A ceiling to floor curtain has been installed which can be pulled to shut off the nine vats from the rest of the winery. When necessary, a heater warms the air around the vats to help both fermentations to start. The wine is then given its appropriate sulphur dioxide dose and cooled by opening doors and windows. It remains in the vats for a

The château building and the nearby vineyards are depicted on the Ausone label. The estate only has seven hectares of vines.

period of a month or slightly more, before being put into new oak barrels.

The wine is put into barrels in February, and the bung-hole stays on top for three months. The first racking takes place towards the end of April, the second in July, and the third just before the new harvest so as not to intrude on the making of the new wine. A pre-fining racking is done in December and post-fining racking takes place in late January or February. There are one or two more rackings before bottling in late spring or early summer, depending on the quality of the vintage. Approximately seven rackings in all are recommended because more may add too much sulphur, which tends to dry out or fatigue the wine. Blending—10 barrels at one time—takes place three times: the first when the first devatting takes place and the press wine is added; the second blending takes place at the racking before fining; and the last just before bottling.

The Ausone *chais* is dug out of the rock under one of the St Emilion cemeteries. The temperature inside varies only from 9°C to 12°C, but the *chais* suffers from humidity.

For Pascal Delbeck, the qualities of well-aged Château Ausone are that the wines possess a powerful nose and complexity. They are both delicate and well-developed.

Château Ausone is one of the main St Emilion estates. The estates in this area tend to be smaller than those on the other side of the Gironde.

The vineyard in front of the château shows the well exposed slope on the hill. This excellent exposition to the sun is a factor in the quality of Ausone's wines (LEFT).

Château Ausone has a power-ful nose and is a complex wine, which is eminently suited to lengthy ageing. These bottles of the 1975 vintage (BELOW) are still in the cellars.

The entrance to the chais (RIGHT) is hewn out of the rock. This is typical of this part of Saint Emilion. The cellars are very damp. Some dampness helps to stop the wine drying out, but an excess can be a prob-lem as fungus can thrive. Great care, therefore, is taken about cleanliness at Ausone.

Wooden fermentation vats (RIGHT) are still used at Château Ausone. There are nine in all, each holding 55 hectolitres.

Opening the windows and doors in the cellar (RIGHT) is a natural way to provide good air circulation. This helps to counteract any damp problem and, especially during fer-mentation, to control the temperature.

The Ausone vineyard is small and compact, so it is still economical to use a horse rather than tractor to work the soil, which is regularly tilled especially in the growing season. The vines are planted horizontal to the slope, rather than vertically (RIGHT).

CHATEAU CHEVAL BLANC

Château Cheval Blanc is always marked by an intense but elegant flavour and great depth of fruit. The bouquet nearly always has a large, ripe character. The wines have the structure and composition to age magnificently, and in overripe years such as 1947 produce an amazing breadth of taste.

The ownership of Cheval Blanc by the Fourcaud-Laussac family goes back to 1850 when a daughter of the proprietor took a portion as dowry when she married into the family. The rest was added in the next 10 years and there have been few changes since that time.

The domain covers a total of 41 hectares, 35 of which are planted with vines. The vines are 33 per cent Merlot, 66 per cent Cabernet Franc and one per cent Malbec. There was probably a little more Malbec in the past, but it was never grown in large proportions. In fact, there is very little left in the region, and when the vines at Cheval Blanc stop producing, they will probably be replaced by the other two grape varieties.

The soil is a mixture of gravel, sand and clay, but dominated by gravel. Being next to the Pomerol area, the soil reflects the typical composition of that region more than it does the standard St Emilion mixture. It even has the solid layer of iron oxide and stone found at Pétrus, although here it lies considerably deeper.

At harvest time some plots ripen sooner than others. Usually the Merlot from the most gravelly plots reaches maturity first, followed by Merlot from the more clay-like soil. According to the owner manager Monsieur Hebrard, they delay picking as long as possible to obtain grapes of maximum ripeness. Cheval Blanc is never among the first to begin and always among the last to finish. The management is fortunate in having several 'regular' families willing to come from the south at a moment's notice to harvest the grapes.

The property's drainage system was installed at least a century ago. The exact date is not known, but at that period it was one of very few vineyards to have any drainage at all. New pockets of underground water have been discovered by simply noting which patches of vines are less developed than their neighbours, and the necessary drainage improvements have then been made.

The average age of the vines is 35 years, which may seem to contradict the well-known fact that Cheval Blanc suffered more than most from the 1956 frost. In fact, this freeze did not kill the roots and the upper frozen part of the vines was sawn off above the former graft and replaced by grafting on fairly advanced branches of *vitis vinifera*. This technique enabled the vines to recover their yield in two years. The estate manager Monsieur Hebrard believes that the vintages of 1958 and 1959 were well up to standard in quality as well as quantity.

The present rotation policy is to replace a plot every two years. The procedure is to uproot the vines, plough up the earth, lay on a mixture of manure and humus and then apply disinfectants. After a stipulated time, an analysis is carried out which determines what may still be lacking and any necessary adjustments are made. The land rests for a total of two years between uprooting and replanting.

Harvested grapes pass into the crusher-destemmer and are completely destalked before being pumped into the vats. There are 18 concrete vats in the cellar with an average capacity of 100 hectolitres. One large vat with a capacity of 240 hectolitres is used for the *assemblage*.

The vatting period, including alcoholic fermentation and maceration, ranges from two to three weeks depending on the state of maturity of the grapes. Generally, the last few vats are kept back for a few days before being drawn off the *marc*. Pumping over the *marc* is carried out every two or three days during fermentation to supply oxygen and help to cool the must. An arrangement of pipes keeps the fermenting temperature below 30°C.

There are two pressings of the *marc*, but the first is gentle with the resulting juice added completely to the free-run wine. The second, which presses out wine much more concentrated with extract, is in fact little used in the Château wine. It is blended with the generic St Emilion or used by the family and staff.

After the wine is run off the *marc*, it is pumped into clean vats for malolactic fermentation. According to the cellar master, there have never been any hitches at this stage, and it is generally finished in two weeks, before the end of November.

The malolactic fermentation is followed by blending various lots. A more important blending then takes place just before the fining the following year. The aim of the later blending is to equalize the taste of the wood which sometimes varies from one cask to another. Just before bottling, a final blending of 20 casks is carried out. This involves 6,000 bottles, or one day's bottling. This wine is not pumped into the special blending vat, but forced in by compressed air. The vat is raised more than two metres above floor level so that the wine can move down to the bottling line through the force of gravity.

Except for light vintages, all the barrels used are made of new oak. The winemakers believe that a robust vintage can tolerate all new wood, whereas wine in years with less tannin would acquire a dominant woody taste. In such years, a proportion of barrels from a preceding year is often used, depending on the tasting. However, no barrels are as a rule used more than twice.

For the initial barrel-ageing, standard Bordeaux practice is followed. Racking is carried out every three months, topping up takes place several times a week, and fining with egg white. However, it is a practice peculiar to Cheval Blanc to leave the bung-hole uppermost for an entire year. A glass stopper is used for six months, followed by a wooden fining plug for the second half of the year. The winemakers believe their wine needs a fair amount of

1975 was an excellent vintage in the Bordeaux area. The label shows the area (St Emilion) and the names of the family who have owned the estate since 1850.

oxygen during the first year, and so do not follow the more economical procedure whereby the bung-hole is rolled to the side after four to six months.

Bottling depends on the quality of the vintage and the length of time it can age in cask. At Cheval Blanc this month-long operation is performed later than most, sometimes before the harvest, sometimes after. It has been performed as late as December, well over two years after the grapes were harvested.

The St Emilion–Pomerol regional climate is much more continental than that of the Médoc and Graves. In place of the Atlantic Ocean and Gironde-Garonne rivers, the major influences are the rivers Dordogne and l'Isle. This means that the climate tends to be less humid than that of the coastal regions, and there is less likelihood of a low yield through too much rain.

The particular geographical position of Cheval Blanc makes it more open to wind from the north than some domains on the hills around the town of St Emilion. It is also less protected than many domains in the *graves* plain which are too low to feel the full force of the icy northern air. This situation makes Cheval Blanc and some of its neighbours much more susceptible to frosts such as those which occurred during the winter of 1956.

Château Cheval Blanc produces no second wine, and

Château Cheval Blanc is in the St Emilion area of Bordeaux but close to Pomerol. The wine therefore shows a combination of characteristics.

The relatively modest, nineteenth century Château Cheval Blanc (ABOVE) contrasts with the elegant modern chais nearby.

everything of inferior quality is sold without a château name to shippers or kept back for the family and staff. The Château wine is sold entirely to *négociant* shippers.

Château Cheval Blanc is in the north-west corner of the St Emilion *appellation* and borders on Pomerol. This region seems to have a greater influence on the wine's style than St Emilion. Like Pomerol, Cheval Blanc has the advantage of being pleasant to drink relatively soon, while remaining suitable for laying down for several decades.

Jacques Hebrard (RIGHT) is Cheval Blanc's manager and is married to a member of the family which owns the Château. The Cabernet Franc vine beside M Hebrard is about 50 years old. This grape variety makes up 66 per cent of the estate's production.

The chef de culture is Guy de Haurut (RIGHT). The vineyards show the flatness of the plateau on which Cheval Blanc stands. The estate adjoins Pomerol.

This part of the modern chais at Cheval Blanc (RIGHT) is used for ageing the first year wine. New oak casks are used and, un-usually, the bung hole is kept uppermost for a whole year. The red staining on the casks comes from the frequent top- *ping up which is necessary when the wine is kept in this way. The maître de chais Gaston Vaissire (INSET) is using a pipette to take wine from a cask for tasting.*

CHATEAU FIGEAC

In St Emilion, Château Figeac displays, in a forceful way, its own unique character. The great individuality and strongly marked flavour of its wines derive from an exceptionally gravelly soil and its high proportion of Cabernet Sauvignon vines for the region, as well as the meticulous care in winemaking. The wine always has immense structure which is combined with dense fruit and a penetrating bouquet.

Like that of Château Ausone, the name Château Figeac goes back to the Roman occupation of France. It is named after Figeacus, who lived in a villa on the site of the vineyard. It became a small fortified château in the Middle Ages and garrisoned some of the troops of Henri de Navarre, later Henry IV of France. When he left for Paris and became a Catholic, the château was destroyed by Protestant Huguenots. It was then rebuilt under Henri's orders.

The family of the Duc Decazes acquired the 500 hectare property around 1450, and, in 1595, the central part of the present château was constructed. It passed to the Comte de Carles in 1654 through his marriage to Marie Decazes and remained with this family until the first half of the nineteenth century.

A plot of 13 hectares was sold to a gentleman called Ducasse in 1838 who also bought another larger plot in 1840. There had long been vineyards on the property and the wine had always been known as Figeac. Ducasse renamed the wine from his portion of the vineyards Cheval Blanc and purchased a third and final piece of land in 1860.

Other parts were sold off after 1838, including some small farms which had their own names. When their proprietors decided to plant vines, they added the name of the original Figeac domain. This is why so many other wine-producing properties in this area appear with the name Figeac attached.

Following the period of de Carles family ownership, there was a succession of six or more proprietors. This continued until Henri de Chevremont and his English-born wife Elizabeth Drake purchased the property as an investment for their eldest daughter in 1892.

The present owner, Monsieur Manoncourt, is grandson of that daughter. After her death, his father, who had never taken an interest in the estate, asked him to take charge. He has made the wine since 1943, but did not manage the estate fully until 1947. The vineyards were yielding only 10 to 17 hectolitres per hectare, so the domain had to be restored to profitability.

The geography of St Emilion is popularly divided into two categories: the *côtes*, which is composed of the slopes surrounding the town, and the *graves* which includes everything else. The first category has primarily limestone soil, but the second consists of *graves* or gravel.

What gravel there is is almost identical to that found in the vineyards of the best classed growths of the Médoc. It was brought here long ago by the river Isle, a tributary of the Dordogne, and covers only 60 hectares of the entire 5,000 hectare region of St Emilion. Figeac has 30 hectares, Cheval Blanc 25, the rest being shared by La Tour Figeac, La Tour du Pin Figeac,

Grand Barrail-Lamarzelle-Figeac and a few others. The soil in the rest of this *graves* region is basically sand and contains no gravel at all. In total, Figeac has 40 hectares of vine, so over 30 per cent of its vineyards are planted in this sandy soil. Cheval Blanc has about the same proportion.

The knolls or gradually rising slopes of the gravel bank resemble the classed growth terrain on the left bank of the Gironde. A series of these knolls begins at Figeac, which has three, continues in a straight line with two in the domain of Cheval Blanc, and ends at Château Pétrus in the Pomerol region.

The harvest, which takes place as late as possible, is performed by a mixture of locals, a regular Spanish family, and a group of gypsies. Grape variety is 35 per cent each for Cabernet Sauvignon and Cabernet Franc and 30 per cent Merlot. According to Manoncourt, Figeac has the highest percentage of Cabernet of any vineyard on the right bank of the Gironde. Average yield is approximately 30 hectolitres per hectare.

There are no Malbec vines because the conditions that made them necessary no longer apply. In the seventeenth and eighteenth centuries, when the English fleet came to Bordeaux in pursuit of wine for the home market, the buyers paid more dearly for the new wine than for that more than a year old. Neither the cork nor the notion of ageing wines in bottle had become accepted and the buyers knew that anything not still *primeur* was in decline. Young Cabernets have a certain amount of tartness or bitterness—particularly in St Emilion—and the Malbec subdued this characteristic. In itself, it has little character and has therefore been almost completely replaced in this region by the Merlot.

The cellars contain 10 oak vats and 10 stainless steel ones. The wooden vats are used first each year because if they are left idle for one or two years there is a danger they will be invaded by insects and tiny woodworm. The number of stainless steel vats used depends on the size of the harvest, then, but they do serve to cool the wine if overheating occurs during fermentation. Wine from wooden vats can be pumped into them, cooled by running water down the side of the vat, and pumped back to rejoin its *marc*.

The grapes are completely removed from the stems before being pumped into the vats. As few yeasts are present, fermentation in the first vat is slow in starting and takes about 10 days. Subsequent vats need only five or six days. Malolactic fermentation takes place in the wooden vats after separation from the *marc*. Wood has the advantage of retaining the warmth of the wine following the alcoholic fermentation, and so speeding up the malolactic fermentation.

Figeac is aged in new oak barrels every year. It is often a problem to resell them

CHATEAU-FIGEAC
PREMIER GRAND CRÛ CLASSÉ

St ÉMILION
Appellation St Emilion 1ᵉʳ Grand Cru Classé Contrôlée

1975

MIS EN BOUTEILLES AU CHÂTEAU
A. MANONCOURT PROPᵗᵉ A St ÉMILION · FRANCE ·

The wine of Château Figeac needs about 10 years before it develops its fine flavour and bouquet. The 1975 vintage produced some classic wines.

Thierry Manoncourt (BELOW)
is the owner of Château
Figeac. The Manoncourt
family have owned the property since the last century.
M Manoncourt personally
directs every aspect of the
winemaking.

Château Figeac is magnificently situated on the St
Emilion plateau on a series of
gravel ridges very well suited
for growing vines. Gravel is
always well drained.

because many growers in the region keep their wine partially or totally in the vat until bottling.

The *marc* is pressed twice and the first *vin de presse* is kept aside then blended with the *vin de goutte* if it is good enough. The first blending is carried out in January before the wine is put in barrel. The bung-hole is rolled to one side just before the first summer. Four rackings are performed in the first year and three in the second, including the bottling in June and July. In years when some lots do not have the character or taste required for Figeac, a second wine called Château Grangeneuve is made mainly from grapes of vines between five and 10 years old. All other wine made on the property is either kept for local consumption or sold as generic St Emilion.

Like those of the neighbouring Cheval Blanc, the vine-yards here were badly affected by the frost of 1956. The entire yield that year amounted to one and a half barrels. In some parts of the vineyard up to 50 per cent of the vines had to be replaced either entirely or partly, by grafting on new plants.

In most years the wine of Château Figeac is fairly powerful and long-lasting, and has some characteristics of Pomerol on which the estate borders. It has its own savour and intensity of flavour and is characterized by a certain finesse and breeding. In its younger years it is usually hard, tannic and closed-in. It needs a full 10 years to develop its real bouquet and fine St Emilion flavour.

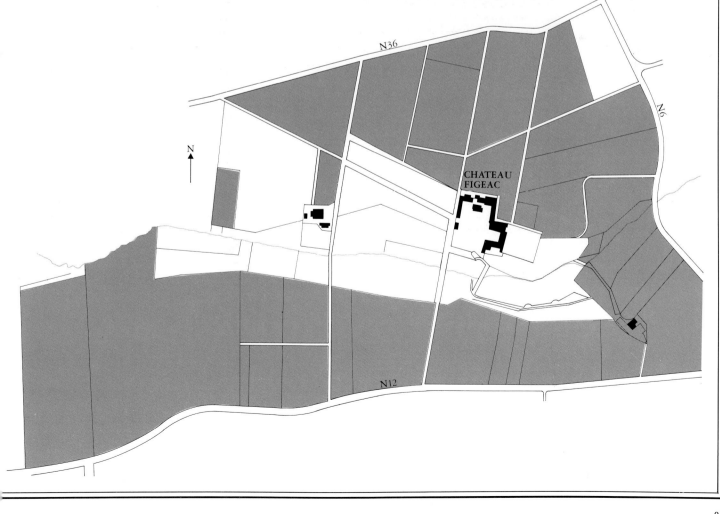

cellerman are racking the wine (RIGHT). In racking, a hose is normally used to run the wine from one cask to another often using gravity. It is vital to monitor the level of the wine in the first cask carefully and to stop the process when it reaches the lees.

In the fifteenth century cellars, the owner and principal

The owner Thierry Manoncourt (RIGHT) is standing in the vineyards on one of the gravel knolls which contribute so much to the quality of the wine.

This Roman built drain (RIGHT) is a reminder of the estate's origins during the Roman occupation of France.

Clément Brochard the chef de culture (FAR RIGHT) is holding some of the large quartz pebbles found on the gravel ridges at Château Figeac.

Unusually for St Emilion, Château Figeac grows a large proportion of Cabernet Sauvignon grapes which are well suited to the estate's gravelly soil. Only perfectly healthy grapes are picked. Those considered unsuitable are left on the vine (ABOVE).

CHATEAU PETRUS

This amazing small property makes some of the most exciting wines in the whole of the Bordeaux area. The taste is exotic, with dimensions and overtones of flavour which almost exceed what one could expect from a wine. Always generous and full, the wine blossoms out to a many-dimensioned complexity as the years go by.

In view of the several hundred years' fine wine history that surrounds most of the Bordeaux region, the area of Pomerol is a relative newcomer on the scene. Prior to the First World War it had little, if any, international reputation, and in France the region was known simply for its pleasant Bordeaux *rouge*. Château Pétrus won a gold medal at the Paris Exhibition of 1878, which was an impressive first for Pomerol and a great achievement. Château Pétrus is also mentioned briefly in the first Féret encyclopedia in 1893. But it is not represented here as an exceptional growth and there is no other nineteenth century classification. It then belonged to a family by the name of Arnaud and the wine was known as Château Pétrus-Arnaud.

Some time during the 1920s, the legendary Madame Loubat and her husband purchased a share in the property. Little by little she increased the size of her share until in 1946 she became sole owner. There were no direct descendants and the only heirs were a niece and nephew. Jean-Pierre Moueix, a good friend of the family, was asked to take the nephew's share, and in 1964, three years after the death of the proprietress, he became co-proprietor with her niece.

The fame of Pétrus owes at least its beginning to the prestige and charm of Madame Loubat, who sincerely believed it to be the best wine in the world. With much style and elegance, she introduced it into finer social circles so that its reputation grew rapidly during the inter-war years and after. Its quality and fame have been even further increased through the efforts of Jean-Pierre Moueix, and today it commands prices equal and often superior to the top classed growths of the Médoc.

The house itself, built by the Arnaud family, cannot be called a château. Indeed, there are few châteaux of the Médocain size or style in the whole commune of Pomerol, although Château Beauregard, Château de Sales and Vieux Château Certan are handsome exceptions. Madame Loubat lived at Château Pétrus for a time, but it has not been inhabited since her death. The building, interior and furnishings are kept in good order, much as they were in her era. However, the habitable portion forms only an addition to what is really the *cuvier* and the storage area for the wine which is produced by Château Pétrus.

Pétrus consists of one vineyard which makes a wide 'L' shape and, since the acquisition of four hectares by Moueix, covers just under 11.5 hectares. Even though the entire region is relatively flat, the ground gradually rises from all points to a plateau in the south-east section of Pomerol. The vineyard occupies an excellent position on the plateau, which gives it good drainage in all directions.

Soil composition is approximately 80 per cent clay mixed with 20 per cent gravel of the type found in the Médoc. Some say it is not real clay, being too coarse, but for the owners it remains that until proved otherwise. Further down, at about 70 centimetres, the clay turns bluish until it reaches the *crasse de fer* at a depth of one metre or so. This is a layer of iron oxide mixed with stone which creates a barrier so solid that not even the vineroots can penetrate it. Obviously, this contributes a great deal to the richness of the wine.

To improve drainage of this soil further, a 500 metre ditch was dug in 1971. Underground drainage pipes branch away from it into some parts of the vineyard to help drain off any water.

Merlot is planted in 11 hectares of the total vineyard area. This means that about 95 per cent of the wine is made from it. The rest is Bouchet, known elsewhere as the Cabernet Franc. The latter's state of maturity at harvest time determines the percentage of the blend and, in some vintages, only the earlier-ripening Merlot is used. The high percentage of this grape is the result of owners' and winemakers' conviction that it makes a perfect marriage with this type and structure of soil.

The severe frosts of 1956, which devastated several Bordeaux vineyards, also took their toll at Pétrus. The vines should have been torn up and replaced, but the ageing owner was afraid she would not see another vintage. As only the upper vines and not the roots were damaged, the local agricultural centre advised grafting new vines onto the existing roots, but predicted that they would only last 10 years. This was done, and most of the vines are still producing well 25 years afterwards. Pétrus may be one of the few châteaux where the actual age of the vineroots is unknown and impossible to know.

Pétrus has the enormous advantage of belonging to owners with several properties to harvest. Each year a small army of 180 pickers is brought in, enabling one hectare to be harvested per hour. It therefore takes only 12 hours to harvest the vineyard, and the best possible weather conditions can be chosen. Picking never begins before 10 am and rarely goes beyond 4 pm, so that morning and evening mists can be avoided. Picking under dry conditions can increase the alcohol yield of the undiluted grapes by as much as half a per cent.

The vineyard contains 6,269 vines to the hectare.

Pruning is carried out according to the simple Guyot method which leaves two shoots, and each plant is limited to six or eight buds. The use of chemical fertilizer is limited, but manuring is fairly important. With 14 different properties to manage, capital investment in equipment is substantial, and as many as six tractors can be put to work in the vineyard at one time.

The oenologist and wine-maker Jean-Claude Berrouet basically determines the style and structure of the wine. It is his

The wines of Pomerol tend to be ready to drink sooner than those of other parts of Bordeaux. 1975 was a very good vintage, with excellent keeping qualities.

practice to leave 10 to 30 per cent of the grapes on the stems, depending on the degree of tannin needed. After crushing, the must undergoes alcoholic fermentation and maceration for a period of two or three weeks. Fermentation is controlled at 29° to 30°C, and the must is pumped over several times a day to soak the *marc* and give the fermenting yeasts sufficient oxygen. This helps with the extraction of colour and tannin. Some press wine is added to the free-run wine before the malolactic fermentation begins to take place in the vats.

Following the malolactic fermentation, the wine is pumped into barrels. These are new unless the tannin content of the grapes is unusually low. According to Berrouet, there must be enough natural tannin to counterbalance the tannin from the wood. Otherwise the finished wine has too much oak or wood in the taste. To counteract this, older barrels can be used.

Christian Moueix, *directeur de vignoble*, is rigorous in his selection of the Pétrus wine. Each vat is as homogeneous as possible and identified according to vineyard plot, conditions at time of picking, and other unique attributes. Three months after the harvest, a blind tasting decides which vats have the necessary quality. Each year, out of eight or nine fairly small-sized vats, one or two are

eliminated. This amounts to 15 to 30 per cent of the vintage, or 60 to 120 hectolitres out of an average yield of 400 hectolitres in total.

In the view of the Pétrus winemakers, the wine gains its character from the constitution of the soil, the local climate, the grape variety and its adaptability, the age of the vines, and—not least—from the people who cultivate the vineyard. The single largest influence on the climate is the river Dordogne which snakes past Libourne, four kilometres away. Several small streams meander towards the Dordogne, so there is no lack of moisture in the atmosphere.

The Merlot is the perfect match for this clay soil and gently sloping plateau. It is a characteristic of Pomerol wines produced in this area that the nearer the vineyard is to the summit, the darker the wines become. This is particularly noticeable in the case of Pétrus.

In the opinion of the winemakers and owners, the wine of Pétrus is easily accessible for its pleasure and sumptuousness, warmth and tenderness. Finally, and like all Pomerol wines, it is ready much sooner than other Bordeaux growths, acquiring its full personality at a relatively early age, but keeping it for a considerable time, as old bottles of Pétrus constantly prove.

Château Pétrus is on the Pomerol plateau near the border with the graves area of St Emilion. Pétrus is noted for the high proportion of clay in its soil.

These harvesting vehicles (LEFT) carry the grapes from the vineyards into the cellars. The covers can be put on to the trucks to protect the grapes against possible showers of rain. It is bad for the grapes to get wet because of the diluted wine which is then produced.

This cartoon hanging in the cellar at Château Pétrus (ABOVE) depicts a maitre de chais holding a tastevin. This type of tasting vessel is in fact much more common in Burgundy than it is here in Bordeaux.

Vintage baskets (LEFT) are traditionally made of wood. Metal can contaminate the must. Cleanliness is extremely important even at this early stage in the winemaking process, so the baskets must be cleaned before the harvest and at the end of each day's picking.

CHATEAU TROTANOY

Château Trotanoy is a beautiful, glossy wine of intense fruit. Its full rich character gives it a lusciousness that is tempting both in youth and in maturity, while the soil provides the firm backbone which gives the wine its structure.

The name Trotanoy is derived from the old French words *trop ennoy* meaning too tedious. It applied to the terrain of the property which is excessively difficult to cultivate. The soil is a mixture of clay and gravel, and when it dries after rain, it solidifies into a concrete-like mass. There used to be a horse and ox to pull the plough, but even today using a tractor there are times when the soil is almost impossible to cultivate.

Trotanoy belonged to a family named Giraud until the Second World War, when they sold it to a gentleman called Pecresse. Like the Giraud family, he was an important property owner and sold the domain to his friend, Jean-Pierre Moueix in 1953. It is now a *société civile* entirely owned by the Moueix family.

Since its purchase by Jean-Pierre Moueix, there has been no change in the vineyard area of nine hectares. However, there has been some problem with rotation because of the fact that all the vines were of the same age. During the 30 years since the present owners bought the property, they have only been able to replant two hectares, so that at present there are seven hectares of very old vines and two relatively young. Christian Moueix, son of Jean-Pierre Moueix and *directeur de vignoble*, maintains that the choice of what to replant is difficult because, depending on the year, different groups of vines will produce different quantities. Furthermore, it is impractical to replant several small plots at one time. Production averages 30 hectolitres per hectare.

The vineyards are located on gentle slopes and are composed of gravel at the highest elevations, clay and gravel around the middle with pure clay at the lowest levels. In general, the soil is very dense and tends to be dry at the summit and often very damp at the bottom.

The whole vineyard faces west. Trotanoy was the only Pomerol vineyard to have a decent harvest in 1956, a year in which the February frosts killed many vines in the region and resulted in a meagre yield of only one or two barrels per domain. The vineyard's unusual position was partly responsible for this, but it was also due to the fact that the vines were a healthy age of around 30 years and thus able to withstand the shock of the extremely low temperatures which occurred that winter.

There are eight hectares of Merlot, one-half of Bouchet (Cabernet Franc) and another half hectare of various varieties, including a few rows of Pressac. Except for some slight clonal difference, this variety is identical to Malbec. It produces a large grape, quite fragile and full of juice so that it has always contributed more to volume than to alcoholic degree or quality. In the past it was used more extensively in the region than it is today, helping to mellow the bitterness of young wines by diluting them with juice which is lower in tannin.

Trotanoy can be picked in two afternoons by the large team that harvests all the Moueix-owned properties. There is a pyramid-like system, with Pétrus receiving top priority, Trotanoy and Magdelaine the next best, and so on.

It would be an enormous and unnecessary expense to install a crusher-destemmer at each property, so there is a new model mounted on a rolling chassis that is towed to each property for use as needed. This machine can be adjusted to different speeds as required by different grape varieties and maturity. It can also separate undeveloped or immature grapes because they are more firmly attached to the stalks.

Christian Moueix and the oenologist Jean-Claude Berrouet are responsible for making the wine. They believe in the concrete vats with their tartrate lining and do not find it inconvenient to use them, except that it is impossible to heat them. This is rarely necessary, but in 1974 it was so cold during the harvest that hot water had to be run through pipes immersed in the must to make the yeasts begin to work.

The ideal *cuvaison* period is two or three days of waiting for the fermentation to begin, eight to 10 days for the first alcoholic fermentation, and one week of maceration. Ideal fermentation temperature is between 27°C and 29°C, and it is very important to keep it from exceeding 32°C, which can destroy natural aroma and stop the fermentation before it is completed. No yeasts are added as the winemakers believe the indigenous yeasts are better for the wine's aroma. Malolactic fermentation takes place before bottling.

The first partial blending takes place after this secondary fermentation is completed and the wine run off. At this point, the press wine is almost always added in entirety to the free-run wine. In general, the winemakers use about 60 per cent—and sometimes as much as 100 per cent—new oak barrels. But this depends on the year, with powerful concentrated vintages meriting all new wood. Racking is carried out systematically every three months until bottling takes place which is usually in June, July or September of the second year.

The originality of Trotanoy's wine is attributable to its location and relatively poor soil. The result is always juice-rich, well-concentrated and powerful. The wines are never too acid or angular; the tannin is mature and never aggressive. The colour of Château Trotanoy is an intense dark purple—a result of the high proportion of clay in the soil and a characteristic of Pomerol wines.

The aroma and taste are superb, redolent of grilled almonds, roasted coffee, cocoa, and even producing a pleasant cooked or tar like odour. These are the recognized characteristics of the wine of Trotanoy, but it is the winemakers whose task is to try to exploit these properties to the fullest possible extent.

Château Trotanoy is in Pomerol. The estate has only nine hectares of vineyard.

APPELLATION POMEROL CONTROLÉE

CHATEAU TROTANOY
POMEROL
1961
SOCIÉTÉ CIVILE DU CHATEAU TROTANOY
PROPRIÉTAIRE A POMEROL (GIRONDE)
MIS EN BOUTEILLES AU CHATEAU

1961 produced the best Bordeaux vintage since the Second World War. The wines of Château Trotanoy have a rich, full character and a classic profile.

RIGHT *The majority of vines at Trotanoy are Merlot. The modest château building* (INSET RIGHT) *is typical of smaller scale properties in Pomerol. This relief sculpture of* notre dame des vignes (INSET LEFT) *is on the walls of the château.*

CHATEAU DE LA BIZOLIERE

Château de la Bizolière epitomizes all that is subtle and intriguing about the wines of Savennières. Beginning life rather austere and dry, the wines gradually soften and become more fruity with a few years in bottle. With age they can become really luscious, with a full-blown bouquet redolent of peaches and honey.

Savennières is an enclave on the right bank of the Loire, 15 kilometres to the west of Angers, within the more general Anjou *appellation* of Coteaux de la Loire. The Savennières *appellation* itself has two *grands crus* within it: La Roche aux Moines, which covers approximately 50 hectares, about 20 of which are planted, and La Coulée de Serrant, which covers less than five hectares.

These are *coteaux* vineyards, with the vines planted on slopes and benefiting from the extra sun that a south or south-west facing position gives. The rays of the afternoon sun are as direct as any winemaker could wish.

The Savennières vineyard begins at La Pointe, in the commune of Bouchemaine, and ends just before the village of La Possonnière. Four slopes, lying perpendicular to the river, succeed each other over a distance of about six kilometres, but the vineyards are never more than about 1,500 metres wide. The total Savennières vineyard area is about 360 hectares with approximately 100 hectares actually planted.

These rocky, schistous spurs jut out towards the river Loire. The roots of the vines penetrate deeply into the meagre, pebbly soil, and the schist undoubtedly reflects light and warmth. There are traces of fragmented blue volcanic rock, and these certainly have some influence on the wine. The vines on the steep slopes are planted vertically, they are trained on three wires at the highest points and on two elsewhere. In a hot summer, the thin soil can become very dry.

The Vignoble de la Bizolière has belonged to the Brincard family for more than a century and is today managed by Baron Marc Brincard. The nineteenth century château stands in a park of beautiful mixed trees and lawn. Baron Brincard is deeply involved in the running of the domain, and is ably assisted by his winemaker and *maître de chais*, Albert Giraud.

The Domaine de la Bizolière occupies 18 hectares of vineyard, mostly in small plots. The largest part of the wine produced on the estate is Château de la Bizolière AC Savennières, but there is a second white wine, Clos des Fougeraies, accorded *appellation* Anjou. The estate also includes a five hectare vineyard in La Roche aux Moines. This part was replanted with Chenin (or Pineau de la Loire, as it is known locally) in 1967 and 1968, after the soil had been rested for five years. The Roche aux Moines is more fertile than Savennières Clos du Papillon, another site along this slope, but it can become very dry in summer.

Other wines of the domain include two reds, Anjou Gamay, and Cabernet d'Anjou. Marc Brincard would prefer to be able to use the Coteaux de la Loire *appellation* for his Cabernet, as he believes this has more appeal for his customers. He also makes some white Sauvignon. This has no right to the Savennières AC, which is reserved solely for the Chenin, and so it is sold as Blanc de Blancs *vin de pays*. Sparkling wine is also offered for sale under the name Château la Bizolière, with the AC Anjou.

Anyone who makes Savennières has to be content with low yields. The maximum allowed is 25 hectolitres to the hectare, laid down by the decree giving the *appellation* in 1952. The possibility of annual adjustments upwards was allowed for, but the average yield remains somewhere around 20 hectolitres to the hectare. Frost can be a great menace to the area. In 1945, only one *barrique* of wine was made at the domain, and the 1977 growth was also badly hit, with the result that only one third of a normal crop was gathered.

The age of the vines varies. The domain vines in the Clos du Papillon, Savennières, are about 35 years old, and the balance between quality and quantity is good. Some of the Cabernet vines are about 30 years old, and the rest were planted some 10 years ago.

The white wines of Anjou are often spoilt by too liberal doses of sulphur, added both for stability and, in a good many cases, to stop the fermentation before the wine has fermented out so that some residual sugar is retained. Bizolière are determined not to fall into this trap, and their sulphur levels for Savennières do not exceed 40 milligrams per litre free, and 190 milligrams per litre total.

The white wines are pressed in a 50 hectolitre Vaslin press and fermented in barrels at a temperature of between 19°C and 20°C. Almost every year the cellar has to be heated to bring up the temperature, and added yeast is used. Bentonite is added to the must before fermentation for the *débourbage* or cleansing and clarification process. This helps the bouquet of the finished wine. The wines are filtered and, normally, bottled in the spring following the vintage, taking the Sauvignon first, followed by Savennières and La Roche aux Moines. The red wines are fermented in vats. The Cabernet receives two rackings and is filtered. The cold conditions in the cellar help to clarify the wine but do not enable it to fall bright—or clarify—completely.

A Savennières wine has to have a minimum of 12.5 per cent alcohol, and in the richest years it can exceed this by one or two per cent quite easily. An acidity of about five grams (expressed in sulphuric acid) would give a good acidity balance, and again, this is usually obtained quite easily, thanks to the fairly high natural acidity of the Chenin grape variety. In poor years, this can even be a problem, and it explains why, in the Loire, the Chenin variety is considered a suitable base wine for sparkling or *mousseux* wine.

Savennières wines present a tremendous contrast of tastes. They appear dry when

SAVENNIÈRES
LA ROCHE AUX MOINES
Appellation Savennières Contrôlée
Récolte du Vignoble du DOMAINE DE LA BIZOLIÈRE
à Savennières (Maine-et-Loire) France
MISE EN BOUTEILLES AU CELLIER DU DOMAINE

CHATEAU de la BIZOLIÈRE
APPELLATION SAVENNIÈRES CONTRÔLÉE
Récolte du Vignoble du DOMAINE DE LA BIZOLIÈRE
à Savennières (Maine-et-Loire) France
STILL TABLE WINE
MISE EN BOUTEILLES AU CELLIER DU DOMAINE

La Roche aux Moines (TOP) is planted with the Chenin grape. Most of the wine from the estate is the appellation contrôlée *Savennières, Château de la Bizolière (ABOVE).*

All the vineyards are on slopes with schistous soil above the Loire. In addition to the excellent exposure to the sun, the schist also benefits the grapes by reflecting the sun's light and warmth.

young, but as the wines age in bottle, they get richer and more 'ample', with the dry, nervy backbone fighting with the more liquorous overtones. This contrast is characteristic of the Chenin grape as grown on these schistous, steep slopes of the Loire. The area seems to bring out a steely side to the grape, which co-exists with the more generous elements. The nose has overtones of flowers and honey, and the palate unfolds gradually to complexity.

However, none of this can be seen in a young Château de la Bizolière Savennières. In 1980, the marvellous 1976 vintage is still relatively closed and reserved, almost spiky still, giving little inkling of the richness in reserve. In general, the wines of rich years, such as this, have five to six grams of residual sugar, whereas the drier years would only contain about one and a half to two grams.

In 1977, a 1947 Château de la Bizolière was still fresh and in full splendour. This was one of the great years, along with 1959, 1964 and 1969. The 1976 Savennières *appellation* Bizolière, the Roche aux Moines, both dry and *demi-sec*, and the Clos du Papillon Savennières *demi-sec* are still mixing austerity with sweetness at the start of the 1980s. Gradually, the latter takes over from the former.

The Countess of Serrant, one of the Empress Josephine's ladies, brought the fame of the wines of Savennières to the court of Napoleon. Since then, the wines of Château de la Bizolière have travelled much further than Paris, and their fame has spread well beyond court circles.

The maître de chais *at Domaine de la Bizolière is Albert Giraud* (LEFT). *The vineyards in the background are planted with Chenin grapes. This is one of the less steep parts of the vineyard.*

The château (BELOW) *is set in a beautiful park where sheep help to keep the grass down.*

CHATEAU DU NOZET—LADOUCETTE

The Pouilly Blanc Fumé wines of Ladoucette have all the pungent aroma of the Sauvignon grape, the flinty taste and the instant appeal that attracts devotees of this famous white wine. The crisp impact on the palate is a delight that whets the appetite and makes the second glass even more tempting.

The wines of Ladoucette elevate the Sauvignon grape to a new category. No one could accuse the Sauvignon of being reticent—it is a pungent, four-square type of grape variety, that wears its heart on its sleeve. It reaches its maximum expression in Pouilly-sur-Loire and Sancerre in central France, in the rather isolated area north of Poitiers known as Haut-Poitou, and in the wineries of a few devotees in California, but is otherwise fairly basic quaffing wine. But, tasting the 1976 Baron de 'L' Pouilly Blanc Fumé in 1980, the top *cuvée* from the Ladoucette stable showed that the Sauvignon could actually be described as having finesse, not a term usually applied to this grape variety. Such a wine is not always drunk as much as four years after it is made, but here the bottle age had done nothing but add a certain distinction, even in 1976, a year which did not have much holding acidity.

As with many great vineyards, the Ladoucette empire began round a château. The Château du Nozet has been in the hands of the Ladoucette family for six generations, and now it, and its vastly expanded business, are run by the young Baron Patrick de Ladoucette. Even in the dark days of 1973 and 1974, he took an optimistic view of the future and saw the potential of really well-made white wine marketed round the world. With this faith to back him up, Patrick de Ladoucette spared no expense in equipping the cellars to handle greatly increased volumes of wine.

The château is in the hills above Pouilly-sur-Loire, surrounded by the vineyards that have the right to produce Pouilly Blanc Fumé—Pouilly-sur-Loire itself, Le Bouchot, Soumard, Les Cassiers, Les Berthiers, Les Loges and Bois-Gibault. It is an area of small growers, with about 80 of them sharing 700 hectares of planted vineyard; so the single holding of 50 hectares of planted and producing vines belonging to the Château du Nozet is of considerable importance in the region. The grapes from the estate are augmented by must brought in from other growers of Pouilly Blanc Fumé, and the whole is vinified together. The wines are all produced at the Château du Nozet cellars which are hewn out of the hillside on three levels on the site of the old stable block. However, if this presents a picture of tradition, the new installations in the Château du Nozet cellars bring the winemaking techniques right up to date. The new plant and machinery are all based on Patrick de Ladoucette's conviction that the Sauvignon, unlike the Chardonnay, does not need any wood ageing or the influence of oak. His aim is to allow the Sauvignon to express its character through fruit and freshness, recognizing that its characteristic inimitable liveliness would fade in the barrel.

Care and attention begin in the vineyard, where the selection of grapes covers all the nuances of soil types in the area. Marl and kimmeridge chalk predominate, especially on the slopes of Les Loges and Les Girarmes at Tracy, with dry clay-calcareous soil at Le Bouchot, clay-siliceous soil at Saint-Andelain, and siliceous soil meeting at Tracy, from Bois-Gibault to La Roche. The Sauvignon particularly likes chalky soil. The great natural danger of the area is late spring frosts, which can decimate the vineyard. About two weeks before the vintage, a quantity of leaves are removed to give the grapes maximum exposure to the sun.

Stalks are removed, and the grapes are pressed in two types of press, pneumatic or mechanical, and then rested for 24 hours to clarify at 5°C or 6°C. Fermentations are carried out at low temperatures, about 14°C or 16°C, with each vat able to be cooled from the outside at the press of a button. This system virtually eliminates all risk of oxidation, which is a disaster for the Sauvignon, even if it is only slight. Whether malolactic fermentation is encouraged or not depends on the nature of the vintage. In a low-acidity year such as 1976, for example, the malolactic fermentation was prevented in order to safeguard as much acidity as possible; and there were only two rackings. However, in a high-acid year such as 1977, the greatest possible malolactic fermentation was induced in order to rid the wine of its 'green' malic acidity.

In January, the wines are tasted and blended in large vats which permit Ladoucette to make 250,000 bottles per *cuvée*, allowing the different wines in the blend to add their particular characteristic and enabling the wine of a year to maintain a constant standard. This often also balances acidity between vats, and the whole normally tastes better than the component parts, say the winemakers.

Following the blending, there is a February racking and the wine is kept in large glass-lined concrete storage tanks for two or three months more, until it is filtered and bottled in spring. Thus, the wine is always stored in large quantities and is bottled young, which maintains its freshness well. Normally, the wines leave the property about a year after being harvested, and this enables them to be kept safely for three, four or even five years on the export markets, although usually Pouilly Blanc Fumé is consumed in the second or third year of its life.

Pouilly Blanc Fumé keeps better than neighbouring Sancerre, which should be drunk very young. The soil difference is the reason for this—Pouilly is more chalky than Sancerre, which has a higher proportion of stones in its soil. Otherwise, the Sauvignon grape in both *appellations* makes them difficult to tell apart—perhaps Sancerre is even crisper and more lively, while Pouilly is slightly more solid, more round, perhaps just the more 'serious' wine.

In the production of their Comte Lafond Sancerre wines, the Ladoucettes buy in 100 per cent of their needs from the

Château du Nozet is an extremely important estate in the area around Pouilly-sur-Loire which has the right to produce Pouilly Blanc Fumé.

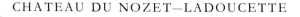

Sancerre growers, in the form of must. There is a separate receiving centre at Sancerre, but all the Comte Lafond Sancerre wines are made at the Château du Nozet.

Such an operation and such modern installations require management by highly trained staff, and the cellar master is Joel André, an oenologist trained at Montpellier. There is still further potential for planting with AC vines around the Château du Nozet, and expansion on the export markets has been at the heart of the success of Ladoucette wines.

The top *cuvée* Baron de 'L' is made from specially selected lots of wine thought to show superior qualities. But perhaps the most pungent Pouilly Blanc Fumé taste is in the straight Ladoucette wine sold round the world. The name *Fumé*, meaning smoke or smoky, comes from the fact that, at the vintage, when the ripe grapes are thrown into the vintage baskets, there is a little dust or 'smoke' which rises from the bloom on the grape skins, but many people still think that the *fumé* element can really be seen in the 'smoky' taste of the wine.

The château and its beautifully maintained gardens (CENTRE) have been in the Ladoucette family for six generations. All the Ladoucette wines are made here. The present owner is Baron Patrick de Ladoucette (FAR LEFT). He is standing by one of the boundary stones around the 50 hectare Ladoucette domain. Rognage (LEFT) is carried out a few weeks before the harvest. Foliage on the top of the vine is removed, allowing the vine to concentrate its strength on the ripening fruit.

These vineyards at Pouilly-sur-Loire (ABOVE) are planted with Sauvignon. The soil is stony and chalky. Pouilly-sur-Loire is one of the villages which makes wine for the Pouilly Blanc Fumé appellation, made entirely from the Sauvignon grape.

Château du Nozet has 50 acres of producing vineyard. The château itself is in the hills above the village of Pouilly-sur-Loire.

DOMAINE CLAIR-DAÜ

The wines of Domaine Clair-Daü are some of the most long-lived in Burgundy, needing time to open out and show their true paces. Even in light years, they have a structure that ensures drinking pleasure over a long period. The old vines give concentration and a tight, rich flavour that becomes more complex and fascinating as the wines mature in bottle.

Although Clair-Daü is the largest domain in private hands on the Côte d'Or, size alone does not imply greatness. It is the quality of the wines and their superb ageing potential which elevates this domain to illustrious heights. Here, even vintage characteristics can be defied, with normally light vintages giving wines of length and depth, while great vintages demand patience, but reward the drinker with splendour. Even a humble Marsannay Rouge can make a 'blind' taster search amongst the big vineyard names when presented with a Bourgogne Rouge de Marsannay 1961.

The sage winemaker behind these great wines is Bernard Clair, a qualified oenologist and a man of immense experience. Bernard Clair has built up the domain to its current size and prestige and does not count the cost in his efforts to make truly fine wine, realizing that short cuts are always found out in the glass.

The list of domain vineyards is impressive. It is headed by its holdings in Marsannay, the northernmost village on the Côte de Nuits below Dijon, where the domain's offices and cellars can also be found. There are 21 hectares planted in red grape varieties, mostly Pinot Noir with a small amount of Gamay. Some 2 or 3 hectares of this go to make Rosé-Marsannay, the colour being obtained from a 48 hour vatting period. The rest makes Bourgogne Rouge de Marsannay and some rather superior Bourgogne Passetoutgrain, which is predominantly Pinot Noir mixed with some Gamay.

At Marsannay, the domain also includes 2 hectares planted in the white grape variety Aligoté, sold almost as soon as it emerges and 1 hectare of Chardonnay—again, eagerly snapped up. The Chardonnay is grown near the top of the slope, where the soil is more chalky. This suits the Chardonnay variety well.

The range in Gevrey-Chambertin *premier cru* is immense, with 4.5 hectares spread over the Combe-aux-Moines, Cazetiers, Estournelles St Jacques, Clos du Fonteny (sole owner) and the great Clos St Jacques, easily, with Les Varoilles, the equal of a *grand cru*. There are 0.4 hectares in Chapelle-Chambertin, just over 1 hectare in Chambertin Clos de Bèze, about 0.3 hectares in Chambolle-Musigny, 2.5 hectares in Bonnes Mares, 0.2 hectares in Chambolle-Musigny Les Amoureuses, the same in Musigny, 0.3 hectares in Clos de Vougeot, 1 hectare in Vosne-Romanée Champs Perdrix and the same in Savigny-Lès-Beaune La Dominode, their only vineyard on the Côte de Beaune. About 1 hectare is being planted in Morey-Saint-Denis, half of which will be, unusually, in white—there is already a tiny proportion of white made in the upper part of the Monts-Luisants.

Fastidious winemaking begins in the vineyard. One of the great viticultural aids of recent years has been the development of spraying against rot, the impetus largely arising from the disaster of the 1975 crop. However, this is expensive (about 1,500 French francs per hectare) and only growers who do not grudge capital outlay insist upon it. In 1980, those growers who economized were caught by the rain, while, in the same year, the Domaine Clair-Daü vines were treated against rot four times. The result is wines far above the average.

When the grapes come in, there is hardly any *foulage* or crushing, as the aim is to preserve the grapes as whole as possible. In 1980, the must came in at a very low 8-9°C (with the vintage being so late in the second half of October, temperatures were naturally low). The temperature was raised to 15-18°C to start the fermentation, even touching 30°C at its peak. In 1979, the grapes came in at 22°C and the temperatures went soaring up; the vintage lasted from 1 to 10 October. Usually, there is a first fermentation stage of two days, followed by a slower process lasting 6 to 8 days. The wine then spends a further 2 or 3 days in the vats—the reds usually in 75 hectolitre wooden vats, while the whites are in enamel-lined ones.

The *chapeau* of skins is pushed down into the fermenting juice very frequently, often as much as three times a day. Chaptalization is done gradually, sometimes in three stages, so that the fermenting juice slowly absorbs the sugar and gains in richness and 'fat.' The policy on adding stalks varies with the year. In the view of the Domaine Clair-Daü a choice has to be made—the stalks take out some colour from the wine, but on the other hand, they add tannin structure.

The casks are oak from the Allier, and about 30 per cent are new each year. In Burgundy, the bung-hole is always on top and the casks are never rolled to the side, as in Bordeaux (*bonde de côté*). Thus, constant topping-up is necessary. The first racking is very late, just before the next vintage.

Clair-Daü consider that they possess the ideal sizes of vat and cask for all stages of the wine's development. While alcoholic fermentation takes place in the 75 hectolitre wooden vats, the malolactic fermentation is usually carried out in 15 hectolitre *foudres* or big barrels, which are imported from Germany. The wine then matures in normal Burgundian *pièces*, or hogsheads, with a capacity of 228 litres. Bernard Clair's wonderfully descriptive phrase to compare the way in which wines develop in different years is to say that vintages with rot 'oxidize,' while good vintages 'oxegenize.' This exactly describes the bad effect that air has on wines that are made with unhealthy grapes and the thoroughly beneficial effect that judiciously controlled air can have on a wine which is made from ripe, healthy grapes.

BONNES-MARES
Appellation Contrôlée
Domaine CLAIR-DAU

Domaine Clair-Daü owns 2.5 hectares in the excellent Bonnes-Mares vineyard.

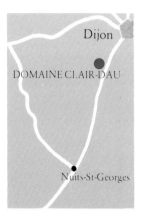

The sign of the domain (BELOW) *is against the creeper-covered house. The domain cellars, offices and about 23 hectares of vineyard are at Marsannay.*

He also has firm attitudes towards the use of fertilizer in the vineyard, believing that it must be put in the ground in November, straight after the vintage, so that it goes down into the earth. If it is put in during April, it is too late in the season. Great care, too, is taken in pruning the vines, with the climatic factors of the previous year being taken into account. In 1977, for example, the vines were pruned short, as they had suffered from the exceptional dryness of the year before.

In common with all other vineyards on the Côte d'Or, the hazards of hail can drastically affect yield. It is hoped that a plan to provide effective protection by using a light aeroplane to break up hail clouds will operate fully from 1981. The hail in 1979 not only had disastrous results in certain vineyards of the Côte de Nuits, but its damage could still be seen on the vines in their 1980 production. An example of how hail can dramatically reduce yield is the hectare of Vosne-Romanée belonging to Clair-Daü. In 1979, it produced only four *pièces* (1,200 bottles) when it would normally be expected to produce 4,655 bottles or even more.

The care and vinification methods of Clair-Daü enable the character of all the vineyards to express itself fully. The one Côte de Beaune wine, the Savigny-Lès-Beaune La Dominode, is greatly influenced by its old vines, almost

The headquarters of Domaine Clair-Daü are in Marsannay, south of Dijon on the Côte de Nuits. The estate includes holdings in Marsannay, Gevrey-Chambertin and Chambolle-Musigny.

Bernard Clair the oenologist owner of Domaine Clair-Daü (LEFT) *is standing in some of the vines at Marsannay during picking. The vines show clearly the effect of* rognage. *This is when, some weeks before the vintage, the top leaves on the vines are trimmed off. It is a frequent practice in Burgundy and Bordeaux. It means that the vine dissipates less energy in the foliage and can concentrate its strength on the ripening grapes.*

Clair-Daü has 4.5 hectares of vineyard in Gevrey-Chambertin premier cru. This section (BELOW) *is at Cazetiers.*

70 years old, which provide the wine with enormous concentration and richness. Chapelle Chambertin as well boasts vines aged about 70 years, while the average age of the vines in the Clos St Jacques is also impressive. Bonnes Mares is usually big structured and round, with Clos de Bèze being the epitome of elegance and breed. The 1962 Bonnes Mares was sheer balance and perfection in 1980, as was Clos Vougeot 1961 drunk in the same year.

All the wines beg for bottle age before being drunk. Even a year like 1973 from this domain is ample and full, only beginning to show its paces in the 1980s. Clair-Daü wines are not for those who lack patience but for those who seek excellence.

The great Clos St Jacques premier cru (BELOW) *is often the equal of a grand* cru. *To use the designation* clos, *the vineyard has to be surrounded by a wall. The gentle slopes and excellent exposure to the sun can be seen during picking* (INSET).

The vineyard at Marsannay (BELOW) *is rather pebbly. Of the 23 hectares, 21 are planted with red grapes and two with white Aligoté. The reds produce Bourgogne Rouge and Rosé de Marsannay.*

This horizontal press (BELOW) *is being filled with white grapes. Whites are pressed before fermentation and red after.*

These grapes (LEFT) *are arriving at the cellars' reception area.*

The horizontal press (LEFT) *is used for pressing the white Chardonnay and Aligoté grapes. Discs at both ends of the press move towards the centre, gradually pressing out the juice. The degree of pressure can be adjusted. White grapes are usually pressed without being de-stemmed. The chains in the press break up the* marc *or solid matter.*

The red wines are fermented in 75 hectolitre wooden vats. Pumping over (LEFT) *in one of the open 75 hectolitre vats where the reds are fermented involves spraying must over the cap of skins to break it up and lower the temperature slightly. This process helps give good colour immersion. If the cap stays on top, the colour is less good because it comes from the skins which should, therefore, be kept in contact with the wine.*

The oak storage barrels for the red wine of Clair-Daü (LEFT) *are 228 litre Burgundy pièces.*

The bubbles in the fermenta-tion vat for red grapes (ABOVE) *show the carbon dioxide produced during the fermentation.*

DOMAINE DES VAROILLES

The Domaine des Varoilles makes some of the most glossy, rich wines on the Côte de Nuits. They have a really full flavour, backed by true vinosity. This combination of charpente or structure and gras or fat ensures an ageing process that is wholly harmonious, although the wines have the felicitous quality of tasting well throughout their lives, both in the great and the lesser years.

The Domaine des Varoilles at Gevrey-Chambertin, like so many *grands domaines* in Burgundy, owns vineyards in a variety of top *appellations*. The Domaine des Varoilles falls into this small, favoured group, with some superb and prestigious plots of land, but also has an enormous asset in being the sole owner of the six hectares of the Clos des Varoilles, which forms the whole of the *premier cru* Gevrey-Chambertin of the same name (sometimes spelt Véroilles). This is the heart of the property, on which the fame and reputation of the whole domain is based. Managed as this is, the Clos des Varoilles is one of the best buys in the whole of the Côte d'Or, for its sheer value and class, a living, breathing example (for wine lives and breathes) of what quality-conscious ownership can do to a particular growth.

Around the Clos des Varoilles, an impressive array ot top vineyards has been built up. These include 1 hectare in Charmes-Chambertin, or Mazoyères-Chambertin, the *grand cru* that is the result of the merging of these two vineyards—the wine can be sold under either name, but the Mazoyères part is regarded as superior. There is also 1 hectare in La Romanée, the *premier cru* (since 1969) in Gevrey-Chambertin adjoining the Clos des Varoilles, 0.6 hectare in another *premier cru*, Les Champonnets, 1 hectare which is in Gevrey-Chambertin Clos du Meix des Ouches, near the *premier cru* Les Fontenys, and some vineyard in the Clos Prieur, only the high part of which is *premier cru*. There is also 0.5 hectare in Bonnes Mares, and 1.5 hectares in Clos Vougeot. There remains a small amount of Gevrey-Chambertin and of Gevrey-Chambertin Clos du Couvent. All these wines are vinified and aged in cellars under the old hospital in the middle of the village of Gevrey-Chambertin, round the corner from the domain offices.

Jean-Pierre Naigeon owns partly and manages the Domaine des Varoilles, and is aided by his sons, Patrice and Pierre. Jean-Pierre Naigeon is a gifted winemaker and a perfectionist, harder with himself than any critic could be. He is quick to recognize intrinsic quality in any wine, and constructively critical when he finds a wine lacking—in short, a taster who is not blinded by his own product. No doubt the fact that he does not have the somewhat insular attitude of some growers is partly attributable to his second profession as a successful *négociant*, but it is more the character of the man himself.

One of the aspects of the Domaine des Varoilles wines that impresses most is the willingness to declassify if, for climatic reasons or owing to the young age of vines, a wine is thought not to live up to its pedigree—that is why some of the more lowly *crus* here are such wonderful buys, while the top *crus* are really first class. The straight Gevrey-Chambertin 1976, for example, has in the blend the produce of the young vines

of the Clos des Varoilles (a much rarer occurrence in Burgundy than in the top châteaux of Bordeaux where, with the greater quantity, an owner can afford to be more restrictive), while the 1977 has, amazingly, some La Romanée and Mazoyères.

Another aspect of the domain is the age of the vines. Jean-Pierre Naigeon describes vines of up to 18 years of age as *jeunes vignes*. The Clos des Varoilles has some vines nearly sixty years old, those at Clos Prieur are about 50, and at Champonnets over 50, and Bonnes Mares has even older vines. La Romanée is considered as youthful, as it was planted in 1964. Some of the vines in the Clos du Meix were planted in 1902.

There are fascinating differences between the vineyards. Champonnets always has good natural acidity, very necessary nowadays in Burgundy where it has been consistently rather too low, because the site faces north to north-east, while the Clos du Meix opposite is differently situated. The Clos des Varoilles has a microclimate which enables picking to take place up to eight days later than elsewhere. Two-thirds of the Clos is surrounded by a wall —as it is over 100 years old this is acceptable, although a Clos created today would have to be entirely enclosed.

It is Naigeon's view that the character of the year dominates over the yield achieved. Vinification of Varoilles is classic, with some modifications. One of these is the way in which the *chapeau* of the fermenting must is kept in contact with the juice, by means of cables across the vats which, when pulled twice a day, break up the cap and mix it with the liquid. Flexibility is shown with regard to whether the grapes are destalked or not; for instance in 1980 no stalks were added. The same policy is applied to how long a wine rests in barrel and whether this is in new or older oak. In principle, each barrel is replaced every four years. Some wine is judged to need new oak, others not— an example of the latter would be the Bonnes Mares 1978, as the berries were small with little juice and a great deal of tannin, and so did not need further concentration. The press wine is always kept apart, and at a later stage Naigeon judges whether a proportion should be added. A first racking takes place when the malolactic fermentation is completed, usually in the spring after the vintage, a second in September, and a third before bottling. Most wines are bottled in the January or February of the second year.

Winemaking is always adapted to the year. The Clos des Varoilles 1976 is big and tannic, with a spicy nose—this wine was only vatted for a week as, given the conditions of the year, a normal vatting would have produced altogether too much tannin. Final selection is remorseless; for example, there was no Bonnes Mares 1979 from the domain as the July hail prevented all hope of good quality.

SOCIÉTÉ CIVILE DU DOMAINE DES VAROILLES
PROPRIÉTAIRE A GEVREY-CHAMBERTIN (CÔTE-D'OR)

GEVREY-CHAMBERTIN
CLOS DU MEIX DES OUCHES
APPELLATION GEVREY-CHAMBERTIN CONTROLÉE

Mis en bouteille   Bouteille N° 000674
au Domaine

PRODUCE OF FRANCE

SOCIÉTÉ CIVILE DU DOMAINE DES VAROILLES
PROPRIÉTAIRE A GEVREY-CHAMBERTIN (CÔTE-D'OR)

GEVREY-CHAMBERTIN
CLOS DES VAROILLES
1er Cru
APPELLATION CONTROLÉE

Mis en bouteille   Bouteille N° 013751
au Domaine

PRODUCE OF FRANCE

Domaine des Varoilles is the sole owner of the premier cru Clos des Varoilles. Some very old vines still grow the Clos de Meix des Ouches.

The outstanding quality of the domain's Clos-Vougeot has shown itself as the amount of vineyard has increased which shows that it is very difficult to make top quality wine without a certain mass. Very tiny parcels of vineyard almost always produce slightly 'edgy' wines. The Clos-Vougeot as made by Jean-Pierre Naigeon shows exactly the glossy depth of fruit and rich structure that this *appellation* should have.

The Domaine des Varoilles always gathers a high proportion of *Tasteviné* labels, a major accolade. This is particularly impressive in years that are not considered the greatest. In 1977, for example, the straight Gevrey-Chambertin, the Gevrey-Chambertin Clos Prieur, the *premier cru* Champonnets and the Clos-Vougeot were all *Tasteviné*. In 1976 both Champonnets and La Romanée received this distinction.

Jean-Pierre Naigeon (BELOW) is part owner of and wine-maker at the Domaine des Varoilles. He is seen here with his son Pierre, in the cellar at Gevrey-Chambertin.

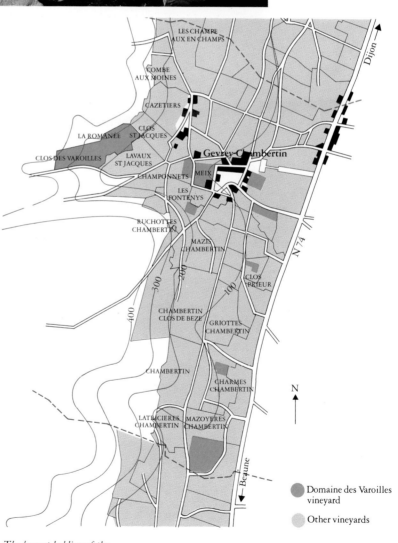

Domaine des Varoilles vineyard

Other vineyards

This part of the Clos Vougeot grand cru (ABOVE) is seen during autumn pruning. The pruned branches of the vines are not burned, as is the practice elsewhere, but are cut into shorter pieces by machine and mixed with the topsoil.

The largest holding of the Domaine des Varoilles is the six hectares Clos des Varoilles of which the domain is sole owner. The domain also includes other significant plots of land in grands and premiers crus within the commune of Gevrey-Chambertin.

*The Clos du Couvent (*INSET LEFT*) is a former seventeenth-century nunnery which is now a cuverie. The Naigeon family live in a beautiful old house (*INSET RIGHT*). The estate is run by the Naigeon family.*

*The Clos des Varoilles, basking in late autumn sunlight (*BELOW*), has a very good microclimate as it is sheltered and receives the last sun of the day. This contributes to maximum ripeness in the grapes.*

The cellars of the domain (BELOW) *are beneath the former convent. Jean-Pierre Naigeon tastes his wine regularly to follow its progress. Cables are stretched across the fermenting vats* (BELOW RIGHT) *which, when pulled, break up the* chapeau *or* cap *and mix it with the fermenting must. This is an innovation unique to the Domaine des Varoilles.*

DOMAINE ARMAND ROUSSEAU

The Domaine Armand Rousseau makes some of the most distinguished wines of the Côte de Nuits. Whether it be the famed Chambertin, with its enormous, all-enveloping bouquet and long, lingering taste, the concentration of the Clos St Jacques or the majestic, long-maturing Clos de Bèze, the wines are stamped with class and intense fruit. These are the heights to which Pinot Noir can aspire.

It would be difficult to find a winemaker more enthusiastic or more willing to share his views and experience than Charles Rousseau, owner of the Domaine Armand Rousseau at Gevrey-Chambertin. His beautiful house and enviable cellars are on the outskirts of the village, opposite a truly Burgundian church on the road to the Combe de Lavaux. The Domaine Armand Rousseau was one of the first properties to sell its wines *mise en bouteille au domaine*, and it has not sold wine to *négociants* for 25 years. This early desire to be entirely responsible for the quality of the wines from the property shows how closely the Rousseau family guard their high reputation and fame. Charles Rousseau has been making wine at the domain since 1946.

The vineyards are some of the noblest in Burgundy. Domaine Armand Rousseau owns a little more than 1.5 hectares of Le Chambertin, less than 1 hectare in Clos de Bèze, 1.5 hectares in Charmes and Mazoyères-Chambertin, 1 hectare in Mazis-Chambertin, just over 1 hectare in the Clos de Ruchottes-Chambertin (the domain is sole owner of the Clos, having bought it in 1977 from Thomas-Bassot), 2.5 hectares of the *premier cru* Clos St Jacques, 1.5 hectares in Clos de la Roche (Grand Cru, Morey-Saint-Denis), a little more than 1 hectare in Gevrey-Chambertin 1er cru and 2.25 hectares of Gevrey-Chambertin straight 'village' wine.

In the vineyard, M Rousseau believes in using the minimum of fertilizer, and no potassium has been added to the soil for 20 years. The soil itself varies with the *cru*—Chambertin is less high than Clos de Bèze and there is more soil, giving power to the wine as the roots do not go down to the rock, as at Clos de Bèze. This area is higher and the vines grow on a calcareous base; Clos de la Roche is on rock, which gives the wine finesse, while Ruchottes has thin soil.

Selection begins at the time of the vintage, when any rotten grapes are taken out of the baskets. It is Charles Rousseau's opinion that the man in charge must be in the vines when picking is taking place, to oversee personally what is happening, and therefore the picking team cannot be too big as this would make direct control impossible.

On the ground level of the buildings, there is the *cuverie* with the presses. The maturing cellars are underneath. The fermenting vats are open stainless-steel and are astonishingly modern for Burgundy. Temperature is controlled by means of a serpent pipe in the middle of the vat which can be used for both cooling and heating. The *chapeau* of skins is kept down in the juice by means of wooden poles called *piges*. Rousseau is not a great believer in adding the stalks, as he considers that they have disagreeable tannin, acidity and water, but sometimes a judicious proportion—between 10 and 15 per cent—is added purely as a means to add some backbone to a wine that needs this.

A small amount of the must is heated in copper pans and then tipped into the mass to start off the fermentation, which usually reaches a maximum of 27°C or 28°C. Fermentation lasts about 15 days, followed immediately afterwards by the malolactic fermentation which takes place in barrels. When Charles Rousseau is sure of the health of the vintage, he leaves the wine on the lees until September, otherwise he will add sulphur dioxide at an earlier stage. He now adjusts the free sulphur dioxide to 20 milligrams per litre before bottling, in order to avoid the necessity of sterile filtering. When a vintage is small, such as in 1978, he only does one bottling, but a larger vintage will be bottled in two stages, in April or May, and then in September, finishing with Chambertin, Clos de Bèze and Clos St Jacques which mature more slowly.

Chaptalization is carried out when the fermentation temperature is at its peak, so that the sugar is easily absorbed. Egg whites are used for fining which is done at the second racking. The *grands crus* are all kept in new wood.

All the usual variations in yield are to be seen at the domain—in 1978 the Clos de Ruchottes produced only 11 hectolitres to the hectare. The Clos St Jacques has vines which are 50 years old, whereas at Ruchottes they are younger, but the wine is just as good. Charmes has both young and old vines, while the Clos de la Roche is relatively young. Gradually, some of the vines at Mazis are being pulled out and will be replaced. The last vines to be picked are usually in the Clos St Jacques. At the Domaine Armand Rousseau, an unusual code is used to identify the wines—oo is Clos de Bèze, 01 Chambertin, 02 Clos St Jacques, 03 Clos des Ruchottes, 04 Clos de la Roche and so on.

Charles Rousseau considers a pH level of 3.3 or 3.4 ideal. He is convinced that when there is intensity of colour in a wine, colouring matter is concentrated, there is tannin and dry extract, and the vintage cannot fail to be good. He felt that about the 1972 wines, which were very astringent to begin with, but always had excellent colour, and the wines have developed into great ones. Faith, tempered with knowledge and experience, rarely makes mistakes of judgement in winemaking.

Purely for family consumption, Charles Rousseau makes an intriguing white wine, in tiny quantity, from the Aligoté grape, together with Chardonnay and Pinot Blanc. In 1980 there was almost no wine produced from the Aligoté variety.

The wines of the Domaine Armand Rousseau are famed for their penetrating, perfumed bouquet, the very essence of Pinot Noir. They have elegance and breed, with the stamp of class. The Clos des Ruchottes is always typified by marvellous, intense fruit, while Clos St Jacques has

Domaine Armand Rousseau was one of the first properties to sell its wines as mise en bouteille au château. *Armand Rousseau own 2.5 hectares of the* premier cru Clos St Jacques.

concentration and depth, Clos de Bèze 'attacks' the palate less than Clos St Jacques, takes longer to taste at its best, but has great finesse and length, while Le Chambertin itself is all power and enveloping colour, combined with taste and texture, this epitomizes what a Côte de Nuits Grand Cru should be. In 1980 the domain's Chambertin 1962 and 1961 were compared. The 1961 was slightly more closed, big and rich with hard-packed flavour; the 1962 was one of the greatest Burgundies of recent years. Its nose was of such intensity and beauty that it seemed to linger in the air around the glass while the taste combined an almost lacy elegance, the very epitome of breed in a wine, with a great, ever-opening array of palate sensations.

Charles Rousseau, the owner of Domaine Armand Rousseau (LEFT), is here seen standing in front of his open stainless steel fermenting vats. A horizontal press can be seen on the left.

This view of the Clos St Jacques shows the surrounding wall (ABOVE). By the end of November, the vines have been pruned and a light dusting of snow has already fallen.

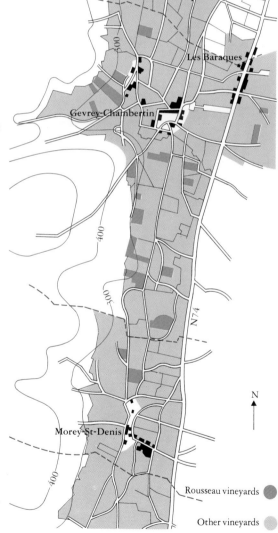

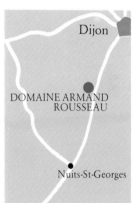

The domain includes parcels of some of the most prestigious climats in the Côte de Nuits which are spread over the communes of Gevrey-Chambertin and Morey-Saint-Denis.

Rousseau vineyards

Other vineyards

The entrance to the cellars of the Domaine Rousseau is seen with the offices to the right (BELOW). *The wooden door leads to the* cuverie *and press house. The maturing cellars are beneath.*

Le Chambertin (BELOW) *is one of the most important holdings in the Domaine Rousseau's grands crus.*

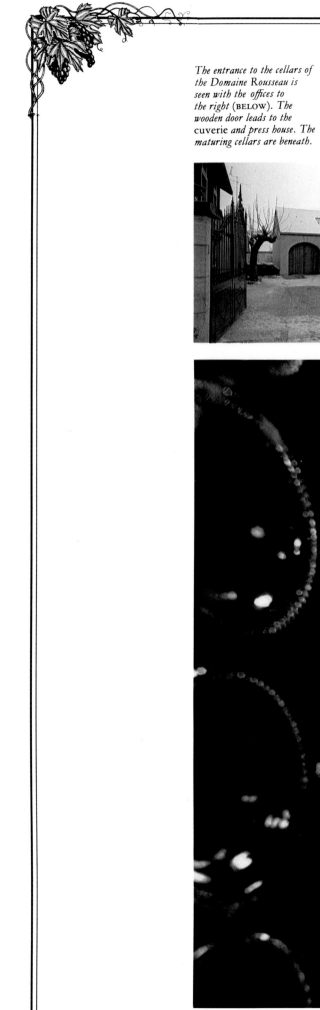

Inspection of the vineyard (BELOW) *at Le Chambertin shows the soil which helps give the wine power. The topsoil includes a fair amount of broken limestone rock.*

Here Charles Rousseau is showing the solid matter which has fallen to the bottom of the red wine vats behind him (BELOW).

A heater is placed in the cellar to encourage the steady progress of the malolactic fermentation in barrels (BELOW). *Warmth creates an atmosphere in which this fermentation can take place; cold will often prevent it from starting.*

The immaculate cellars of the Domaine Rousseau (LEFT) *are a good example of perfect bottle stacking.*

DOMAINE DUJAC

Domaine Dujac made its mark with astonishing speed amongst the group of top winemakers in Burgundy. Combining modern technical knowledge with tradition, the domain presents wines with a classic profile, beautifully designed for ageing, conserving all the body and fruit of Pinot Noir treated with the utmost care and with the minimum of handling.

Domaine Dujac, perhaps more than any other, shows what peaks can be reached when tradition and experience are married to the modern, questing mind and technical advances. The owner, Jacques Seysses, brings with him respect for what Burgundy can produce at its very best coupled with a willingness to borrow any new technology that areas such as California have developed and found to be good.

He is a comparatively new wine domain owner, coming, like many of the top Californian producers, from another profession, and he is fired with the perfectionism that this often brings. After a career in the baking industry, he and his father bought the Domaine in 1968 and have built it up to its present size and reputation.

Domaine Dujac consists of 11 hectares in the midst of some of the finest land on the Côte de Nuits. It has 10 *appellations*, including five *grands crus*. The Domaine holdings are in *village* or *commune* wines Morey-Saint-Denis and Chambolle-Musigny; for *premiers crus*: Morey-Saint-Denis *premier cru* and Chambolle-Musigny *premier cru*, both of them situated just below Bonnes-Mares; Gevrey-Chambertin *premier cru* Aux Combottes, magnificently situated, with *grands crus* plots on most sides. The *grands crus* are Charmes-Chambertin, Clos St-Denis, Clos de la Roche, Echézeaux, and Bonnes-Mares.

The domain house and cellars are in the village of Morey-Saint-Denis. The modest village of Morey added Saint-Denis to its name, after its famous Clos, following the example of Gevrey with Chambertin, Nuits with St Georges and Chambolle with Musigny. There are four *grands crus* in the commune of Morey-Saint-Denis, and Jacques Seysses has holdings in three of them—Clos Saint-Denis, Clos de la Roche, and Bonnes-Mares. Only a small part of Bonnes Mares lies in Morey; by far the greater part is in the commune of Chambolle-Musigny. Clos de la Roche, along with Clos de Tart, the other *grand cru* of the commune, tends to be the biggest wine of Morey. The Clos Saint-Denis combines charm and *mâche*, while Bonnes-Mares has much of the character and breed of the neighbouring commune.

Pruning is short, to limit quantity. This is a key to quality with the Pinot Noir. The desired yield at Dujac is under five tons per hectare. Dujac uses no chemical fertilizer, only manure and organic fertilizer such as guano. Obviously, with small plots of vineyard in 10 different *appellations*, a high degree of organization is required if the wines are to be made well. The order of picking has to be well-defined, and Domaine policy is to start with the young vines. The entire 11 hectares of the Domaine is vintaged in a week, thereby enabling all the different plots in the vineyard to be picked at more or less the optimum moment.

Jacques Seysses, who qualified as an oenologist in 1980, is working closely with colleagues at the University of Dijon. He is very flexible in his views on vinification, perhaps surprisingly so in this area of well-entrenched methods. He is reviewing the possibility of letting the must have a cold maceration for four or six days before starting the fermentation, with the aim of gaining colour at low temperature, and giving stability and complexity. The temperature would then be raised to 15°C to enable the fermentation to start, rise to 27 to 28°C, and then gradually be brought down again. No yeasts are added to the wine.

Unusually for Burgundy, Jacques Seysses neither de-stalks nor crushes his grapes before fermentation, thus the wines have firm body for lengthy ageing. *Vin de goutte* (free-run wine) and *vin de presse* (press wine) are mixed. Seysses prefers to chaptalize gradually near the end of the fermentation, as he sees the fermentation develop, rather than doing it all at once at the beginning of the process which tends to be the more usual method. It has to be admitted that the practice used at the Domaine could only be contemplated by very technically competent winemakers, as it demands extremely careful monitoring if no residual sugar is to be left in the wine.

Chaptalization is carried out with pure cane sugar. Naturally, in some years, it is barely, if at all, necessary. 1976 was such a year. Jacques Seysses is not in favour of wines that are too alcoholic.

Wine is not filtered at Dujac, so that there is a resultant natural deposit. Sometimes it is not fined (1976, 1978), and cold and natural gravity are relied on for clarification. Bottling tends to be quite early. In November 1979, the Morey Saint-Denis *premier cru* 1978 had already been bottled, but other wines receive longer ageing in the traditional Burgundian casks or *pièces*. Jacques Seysses is experimenting here too, as he feels he does not yet know the ideal bottling time for his wine. For example, with the Morey 1976, two barrels were bottled after 12 months, the main part of the crop after 17 months, and one barrel after 24 months. Time will tell if the wines develop differently.

At this domain, bottling is done entirely by gravity, as Seysses is convinced that pumping oxidizes and tires the wine. When in vat, the wine is kept under nitrogen gas. This certainly helps the keeping properties of the wine, but, if it is to be consumed young, it is better to decant in order to give maximum aeration and the chance to 'open up'. The deposit can also be left behind at the same time.

Jacques Seysses is convinced that the oenologists at the University and the winegrowers should have more and closer contact, carrying out research together and studying results. He does much active work to bring the two groups together. If the balance, restrained power and length of the

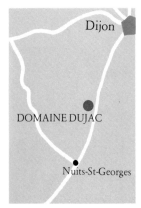

Dijon

DOMAINE DUJAC

Nuits-St-Georges

BONNES MARES
1976
DOMAINE DUJAC

CLOS SAINT-DENIS
1978
DOMAINE DUJAC

Bonnes Mares and Clos St-Denis are grands crus parts of which are in the Dujac domain. Both 1976 and 1978 were very good vintages in the Côte de Nuits.

The owner of Domaine Dujac
is Jacques Seysses, who stands
with his American wife outside their home and the cellars
in Morey-Saint-Denis
(BELOW). He is a qualified
oenologist who has built the
domain up to its present
reputation for quality wines.

Domaine Dujac wines are taken as an example of what
technical and traditional co-operation can achieve, many
more people should be persuaded to co-ordinate their
work and ideas.

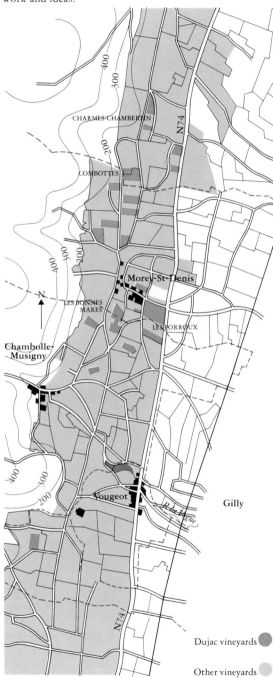

CHARMES-CHAMBERTIN

COMBOTTES

Morey-St-Denis

N

LES BONNES
MARES

LES PORROUX

Chambolle-
Musigny

Vougeot

Gilly

Dujac vineyards

Other vineyards

The 11 hectares of the
Domaine Dujac are spread
over 10 appellations in the
communes of Morey-Saint-
Denis, Chambolle-Musigny,
Gevrey-Chambertin and
Bonnes-Mares. This is a good
example of how some top
Burgundian domains are
fragmented.

The vineyard in front of the domain house and cellars is pruned in November (RIGHT). Small branches are burned in iron barrows that are brought into the vineyard.

In a part of the grand cru *Clos de la Roche* (RIGHT) vines have not yet been pruned and the very thin, stony soil is clearly visible.

Jacques Seysses is a collector of many types of wine (RIGHT) which reflects his wide interest in the subject.

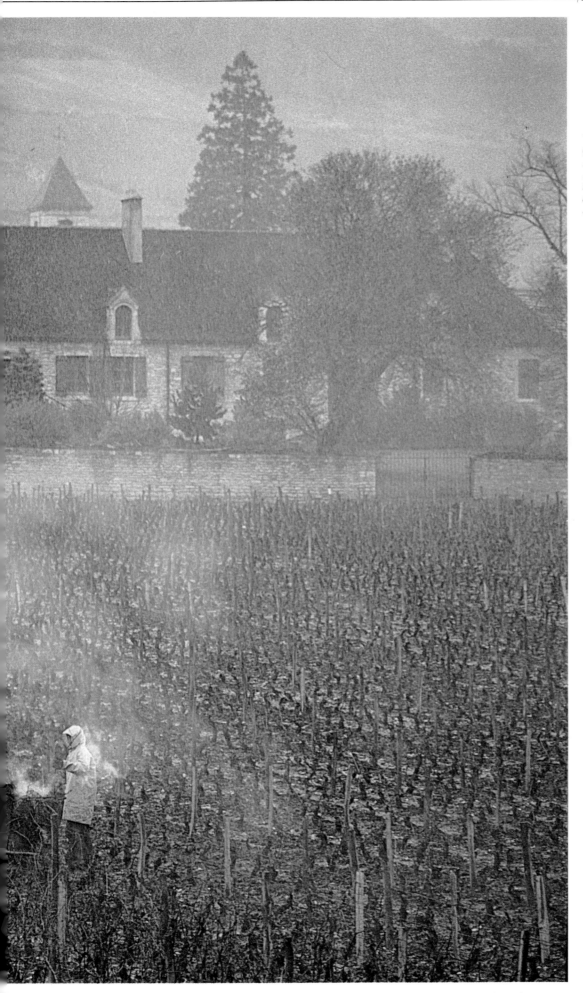

This view of the vineyard (LEFT) *close to the Domaine Dujac buildings shows the cold November work of preparatory pruning. The house has been carefully restored by the Seysses family.*

DOMAINE COMTE DE VOGUE

The wines from the Domaine Comte de Vogüé epitomize breed in some of the best growths of the Côte de Nuits. Their alluring scent, finesse and many-dimensioned complexity are worth waiting for, as these wines have the perfect balance and aristocratic origins in soil and site to develop in the greatest style and with the maximum of elegance. The wines have a real 'heart' to them.

The origin of the domain which today carries the name of Domaine de Vogüé goes back to 1450, and parts of this original holding have passed to the present owner, Comte Georges de Vogüé, through inheritance or marriage. Naturally, the composition of the estate has been modified and enlarged by exchanges or acquisitions carried out over more than five centuries by 17 generations of owners. From 1766, when Cerice François Melchior de Vogüé married Cathérine Bouhier de Versalieu, the de Vogüé family have been the proprietors of some of the most noble vineyards of the village of Chambolle. For over two centuries, five generations of the de Vogüé family have improved the composition of the domain, polishing the jewel which it undoubtedly is in the crown of the Côte de Nuits.

The vineyard holdings are formidable. The domain is the largest owner of the great *grand cru* Musigny, with over 7 hectares out of the total 10.71 hectare area. They also have over 2.5 hectares of Bonnes Mares, and 0.5 hectare in the *premier cru* Chambolle-Musigny Les Amoureuses. Together with the 1.80 hectares in Chambolle Musigny, the domain totals over 12 hectares in some of the most aristocratic vineyard land in the world.

The Domaine Comte de Vogüé is managed by one of the most able of the profession's *régisseurs*, Alain Roumier. He is steeped in Burgundian tradition and experience, as he comes from one of the great vine-growing families of the Côte de Nuits—the Domaine Roumier are worthy owners in Chambolle Musigny. There is no doubt that the cross-fertilization of ideas and knowledge is beneficial to both domains. With regard to the domain's policy on replanting, between 30 and 40 ares of vines are pulled out each year, giving their vines an overall age of about 30 years. Six or seven men work in the vines and cellars, a number which is altogether necessary in an area and with winemaking techniques which are highly labour-intensive.

Fermentation takes place in large wooden vats, of which the domain has an impressive array. Alain Roumier thinks that stainless steel vats would be much colder, and the must would often have to be heated, since the vats in Burgundy do not have a double lining to them to enable the temperature to be controlled easily. The vats are open, with the *chapeau* of skins floating on top. This is submerged regularly by means of *remontages* through a perforated pipe plunged into the middle of the vat, and also the use of poles to push the cap down into the fermenting juice, in order to give maximum colour and extract. Before fermentation, the grapes are crushed and mostly destalked. Between 5 and 10 per cent of the stalks are retained to give tannin, although too many stalks are thought to give bitterness to the wine. Chaptalization is carried out gradually, especially in warm years, when precautions have to

be taken to safeguard the wine against sudden rises in temperature during fermentation. The presses are Vaslin Véritas 22.

The domain carries out three rackings of the wine, mainly to eliminate the remaining gas in it—the first is after the malolactic fermentation, in the September when the wine is one year old, and before bottling. Great emphasis is given to cleanliness, and there is sterile filtering. There are small vats for assembling the wine before bottling, where the wine is kept under a layer of inert gas.

The de Vogüé domain has firm views on guarding its remarkable reputation, and does not hesitate to declassify when it considers that, through poor climatic conditions, it is not possible to make wines worthy of their illustrious *appellations*. Thus, in 1975, a year with much rot, the wines were completely declassified to the straight *village appellation* of Chambolle-Musigny. Even in 1977, a much better year, a few wines were demoted.

The domain's Musigny Cuvée Vieilles Vignes is justly famous; it is a wine of immense finesse and perfume. The meagre earth on this *grand cru* ensures that the vine has to struggle and gives the resulting wine a breed and astonishing blue-bloodedness. There are some young vines at Musigny, which are not included in this top *cuvée* but which provide the continuity necessary to ensure the future of the Cuvée Vieilles Vignes, for, if there was never any replacement of vines, there would come a time when the whole vineyard would have to be uprooted more or less at once. Musigny is always a very complete wine, without being hard; however, it develops only slowly to its full splendour.

The *grand cru* of Bonnes Mares is mostly in Chambolle, with a small part lying in Morey-Saint-Denis where the soil is more marly. With hard calcareous rock under the soil, vineyard work can require the use of dynamite. The Bonnes Mares of the Domaine de Vogüé is always a wine of substance, rich, very long on the palate, with backbone and structure.

Alain Roumier considers that a year with high sugar content and grape maturity is not enough in itself to produce great wine, proving that alcohol is not all. For instance, 1971, a year with hardly a rival for natural ripeness and sugar, is not the year with the most balance and longevity. The wines that may seem less stunning in youth, such as the 1972, are in fact those which can ultimately produce the finest bottles.

The rare white wine de Vogüé Musigny Blanc is produced from a small area—only 30 ares—of Chardonnay, and a little Pinot Blanc. This wine is an experience to be treasured by any drinker. However, the great red wines of the domain leave the drinker with the sensation that life has been truly enhanced.

CHAMBOLLE·MUSIGNY LES AMOUREUSES
APPELLATION CONTROLÉE
Domaine Comte Georges de VOGÜÉ
CHAMBOLLE - MUSIGNY (CÔTE-D'OR)
PRODUCE OF FRANCE
1959
Mis en bouteilles au domaine

BONNES·MARES
APPELLATION CONTROLÉE
Domaine Comte Georges de VOGÜÉ
CHAMBOLLE - MUSIGNY (CÔTE-D'OR)
1961
Mis en bouteilles au domaine
PRODUCE OF FRANCE

Domaine de Vogüé owns more than 2 hectares of the grand cru *Bonnes-Mares and a small area of the* premier cru *Chambolle-Musigny Les Amoureuses.*

An early fall of snow in November has covered the tiny Musigny Blanc vineyard (BELOW). Musigny Blanc is made mainly from Chardonnay but with a small amount of Pinot Blanc.

Alain Roumier (BELOW) is the régisseur of the de Vogüé domain. He comes from a highly respected Burgundy wine-producing family. He is standing in the maturing cellars with the Burgundian pièces stacked behind him.

A sample of a very young wine is being taken from an oak foudre with a pipette (CENTRE). The young wine remains in these large barrels for a short time after fermentation before being put into the smaller pièces. A sample is being taken from the stainless steel vat where the Musigny Blanc has been fermenting (LEFT). This wine is fermented in steel rather than wood to avoid the risk of oxidation.

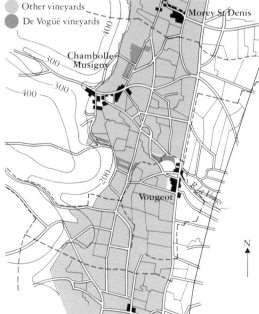

Other vineyards
De Vogüé vineyards

Morey St Denis

Chambolle Musigny

Vougeot

N

Dijon

DOMAINE COMTE DE VOGUE

Nuits-St-Georges

The de Vogüé family own the most prestigous holding in Chambolle Musigny, being the largest owner of Le Musigny with plots in some of the other renowned crus of the commune.

DOMAINE HENRI GOUGES

The Domaine Henri Gouges has always been renowned for the ability to bring out all that is magical in the taste of Nuits-St-Georges. The great splendour of the finest growths of this famous commune is the richness of flavour, the almost heady bouquet of concentrated fruit and the fine structure of the wines. They have an almost earthy intensity about them, coupled with a mouth-filling depth which is memorable.

No keen wine-drinker should pass up a chance to see exactly what constitutes that unique, earthy, individual taste that is Nuits-Saint-Georges, and there is no better way to learn than with a bottle from the Domaine Henri Gouges. With the domain's entire holdings in Nuits-Saint-Georges itself, there are no greater specialists of this fascinating commune, and no wine-makers more experienced in bringing out its intense, infinitely voluptuous taste.

The late Henri Gouges, the father of the present owners of the domain, Michel and Henri Gouges, built up the great reputation of this domain, a reputation which has continued to command the highest respect. Henri Gouges *père* was one of the first in Burgundy to institute domain-bottling, together with other reputed estates such as the Domaine Armand Rousseau and that of the Marquis d'Angerville at Volnay. Domaine Henri Gouges consists of just under 11 hectares of some of the best vineyard land in Nuits-Saint-Georges. This includes 1 hectare of the *premier cru* which attached its name to the village of Nuits, Les Saint-Georges. There is also nearly a hectare in *premier cru* Les Vaucrains, over 3.5 hectares in the *premier cru* Les Porrets (Gouges are the sole owners of the Clos des Porrets), just under 2 hectares in *premier cru* Les Pruliers, about 0.5 hectare in *premier cru* Les Chaignots, and under 0.5 hectare in La Perrière planted in a white mutation of the Pinot Noir which is very rare in the Côte de Nuits.

There are no *grands crus* in Nuits-Saint Georges, although Les Saint-Georges and Vaucrains would be candidates in the hands of the Gouges brothers. These, with Les Porrets, Les Pruliers and La Perrière are all on the south side of the village of Nuits, with Chaignots to the north. The soil at Nuits is basically marl, which would be too rich on its own for the production of top quality wine, but is ideal when mixed with sand and pebbles which have come away from the hard calcareous summit of the *côte*, or slope. Calcareous soil, on its own, would be too poor for the production of this kind of rich red wine. There are differences and nuances between the soils of the different *premiers crus*—the Clos des Porrets has some clay and heavier soil than most, Les Vaucrains has both types of soil, some lighter with sand and some with clay, while Les Saint-Georges has more sand in the mixture, giving it its extreme finesse. As a general guideline, the vineyards to the north of the village produce wines that veer, naturally, in character towards Vosne and are more supple and mature earlier than the great *climats* to the south of the village of Nuits.

The rootstocks used are the 16149, which gives grapes that mature well, and now a good deal of Téléki 5BB, which has given renewed vigour to the vines. Hail can be a problem at Nuits, and the Gouges brothers would like to see the effective organization of aeroplanes to break up the storm clouds and cause only rain, and not hail, to fall. In 1979, hail hit Nuits hard, especially the small parcels of land near the village, where the holdings are often tiny.

When the Pinot Noir grapes come in at the Domaine Henri Gouges they are nearly always destalked, as often the stalks add more 'greenness' than beneficial properties. The grapes are only lightly crushed, to separate stalk from berry. Fermentation takes place in closed concrete vats, and, as the grapes are virtually whole, it is partly a *macération carbonique* process, with the carbon dioxide gas present in the closed vat preventing the temperature from rising too high. On the other hand, the length of the fermentation is traditional. The inside of the vats is washed with hot water and tartaric acid before fermentation in order to destroy any acidity that might be in the concrete, as well as to clean the vats.

Chaptalization takes place at the beginning of fermentation. The Gouges are wary of too much new oak and its influence on their wine. They usually buy about 20 new barrels a year, and the wine only remains in them until the first racking. On the whole, the Gouges consider that new oak hogsheads or *pièces* contribute too much tannin to the kind of wine they intend to make, and the new barrels must always be *déboisés*.

Fining is done with albumen. Their policy towards racking is that it should not be a fixed 'routine', and that it is sometimes done more than is necessary. As a general rule, racking would take place three times during the life of the wine in barrel—in the spring after the vintage, before the following vintage, and before bottling. Perhaps it was done more frequently in the past, sometimes drying or oxidizing the wines, and they consider that if the harvest gives really healthy grapes, rackings are not so necessary. Before bottling, the wines receive a light filtering. As *pièces* differ one from another, and some homogeneous quality within the same *climat* is desirable, four or five hogsheads are blended together for a bottling.

The Gouges domain's white Nuits-Saint-Georges is a rarity and not at all in the classic White Burgundy mould of the Côte de Beaune. It is not made from the Chardonnay, but a white mutation of the Pinot Noir, which is not at all the same grape variety. These bone dry wines are strong and forceful, bottled a year before the reds. The 1978, a concentrated year, was very big, with 12.8 per cent alcohol, with a somewhat spicy nose which is not at all what one would expect from Côte de Beaune Burgundy from the Chardonnay. The 1977 had more acidity. They are wines of great individuality, but the fame of Domaine Henri Gouges rests on the noble red wines they produce.

A tasting *chez* Gouges of the magnificent

Both Les Pruliers and Clos des Porrets are south of the village of Nuits. There are no grands crus in Nuits-Saint-Georges, but superb premiers crus.

Domaine Gouges have planted the La Perrière premier cru vineyard (BELOW) in Nuits-St-Georges in a white clone of the Pinot Noir from which Gouges make one of the very rare whites from the Côte de Nuits. Gouges are the sole owners of Clos des Porrets

(BOTTOM) which is an enclave in the premier cru of Les Porrets. The vines have been uprooted from the strip of fallow land. When vines become too old, they must be taken up and the land allowed to rest and then be disinfected before replanting.

The vineyards of Domaine Gouges are entirely in Nuits-St-Georges where this estate owns a good proportion of the best premiers crus. In Burgundy domains often include vineyards in more than one appellation.

1978 vintage showed the immense variety of tastes within the range of *premier cru* vineyards, but all of them imprinted with that inimitable Nuits-Saint-Georges earthy, somewhat vegetal richness, with a great irony backbone running through them. Les Pruliers had a rich and fruity nose, with a wonderfully long and sweet, ripe finish. Clos des Porrets was truly stamped with that Nuits *goût de terroir*—earthy flavour—and was very flattering and all-enveloping. Vaucrains was the most tannic, the most closed at this young stage (when tasting from cask at 15 months old), with excellent depth and body—it is the most solid of all the *premiers crus* in Nuits, and the slowest to mature. Finally, Les Saint-Georges had a bouquet that was the most *grand vin*, complete, with all the ingredient parts utterly complementary. Underneath the power, there is finesse—in other words, it has everything. Gouges wines always mature slowly and grandly, although they never go through a 'spiky' adolescent stage. The 1978 vintage will be one of the greatest of all years.

DOMAINES LOUIS LATOUR

The domain wines of Louis Latour are some of the best known Burgundies throughout the world. The reds are big and built for long ageing, and the whites include one of the most distinguished produced anywhere from the Chardonnay grape—Corton Charlemagne. This is a fascinating, complex wine which, given the years it deserves in bottle, can astonish with its diversity of flavours.

The history of the Latour family can be traced back in parochial records to the beginning of the seventeenth century. Successive generations are listed as *vignerons* and coopers, and it was these professions that brought the Latours to Aloxe-Corton before the French Revolution. Through purchase, legacy and good fortune, the family acquired a small-holding of a few hectares of vineyard and was eventually able to establish a brokerage, supplying the merchants in town with wine from the vineyards in the vicinity.

In the mid eighteenth century, Louis Latour managed to expand the family business and established himself in Beaune as *négociant en vins fins*. He rapidly built up the company, creating markets in countries as far away as Russia. A few years later he was able to buy the considerable estate of the Counts of Grancey at Aloxe-Corton. Along with hundreds of others in Burgundy, this domain had been devastated by phylloxera and the problems of re-planting for absentee landlords like the Granceys were too great.

Louis Latour's acquisition consisted of 15 hectares of Corton vineyards, a run-down château and one of the finest *cuveries* in the Côte d'Or. Successive purchases, notably in Chambertin Cuvé Héritiers Latour, Romanée Saint-Vivant Les Quatre Journaux, and Chevalier Montrachet Les Demoiselles together with important vineyards in Beaune, Pommard and Volnay, have brought the Latour domain to its present impressive total of 45 hectares in *grands* and *premiers crus*.

Three-quarters of the Latour domain lies in the commune of Aloxe-Corton. The village is centrally placed in the Côte d'Or and is unique in producing both a red and white *grand cru*. The reds have a strong and robust quality which were much appreciated by Voltaire and Wagner, and the white Corton Charlemagne can rival Le Montrachet in finesse.

The Latour family has done much to improve viticulture in the village. Shortly after the war of 1870, the present Louis Latour's great-grandfather was responsible for planting Chardonnay vines on the south-facing slopes of the Corton hill, which for centuries had been dominated by the Aligoté and Pinot Noir. This pioneering work was instrumental in creating the famous *appellation* Corton Charlemagne. Today, the company owns over nine hectares of this vineyard and has a justifiable pride in the wine produced from the grapes which grow here.

During the phylloxera crisis, Latour's great-grandfather was among the few who strongly advocated the new method of grafting indigenous vines on to American root stock. The obstinate Burgundians took exception to these attempts to anglicize their vines, and hurled stones at him when he arrived at Pommard to demonstrate the

new technique. He persisted, however, and with others helped to bring about the successful replanting of the Côte d'Or vineyards.

All domain wines are vinified in the imposing early eighteenth century *cuverie* which stands in the middle of Corton Perrières opposite Château Corton Grancey. The grapes are crushed and destalked at ground level and the must is taken up to the vats on the first floor in large stainless steel waggons. These waggons or buggies are pushed along an intriguing old railway track which runs around the *cuverie* along the tops of the vats.

The wine is fermented for five to six days in large open-topped oak vats. To achieve a short fermentation, the must needs to be brought up to the temperature at which the yeasts will start acting on the sugars, so that the process does not occur haphazardly. Prior to being put into the vat, part of the must is heated in large double-skinned copper bowls which are mounted on the *cuverie* railway. Steam passed between the double lining of the bowls can heat up to 300 litres of liquid in only a few minutes. In a similar way, sugar is dissolved in a quantity of must before being added to a vat for chaptalization.

During fermentation, the skins rise to the surface of the vats and form a hard skin or cap. These caps are broken every morning by a squad of boys, clad in shorts, who jump into the vats and tread the skins back into the fermenting liquid beneath. This allows the alcohol to extract as much pigment as possible from the skins and prevents the fermentation from stalling.

The must is kept at a maximum temperature of 30°C by submerging large cold-water radiators into the vat. This simple but effective method keeps the must from reaching a temperature which might kill the yeasts and leave unfermented sugar in the wine.

Louis Latour has little time for the controversy over short versus long fermentation times. He follows a method that his family has always used and that has been practised in Burgundy for centuries. He is careful not to place too much importance on the fermentation, maintaining that the surest way to a good wine lies in the selection and care of the vine. Fermentation is simply the process of turning the finest grapes into the greatest wines.

The company is the last firm of *négociants* in Burgundy still to have its own fully operative cooperage. About 200 barrels are made every year from selected Jura oak. The wine spends up to 18 months in wood before being taken to the new bottling plant on the outskirts of Beaune. After careful filtering, the wine is bottled and returned to the cellars at Aloxe-Corton. These cellars lie over 20 metres under the hill of Corton and are a perfect breeding-ground for *racodium cellare*, a prolific white mould which the Burgundians call *confiture de cave*. It covers the

Les Chaillots at Aloxe-Corton (TOP) *is one of the important vineyards in the Latour estate; another is the magnificent Les Quatre Journaux in Romanée St-Vivant* (ABOVE).

Louis Latour (BELOW) is
head of Maison Louis Latour.
He is standing in the courtyard of the offices in front of
an old vertical wooden press.

bottles—and anything else left in the cellars for more than a couple of weeks—in a thick insulating blanket. These are perfect conditions for maturing wine. The mould preserves the corks, absorbs excess alcohol fumes and helps to regulate levels of both temperature and humidity.

Before any bottle leaves these cellars, the sediment is removed using a simple syphoning device invented by Louis Latour's grandfather. The bottles are brought up from the cellar, knocked gently on the ground and allowed to stand upright overnight. The next morning the old cork is removed and a nine inch long copper or silver needle is inserted into the bottle. The deposit, now lying conveniently at the bottom, is syphoned out along with a minute quantity of wine and passes via a rubber tube into another bottle underneath. The cleared wine is topped up and stoppered with a new cork before being sent back to the bottling plant to be cleaned, labelled and packed. Although another deposit is likely to form within five years, this method of removing the lees from the wine leaves the wine clear, and decanting, a process disapproved of by many Burgundians, is rendered unnecessary.

The day-to-day running of the domain is supervised by Louis Latour's young agronomist Denis Fetzman. During his five years with the company he has seen many dramatic scientific developments in viticulture and has been quick to see the benefits of the resulting new products. The effect of their implementation on the domain has been encouraging. The vulgar rot which so badly affected the 1980 vintage was successfully countered in many areas, and the incidence of both mildew and oidium is decreasing.

The Corton hill poses its own problems for the agronomist. The upper slopes are among the steepest in the Côte d'Or and everything from ploughing to spraying has to be done by hand. Erosion is severe and manure is still used as fertilizer, as it helps to bind the soil. The hill dominates the countryside between Beaune and the marble quarries of Corgoloin. Unusually for the Côte d'Or, it has a perfect south-facing exposure which is protected from the cold westerly winds by the hills surrounding the neighbouring village of Pernand Vergelesses. The angle of these southern slopes is ideal and acts as a sun trap which catches even the last rays of the evening sun. The exposure, together with a good, chalky, moisture-retaining soil, provides unusually fine conditions for the Chardonnay variety of vine.

Around the hill to the east the soil has more clay and is rocky and compact. The stones retain the heat and keep the ground warm. Frosts are rare as the cold air tends to fall to the lower slopes around the village. Grown on these eastern slopes, the Pinot Noir gives wines of outstanding bouquet and body. The complicated Corton *appellation* is divided between more than 10 separate vineyards. Louis Latour has holdings in six of these and in good years blends the finest wines to make his exceptional *grand cru*, Château Corton Grancey.

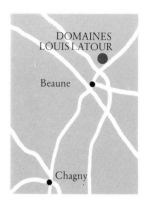

The village of Aloxe-Corton, in which some three-quarters of the Louis Latour vineyards are situated, is near Beaune in the Côte d'Or. The domain also has holdings in other areas, including Chambertin, Beaune and Pommard.

Chambertin soil (FAR LEFT) is reddish coloured and marly. It is less poor than in some areas, and this gives the wine great power. This bunch of perfectly healthy Pinot Noir grapes (LEFT) comes from the vineyard of Romanée St Vivant Les Quatre Journaux.

The Latour *vendages* are always started late to give the grapes as much chance as possible to ripen fully. In one or two years out of 20 this policy fails through extraordinarily severe conditions in late October, but the success rate far outweighs such failures. On a domain that stretches almost the full length of the Côte d'Or, care must be taken to prevent the grapes oxidizing during the journey to the *cuverie*, which can take up to 40 minutes. Traditional wicker baskets, now unhappily seldom seen at harvest time, are always used. The basket stops the bunches of grapes at the bottom from squashing before they reach their destination.

The wines of Louis Latour have always been renowned for their excellence. The family, still engaged in the daily running of the company, remains true to the local, loyal and constant traditions of Burgundy. But, like their ancestors, the Latours are innovators and are fully involved in the research, development and testing of new ideas and products for the benefit of both their own wines and those of Burgundy.

The firm has considerably expanded during the last decade. It is now probably the largest single buyer of top quality wine in the region. The company is still run from offices in Rue des Tonneliers in the centre of Beaune, which have been its commercial base for over a century. But, if its head and shoulders are here, then the heart of the Domaines Louis Latour lies where it has always been—in the village and vineyards of Aloxe-Corton.

These bunches of Pinot Noir (LEFT) have just been picked in the vineyard at Aloxe-Corton Les Chaillots. The grapes are taken from the vineyard to the cellar where they are loaded into the destalker (BELOW). Before picking, it is important to check the sugar content of the grapes. The refractometer (CENTRE) is a handy, portable device for this purpose. A drop is placed on the face of the prism at the base of the instrument and the result read off a calibrated scale.

This is the traditional way of keeping the chapeau in contact with the wine (LEFT). This takes place particularly in the initial stages of fermentation. Treading the wine is a relatively rare practice nowadays.

Density checks (CENTRE) take place regularly on the fermenting must to see what stage the fermentation has reached. As fermentation proceeds, the sugar reading goes down and the alcohol reading up.

In chaptalization (LEFT) the sugar is dissolved in some fermenting must and then returned to the vat and stirred. It is illegal to dissolve the sugar in water.

Chaptalization or the addition of sugar which is turned into alcohol (ABOVE) takes place near the beginning of fermentation. Copper implements do not taint the wine's taste. The froth shows that the fermentation is in progress.

After the wine has been run off from the fermenting vats, the remaining marc is removed prior to pressing (ABOVE). Topping up (ABOVE RIGHT) with a similar quality of wine helps compensate for evaporation.

DOMAINE LOUIS LATOUR

BOUCHARD PÈRE ET FILS

The wines of this great Burgundy firm are some of the most classic of their type in Burgundy, with the characteristics of each <u>cru</u> within an <u>appellation</u> keeping their individuality. The range of Beaune <u>premier cru</u> wines shows just how much one can differ from another, always scented and with a remarkable depth of fruit and rich texture.

Bouchard Père et Fils is not only one of the most renowned *négociants* in Burgundy, but also the largest single vineyard owner. The domain in all its many parts is titled the Domaines du Château de Beaune, since the offices of Bouchard Père are *au château*, in the walls surrounding the city of Beaune.

Bouchard Père have existed since 1731. The present president and managing director Claude Bouchard, represents the eighth generation of the family; and the ninth is already active in the company—one of Claude Bouchard's sons and one of his cousins already work there. Continuity evidently brings its advantages, but this has to be backed up by a willingness to adapt to the times, while refusing to discard the worthwhile traditions of the past.

The extent of the domain is handsome, with 80 hectares, of which 68 hectares are *grands crus* and *premiers crus*. This would be astonishing enough without the very complicated and intricate division of this domain. The staff managing such vineyards—both viticulturally and in the cellar—must have a head for logistics and good, sound training. André Bouchard, a cousin of Claude Bouchard, is responsible for the domain vineyards, the oenologist in the laboratory is M Prost, and M Bailly is *chef de caves*.

The most important vineyard holdings are in Beaune *premier cru*, where, amongst others, there are 8 hectares in Les Aigrots, 4.5 hectares in Les Avaux, 3.5 hectares in the Clos de la Mousse, over 2 hectares in Les Cent Vignes and Les Marconnets, almost 4 hectares in Grèves Vigne de l'Enfant Jésus, nearly 2 hectares in Les Grèves, and 2.5 hectares in Les Teurons. There are 4 hectares at Savigny lès Beaune les Lavières, as well as holdings in Pommard, Rugiens and Combes. Volnay is well represented, with nearly 4 hectares in Caillerets, 1.5 hectares in Fremiets and over 1 hectare in Taillepieds. At Aloxe-Corton, there are nearly 4 hectares of the magnificent red Le Corton and 3 hectares of white Corton Charlemagne. More white vineyard is found at Puligny Montrachet, with 1 hectare of Montrachet itself (about 15 per cent of the total) and 2 hectares of Chevalier Montrachet (about 30 per cent of the total), while there are nearly 1.5 hectares of Meursault Genevrières—the Bouchards own more than 10 per cent of this vineyard. Bouchard Père even have a small holding in Chambertin and a larger one in Chambolle-Musigny on the Côte de Nuits. The family has also put their faith in the lesser wines of Burgundy —they own 5.5 hectares of Bourgogne Aligoté at the village of Bouzeron on the Côte Chalonnaise, the best area for this white grape variety.

Bouchard Père vinify about 3,000 *pièces* of wine a year—a Burgundian *pièce* or hogshead holds 228 litres, or 300 bottles. The first figure includes the grapes from their own domain, plus those they buy in— about the same quantity—in their role as a *négociant-éleveur*. They vinify Nuits-Saint-Georges Clos Saint Marc, for which they are the exclusive distributor, as well as the more modest Château de Mandelot in the Hautes Côtes de Beaune. They have other contracts for the purchase of grapes bought from some of the top *premiers crus* in Nuits-Saint-Georges, and at Chambolle Musigny, Clos de la Roche, Morey-Saint-Denis and Gevrey Chambertin. Wine is bought in from Meursault and Puligny-Montrachet, and must from Chablis. They also distribute the prestigious La Romanée and Vosne-Romanée 'Aux Reignots,' which are vinified at the Château de Vosne-Romanée with technical assistance from Bouchard Père.

There is obviously an order of picking in all these vineyards. This follows a steady course, whatever the character of the vintage. Volnay Caillerets and Beaune Marconnets, for instance, are picked before their neighbours in the same *appellation*, due to their excellent exposure to the sun and subsequent good ripening; the Caillerets also has old vines, which add to its natural characteristic as the most robust and full-bodied of the Volnay *crus*. The Clos de la Mousse is picked before either of these two, while Le Corton is picked last of all.

The red wines are fermented in traditional wooden vats, with the *chapeau* of the skins kept in contact with the juice by means of *remontages*, pumping the must from the base of the vat and spraying the top, treading, or the use of a wooden pole during the tumultuous stage of the fermentation. Finally chaptalization takes place and a wooden lid is placed on the *chapeau*. White wines are fermented in oak casks. With the Pinot Noir, the maximum colouring matter is drawn out of the skins in about 4 days, and the optimum tannin in about six to 10 days, the time it takes for alcoholic fermentation. Bouchard Père are very well equipped to control both temperatures and sugar densities, to warm or to cool the must, and to see that the alcoholic fermentation progresses continuously and steadily.

In most years, the malolactic fermentations follow naturally soon after the alcoholic fermentations; usually at tastings of the new wine in November, some have started and some have not. During this period the cellar is kept at about 17°C. The malolactic fermentation takes far longer to start on the white wines than the reds. When the process is finished, both red and white wines are taken off the lees immediately.

There is a rotation of new and old Limousin oak casks. The new ones have water—never steam—run through them when they arrive, while the casks that have held red wine can be 'decoloured' to hold white. Wines like the Beaune-Grèves Vigne de l'Enfant Jésus and Le Corton spend 50 per cent of their total maturing period in new oak casks and rotate with the older casks at the rackings. There are usually three rackings before bottling—the 1979s

The Corton vineyard, of which Bouchard Père own 4 hectares, produces fine red wine which ages beautifully.

started to be bottled in December 1980, while the incredible red 1978s were bottled in January 1980. The red wines are fined with egg whites, the great white wines with fish glue.

As an indication of yield, the figure of approximately 25 hectolitres to the hectare was achieved for the reds in 1980, against between 35 and 40 in 1979. Chevalier-Montrachet in 1980 produced 40 hectolitres per hectare, but Le Montrachet only produced 38 hectolitres per hectare.

One of the really exhilarating elements of tasting the domain wines of Bouchard Père is their great diversity. This enables the taster to see the differences between the various vineyards within Beaune or Volnay, for example, and the differences between the 'male' character of Pommard against the more 'female' delicacy of Volnay. Although the wines of Beaune present a minefield to the unknowing or inexperienced taster—so often the best that can be hoped for is a pleasant, straightforward taste— at Bouchard Père, all the nuances are there to be discovered. A tasting of Beaune might begin with the Clos de la Mousse, usually light and delicate, with charm and *nerf*, as befits its origins on a rocky base. Beside it Beaune Teurons always seems more brutal in youth since the former evolves far

quicker even in cask, but Teurons is bigger and well-constituted, with a very 'Beaune-like' character. Beaune Marconnets usually has more backbone and often more tannin; it is a deeper wine, perhaps even with a little Côte de Nuits character. Then, finally, there is Beaune-Grèves (derived from *graves*, or gravel, leading to finesse in wine) Vigne de l'Enfant Jésus, all elegance and length and breed.

The same evolution of taste can be seen in the Volnays, with Taillepieds usually soft and forward, Fremiets delicate and scented, and the majestic Caillerets with more *ampleur*, power and richness. Le Corton is perhaps the finest red wine of Bouchard Père. It is the longest-living wine of the Côte de Beaune and, as befits a very *grand cru*, is one-dimensional in youth and matures to absolute complexity.

Beaune du Château is a most successful blend of *premier cru* vineyards, such as Cent Vignes, blended between vintages to obtain a full style. It exists in red and white. The same differences show themselves with the white wines. The Chevalier Montrachet is always more forward than the Montrachet, showing its personality first, while the Montrachet is always fatter.

The Domaines du Château de Beaune is the largest single vineyard owner in Côte d'Or with holdings in all the best climats of Beaune as well as in Volnay, and Pommard, and great white vineyard in Montrachet.

The Montrachet vineyard (LEFT) produces one of the greatest white wines in the world. Bouchard own one hectare in Le Montrachet and make wines that take years of bottle development to show at their most superb.

Claude Bouchard (TOP) is now president and managing director of Bouchard Père et Fils. He has a great knowledge of Burgundy's vineyards and follows all the developments at the Bouchard domain closely.

The sediment is being checked in a bottle of wine (ABOVE LEFT). It is important to keep the bottle horizontal when it has been taken from the bins. The cellars for ageing the wine in bottle are under the Château de Beaune in the town walls.

DOMAINE COMTE ARMAND

The Clos des Epeneaux from the Domaine Comte Armand is a single vineyard of the greatest class. This entirely enclosed plot of vines within the top Pommard <u>premier cru</u> of Les Epenots produces one of the most majestic, rich wines of the Côte de Beaune. The great style and power of this wine is only matched by its projection of bouquet and fruit that makes it distinction itself.

One result of the French Revolution was that large properties, whether they were owned by aristocrats or not, were split up and divided into a mass of smaller farms or vineyard domains. It was rare for a great estate to survive this radical social change, but the magnificent Clos des Epenaux, as it was spelt in 1892, at Pommard remained the property of M le Comte Armand, and is still entirely owned by the present Comte Armand today. He is, in fact, Vicomte Armand, and a barrister.

The Comte Armand is sole owner of the Clos des Epeneaux, the remarkable 5.16 hectare plot of vines within the top Pommard *premier cru* of Les Epenots. This enclave of precious land is enclosed by a wall, as French law now stipulates if a vineyard is to benefit from the name of *clos*. Within the *appellation* of Pommard there are vineyards both to the north and south of the village. Clos des Epeneaux and Les Epenots lie to the north of the village on relatively gentle, south-facing slopes. To the south of the village on much steeper, east-facing slopes lie the other great *premier cru* vineyards of Les Rugiens which produce rich, powerful wines from reddish, irony soil.

Philibert Rossignol has been *régisseur* at the Clos des Epeneaux since 1950. His son, Michel, is in charge of the vines. They both bring a wealth of experience to these responsible posts, also owning vines and making wine in their own right—Rossignol is a well-known name in the area, with 17 different members of the family in and around Pommard.

The soil of the Clos is both pebbly and of a warm, ochre colour, resting on a lava rock base. Michel Rossignol is quick to point out that the soil of the Côte d'Or is tired, having been continuously planted since the Middle Ages, and therefore it must be handled with care and never forced to produce too much. In the middle of the Clos there is a small wooden cabin, and around this are the oldest vines. Otherwise, there is a good rotation of vines of all ages, with the youngest between five and eight or nine years old, and the greater part averaging around 18 or 20 years. When a young vine reaches production in the fourth year, another can be pulled out. Wine from the very young vines is sold to *négociants*. The rootstocks chosen are the Téléki 5BB variety, which gives a good deal of wood, with powerful wines which can, however, lack sugar. The 16149 is more fragile, but with less wood and gives grapes that ripen well, while the SO4 gives wines that do not last long. Weed-killer is used in the vineyard, rather than working the soil. The tools used for this work can harm the base of the vines, and often the scars can be seen on the wood. Michel Rossignol's view of picking machines is that they are bound to come—perhaps in 10 years' time they will be visible all over the Côte d'Or. But he is not training his vines higher to prepare them for this;

his attitude is that the machines must adapt to the height at which the Pinot Noir is traditionally grown in Burgundy.

Fermentation takes place in large wooden barrels, and vinification lasts about 10 days. The policy on addition of stalks is flexible—in 1980, for instance, about 30 per cent were added. Clos des Epeneaux is always blended to make one *cuvée*, there is no second wine, so that all the parts of the Clos, and the vines of different ages can find and complement each other in the blending or *assemblage*.

The wine is matured in the traditional oak *pièces* or hogsheads of 228 litres each. Fining is normally carried out with whites of egg, but in the unusually tannic 1976 vintage, the wines were fined with gelatine to reduce this tannin—25 grams of gelatine were added per hogshead. The right strength of alcohol for the wines is around 12.5 or 12.8 per cent. The policy on chaptalization is that about three kilograms of sugar are added to each hogshead.

Michel Rossignol says that the wall surrounding the Clos does not in practical terms alter in any way the microclimate of this famous vineyard. Working in both Pommard and Volnay, however, he has noticed that the grapes

The wines of Clos des Epeneaux at Pommard, of which Comte Armand is sole owner, have great elegance and distinction.

Philibert Rossignol (BELOW) has been régisseur at the Clos des Epeneaux for over 30 years. His son, Michel, is in charge of the work in the vineyards. They are here standing in the cellars of the Domaine Comte Armand.

of Pommard give more juice than those of Volnay.

The magnificent cellars of the Clos des Epeneaux are in the village of Pommard in sixteenth century buildings. The temperature remains at about 15°C all year due to the thick walls and the airholes which were intelligently thought out by the original builders. Bottling usually takes place when the wine is between 16 and 18 months old—the 1979 wine, for example, was bottled in February 1981. Yields vary with the year. In 1979 it was about 34 hectolitres to the hectare, in 1980 about 29 or 30, while in 1971 only 31 hogsheads were made from the whole Clos. Spraying against rot takes place, and this certainly increases the juice from the grapes, as they remain healthy and also give a very good colour.

Clos des Epeneaux combines a wonderful richness of fruit with great elegance and distinction, always a complete, multi-dimensional wine. The 1979 is wonderfully powerful, in a natural way, coming from the earth and the careful vinification, which is seen less and less nowadays. The 1978 has structure, glycerine and balance while the 1977 has a lovely colour, is scented, fruity and charming.

The Clos des Epeneaux is situated in the top Pommard premier cru of Les Epenots, north of Pommard.

The present owner of the Clos des Epeneaux is M le Comte Armand, here proudly displaying magnums of the great 1976 vintage in his Paris cellars (LEFT).

The soil at the Clos des Epenaux has a stony covering mixed with warm-coloured, heavy soil (LEFT).

The entrance to the cellars of the Domaine Comte Armand in Pommard (ABOVE) reveals a beautifully kept sixteenth century building with typical Burgundian roof and solid walls.

DOMAINE AMPEAU

This domain encompasses some of the most sought-after growths on the Côte de Beaune, and whether this remarkable grower and vinifier is making red or white, the talented hand shows through. The wines are always amazingly youthful, keeping an enticing fruit as they age in bottle, rich, and thoroughly representative of their respective noble origins.

Robert Ampeau of Meursault has always been known for the astonishing quality of his white wines, sought after by all who wish to see the true length and breed of Côte de Beaune Chardonnay, but throughout the late 1960s and the 1970s the red wines of this domain have been rising to new heights. Now, the two are worthy partners, with truly memorable bottles being produced with the apparent ease displayed by all real masters of their craft.

However, the ease is an illusion, as Ampeau is one of those producers of fine wine who is most likely to have the mud from the vineyards on his hands, not just his boots. He and his son, Michel, do practically all the work on the 10 hectare domain themselves, in both vineyard and cellar, with only part-time help in the vines from people in the village. At dusk you can come upon Ampeau father and son finishing their November pruning in the vineyard, with smoke from the burners mingling with the chill air to fade into blueness together. And yet, there is always time to taste and to talk, to compare and to ponder on the future of a wine.

The red wine holdings are at Auxey-Duresses Les Ecusseaux *premier cru*, Savigny-lès-Beaune and Savigny Les Lavières *premier cru*, Beaune and Beaune Clos du Roi *premier cru*, Volnay-Santenots *premier cru*, which is in fact in the commune of Meursault, Pommard and Blagny La Pièce sous le Bois. The red wines from around the hamlet of Blagny, between the communes of Puligny-Montrachet and Meursault, are sold under the name of Blagny. Robert Ampeau owns 1.5 hectares at Blagny, and understands, and can bring out to its full expression, the earthy attractive character of its wines. White wine holdings are straight Meursault, two top *premiers crus*, Meursault-Charmes and Meursault-Perrières, Puligny-Montrachet Les Combettes *premier cru*, and Meursault-Blagny *premier cru* or Meursault La Pièce sous le Bois, which is its new name. In the vineyard, Robert Ampeau's view is that very young vines cannot be asked to produce as much as vines aged 10 years or more, quality would certainly be affected if this was the case. But 10 or 12 year old vines can give excellent wines, with quantity, and be better balanced than either younger or older vines. For him, the site and the direction it faces are very important in deciding the ultimate quality of the wine. He has used a picking machine on an experimental basis for both white and red grapes, and sees no difference in the end result. There is, of course, a danger of oxidation on the whites. In his opinion, one of the reasons why machine picking is not proceeding quickly on the Côte d'Or is the loss of grapes, up to 20 per cent, recorded in this region. White grapes, when picked by hand, are left on their stalks, but the machine-picked grapes are destalked. Robert Ampeau allows for

this at the pressing stage, using Mabille pneumatic presses; when the grapes are destalked, he does not fill the press so full and this allows for better drainage.

Robert Ampeau is totally flexible in his attitude to modern innovations, provided that they are seen to work in the taste of the finished wine. He has a close co-operation with Max Léglise, director of the *Station Oenologique* at Beaune, and they discuss methods and machinery together. Ampeau is convinced that a winemaker must work with very clean must. He used to clarify the white must, a process called *débourbage*, for three days, but now centrifuges after pressing and before fermentation. The white wines are fermented in oak *pièces*.

The red wines are fermented in concrete vats, which are not lined but treated with tartaric acid. It should be remembered that colour is only really extracted during the first three days of the vatting. Ampeau times his vatting to correspond to the state of health of the vintage. In a year when there is some rot, he vats at a quite high temperature to extract the colour as rapidly as possible, not leaving the fermenting juice in contact with less-than-perfect lees for longer than necessary. However, in years when the must is in perfect condition the vatting can be longer, at a high temperature only for the two first days and then lowering it in order that the fermentation can be prolonged.

The red wines only receive one racking in the year, and are filtered before bottling. The reds only receive one year of ageing before being bottled, but it would not be accurate to say that these 'early' bottlings are a recent innovation as Robert Ampeau's father followed the same policy in 1937. Robert Ampeau considers that his wine matures best in bottle, and if this is the secret of their extraordinary youth, there is a lot to be learned from this policy. The whites are also bottled after about a year; the 1979 whites were bottled in September 1980.

Before the malolactic fermentation, the wines have a pH level of about 3.25, which is suitable for this second fermentation to begin, and after it has finished, the pH goes up a little. There is rotation of new and older oak barrels in the cellar, and it is certain that some wines, in some years, take better to the influence of new oak than others—judicious tasting is necessary to determine this.

The wonderful slow, gradual development of both red and white wines from Domaine Ampeau is a feature of their class. Tasters are frequently years off target in attempting to give a vintage to a wine, as they taste so youthful and fresh. Even the very full, sometimes overblown, rich white wines of the 1971 vintage taste balanced and in the first bloom of youth here.

The Blagny wines always have very marked character—definite and emphatic. Blagny reds almost have a slightly untamed air about them, a wonderful fruity, gamey

Both red and white wines from Robert Ampeau develop slowly to show their great class. La Pièce sous le Bois (TOP) and Perrières (ABOVE) are both top white wine holdings.

taste. Volnay-Santenots has tremendous breed and elegance, Auxey-Duresses can have a whiff of strawberries, while Savigny can be more redolent of cherries, with a rustic core to it. In all of these red wines, there is a magnificence of vivid fruit.

The 1978 reds have a glorious opulence, while the 1977 whites are a lesson in style and sheer beauty—the nose of the 1977 Perrières is the nearest thing to grilled almonds that the grape can produce. Robert Ampeau considers that in 1976 the character of the year tends to dominate the character of the individual area, and the 1976 reds are clearly huge, alcoholic wines that will need decades to emerge from their protective covering. In a year like 1974, generally considered as somewhat dull, both the red and white wines shine, while the 1972 whites show great length and breed and none of the coarseness sometimes visible in this vintage. M Ampeau says there are no secrets to his winemaking, but he certainly knows how to tame the most varied of years, and happily he has passed on this knowledge to his son.

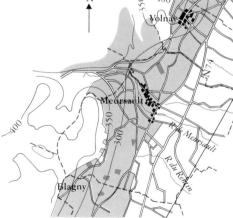

The Ampeau home and cellars are in Meursault with the vineyards chiefly in Meursault and Volnay, although some excellent plots are in other communes.

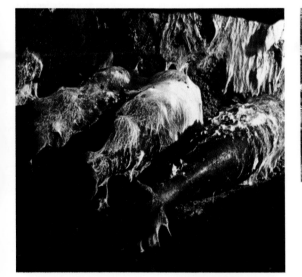

The first pruning after the vintage normally takes place in November. In the Meursault-Perrières vineyard (ABOVE), the vines are being cut back and the wood burnt. The larger branches are taken outside the vineyard and burnt, while smaller ones are burnt in iron braziers in the vineyard. Robert Ampeau's son Michel helps his father run the domain. He is pruning vines with secateurs (ABOVE).

Ampeau vineyards
Other vineyards

Ampeau uses both new oak barrels (ABOVE LEFT) and old oak ones. Robert Ampeau decides personally which wines should go in which type of barrel. He feels that in some years the character of the vintage needs new oak more than others. Ampeau bottles his wines relatively early for a grower on the Côte de Beaune.

Bottles of 1964 Meursault (ABOVE FAR LEFT) are covered in a white, snow-like fungal growth which occurs in one small section of the Ampeau cellars. It is caused by dampness but has no adverse affect on the wine.

DOMAINE LEFLAIVE

Domaine Leflaive without doubt produces some of the world's greatest dry white wines. The immense breed and distinction of their wines come not only from the impeccable pedigree of the sites, but also the skill and care of their winemaking. To breathe in the heady bouquet and taste the amazing complexity of flavours of these wines is to experience the Chardonnay grape in one of its finest manifestations.

Domaine Leflaive is one of the greatest white wine estates in the world. This is a claim easy to justify because the Côte de Beaune is one of the best areas in the world for making dry white wine, the Chardonnay is one of the noblest white wine grape varieties in the world, and this particular domain is owned by the Leflaive family, who possess the skill, experience and professional integrity to continue to make wine up to its fullest potential.

Vincent Leflaive personally controls all operations, aided by his brother Joseph, and a most capable *régisseur*, Jean Virot. The vineyard holdings command respect. Domaine Leflaive owns a little less than 2 hectares of Chevalier Montrachet, the *grand cru* of 7 hectares, marginally smaller than Le Montrachet itself but producing proportionately less. Again, the same amount of land, a little less than 2 hectares, is owned in the *grand cru* Bâtard Montrachet, which has a total area of nearly 12 hectares. Domaine Leflaive owns a little more than 1 hectare of the tiny Bienvenues-Bâtard-Montrachet, the 2.3 hectare *grand cru*, and 3 hectares of the *premier cru* Puligny-Montrachet, Les Pucelles. Other holdings are an impressive 5 hectares in *premier cru* Clavoillon, 0.75 hectare of *premier cru* Les Combettes, and nearly 4 hectares of Puligny-Montrachet. There are 1.5 hectares, producing yearly between 10 and 15 *pièces* or hogsheads of red Blagny Le Dos d'Ane, and 2 hectares of vineyard with the Bourgogne *appellation*. In a normal year, annual production at the domain would be about 450 *pièces*—there were just under 500 in the prolific year of 1979, which was particularly noted for the quantity of wine produced in the white wine *appellations*.

The headquarters of Domaine Leflaive are at Puligny-Montrachet, with the ground-level cellars round a wide and elegant courtyard. Cellars cannot be underground at Puligny because of the relatively high water-table unlike in Meursault, where this is possible. There is also a *cuverie* near the church of Puligny.

At Domaine Leflaive, the Chardonnay vines are pulled out when they are 30 years old. It is felt that the vines give of their best at around 10 years of age, younger than the reds. There is some vine degeneration and the virus *court-noué* or fanleaf among the white vines, so great care must be taken here. When the virus really attacks, uprooting is the only solution. Leflaive think it is important to use a good *pépiniériste* or supplier of vines, and they use Jean Pascal. Weed-killer is not used in the Leflaive vineyards, and when horses are used for tilling around the vines it is felt there is less damage to them than when using a tractor. From about 15 May they spray against mildew, and treatment must sometimes be effected against mites and moths. It is felt that when grapes are treated against rot, the ensuing fermentation of the must is longer, an altogether desirable process.

At Domaine Leflaive, cleanliness is a byword. They themselves admit they are 'fanatical' about it, and everything is regularly and copiously washed with water. Normally, professional tasters spit on the ground in cellars, but at Leflaive tasters have to spit into a bucket. At Domaine Leflaive meticulous care in cleaning everything is stressed above all else.

Vaslin presses are used, and the grapes should be pressed as quickly as possible—after as little as two hours, the grapes are pressed and in vat for the *débourbage* process on the must, which here takes 12 hours. The presses can be filled from the baskets in 15 minutes. Fermentation takes place in wood and in vat. Chaptalization is effected after the *débourbage* but before fermentation. During fermentation, the lees are periodically roused and mixed with the wine to give extra body and flavour. A sign of a poor vintage in white wines is when the lees are a brownish colour. The malolactic fermentation finishes about mid May, and then the wine is racked. Part of the wine is in Limousin oak *pièces* and part in stainless steel vats. The wines are usually fined with casein (a milk protein) and sometimes with bentonite. Experiments with fining using ordinary milk took place, but it was found that the wines took on a taste of milk. There is only a light filtering before bottling takes place.

The wines are bottled at the age of about 18 months, always after the second winter after the vintage. The free sulphur dioxide is usually around 10 milligrams per litre, and always between 5 and 15 milligrams per litre. As free sulphur dioxide guards against oxidation, it could not be as low as this if extraordinary care had not been taken during the vinification and maturation processes.

For these wines to be seen at their best, they have to be treated gently and given time. A wine that has been fermented long at a low temperature and given the barest of filtering will have a great deal of bouquet, body and extract, and, with time, power and concentration appear, even in a white wine. A wine fermented quickly, refrigerated and stabilized to a high degree, cannot have such dimensions of nose and taste, and it can be drunk immediately after bottling. Just as a wine has to be handled carefully at all stages of vinification, it has to be equally cared for when it is in bottle. Great white Burgundy should never be too chilled in a refrigerator, as much of the taste is then removed, and even on warming up, it does not quite recover. Vincent Leflaive says, picturesquely, that his wine 'stays in its shell' if one treats it roughly.

The *premier cru* vineyards near the Montrachet group, such as Les Pucelles and Clavoillon, share some of the characteristics of their neighbours but in slightly less rich

Leflaive have just under two hectares in grand cru Chevalier-Montrachet *(TOP) and three hectares of Puligny-Montrachet Les Pucelles (ABOVE).*

The brothers Joseph and Vincent Leflaive (BELOW), whose family owns Domaine Leflaive, are personally involved in making the wine and running the domain. They are convinced that extreme cleanliness at all stages of the wine's life is vital when producing long lasting white wines. Here the barrels are being cleaned (CENTRE LEFT). This healthy bunch of Chardonnay (CENTRE) is being picked in the premier cru Puligny-Montrachet Les Pucelles.

The Chardonnay grape grows well on the calcareous soil of Puligny-Montrachet.

form. Les Combettes is near the border with Meursault, and is perhaps slightly more mellow. As for the Bienvenues, they can be described as 'honeyed', even in youth, attaining velvety smoothness with a bit of bottle age. Mature Bâtard-Montrachet is rich, round and concentrated, with Chevalier-Montrachet the ultimate in class where Chardonnay is concerned, austere in youth, opening out to utter splendour and length on the palate. The last three vintages of the 1970s—1977, 1978 and 1979—produced lovely white wines, with 1978 perhaps the most tight-knit, 1977 deep and very classic, and 1979 full of charm. In all these wines, there is a centre, a kernel, that lies hidden until it wishes to show itself, and those who quaff these wines thoughtlessly in youth will miss all the variations on a theme.

The grapes are loaded into the trailer before being taken to the cellars (LEFT). Some experiments with mechanical picking have been done on the Côte de Beaune, but machines are not very suitable for picking low vines such as the Chardonnay.

This view of Puligny-Montrachet (LEFT), famed for its white wines, shows the gently sloping vineyards of Domaine Leflaive in the foreground.

DOMAINE DE LA RENARDE

The emergence of the Domaine de la Renarde on the Côte Chalonnaise is one of the finest examples of winemaking skills in a less well-known area. The wonderfully elegant white wines of Rully, with their fruity, lingering scent and incisive taste, and the perfumed, charming reds, the very essence of Pinot Noir, are proof that the Côte Chalonnaise is a vital part of the Burgundian vineyard.

This is one of the newest European domains in this book, the result of the intelligence, vision and enthusiasm of a remarkable man of wine, Jean-François Delorme. Trained in oenology at the University of Dijon, Jean-François Delorme is the son of a grower at Rully, and saw the future of the Côte Chalonnaise, for long the neglected cousin of the Côte d'Or. He set about expanding his vineyards and researching into the best possible methods of making wine from the Pinot Noir and the Chardonnay on the Côte Chalonnaise. At the same time, he also runs the *négociant* business of Delorme-Meulien, which is known particularly for its high-quality sparkling wines of Burgundy, especially Crémant de Bourgogne.

By far the greatest part of the Domaine de la Renarde is at Rully, the first *appellation* reached after Chagny on the Côte Chalonnaise. The domain consists of about 42 hectares at Rully, 22 hectares in Rully Rouge and 20 in Rully Blanc, including all Varot (18 hectares in one parcel), and parts of Monthelon, Les Cloux, La Fosse and Grésigny. At Mercurey, there are 3.7 hectares of red wine production, and at Givry 4.5 hectares, also of red. There are plans in the future for a further 18 hectares of vines to be planted in the next four or five years, all in Rully—4 hectares were planted in 1980; 10,000 vines are planted per hectare. This is considerable, and demands much labour. Delorme uses the first clones that were commercially selected. The most usual root-stocks are the SO4 and the Téléki 5BB, which is older than the SO4 and very strong. The 41B is used on high ground, the 3309 on dry soil, and the Riparia on low ground—the last ripens well, but is not good on calcareous soil which has a tendency to fall to the foot of the slopes.

The soil is basically clay/calcareous, with the clay having more Pinot Noir planted on it, and the chalky plots the Chardonnay. The soil does not vary much between the villages of the Côte Chalonnaise, but the microclimate does. The vines are usually grown at heights of between 300 and 335 metres, higher than the Côte d'Or and therefore usually vintaging three or four days later.

The red grapes are not destalked, as the vines are comparatively young and Jean-François Delorme feels that they need the extra tannin and body that this gives. Fermentations are in completely closed, 80 hectolitre vats. The vats are made of steel and lined with plastified enamel. Temperature is controlled very carefully, and when the grapes are healthy, the temperature can be allowed to go over 32°C without an increase in volatile acidity, as the equipment is so modern and precise. Chaptalization is done gradually and in small amounts. In January, the free-run wine and the press-wine are blended.

The presses are Vaslin horizontal. The policy is to press very slowly and gently so as to avoid any bitterness, but this results in as much as between five and seven per cent loss of juice. Two hours would be a normal pressing time, but at Delorme the length of time taken for a load to be pressed is 3½ hours. The white wines then rest and clarify (*débourbage*) for 24 hours before fermentation. In a year when there might be an element of rot, enzymes—pectins—are used for extra clarification. The white wines are fermented at 18°C or 19°C for between 15 and 21 days. Total acidity is usually around 6 grams per litre, and natural alcohol between 10 and 12 per cent.

Casks are made of oak from the Tronçais and the Châtillonnais, the plateau of the Côte d'Or, north of Dijon at Châtillon-sur-Seine, and they are usually piled three high in the large, airy *chais*. Jean-François Delorme is meticulous about checking for volatile acidity every month in those *pièces* on the top layer; it should be less than 0.3 grams per litre. There is a wide range of casks, with some totally new, some two years, five years and nine years of age. At racking, the wine is rotated between the different casks, giving it a chance to mature in both new and more mature oak. Bottlings are relatively early, in order to capture the maximum fruit. Balance and harmony are essentially what the winemaker at Domaine de la Renarde is seeking. The Côte Chalonnaise wines do not have the body of, say, the *grands crus* of the Côte de Nuits, and too long a period in cask would dry them out and diminish their great charm. The white wines are bottled in the spring after the vintage, while the reds need a few months more and are bottled in the summer.

Yields vary considerably with the year, as in all Burgundy, but it cannot be assumed that they are more important on the Côte Chalonnaise than on the Côte d'Or. In 1979, for example, the white wines on the Côte Chalonnaise usually achieved 55 hectolitres per hectare, while at Puligny the yields sometimes reached an amazing 80 hectolitres per hectare. But in 1978, the white Rully La Chaume was the result of a small, concentrated yield of 24 hectolitres per hectare. Red wine yields can be shown by figures at Rully of 55 hectolitres per hectare for 1979 and 45 hectolitres per hectare for 1978.

There are, clearly, marked differences between the villages or communes of the Côte Chalonnaise, and even between the vineyards within one commune. One example is Monthelon, one of the top areas in Rully, which is on the border between Rully and Mercurey, combining the characteristics of both—it is also high on the hill and very pebbly, with thin soil, which tends to dryness because of good drainage. To confuse the issue slightly, Monthelon will become La Chaume, another *climat*, in the future, with a change in the delimitation of the *climats*. Amongst the *climats* of Rully, La Chaume usually has a

*Delorme owns all of the 18 hectare Varot vineyard in Rully (*TOP*). Delorme produces interesting wines from the Aligoté grape grown in Bouzeron, north of Rully (*ABOVE*).*

The village and church of Rully (BELOW) are set in the rolling countryside of the Côte Chalonnaise. This area was long neglected as a wine growing district, but the Domaine de la Renarde has done much to counter this.

Nearly all of the domain is at Rully. Half is planted with red Pinot Noir and half with white Chardonnay. The entire 18 hectare Varot vineyard is owned by the domain.

great deal of *sève* and attack—at the moment, the vines are youngest here. La Chaume consists of 10 hectares, of which half are white and half red. The red Givry Clos du Cellier aux Moines can have a fascinating nose of toasted bread, at other times it is more *herbes de Provence*, and utterly seductive. Mercurey red wines tend to be bigger-bodied, finishing long on the palate. In the hands of Delorme, even a simple Bourgogne Rouge, La Croix-Lieux is delicate and violet-scented—light, refreshing, completely frank Pinot Noir.

The white Côte Chalonnaise wines, which at the Domaine de la Renarde mean a fascinating range of Rully wines, have a fresh grassy smell, great elegance and breed, sometimes developing a full, almost cinammon flavour with a few months in bottle. Even the Aligoté from the village of Bouzeron, just north of Rully, takes on a new sheen when the *élevage* is *chez* Delorme, although this is not Domaine produced. When Jean-François Delorme buys in the newly-made wine, it is in *pièces*, but he immediately puts it in vat, as he considers that the influence of wood is not needed with this grape variety to draw out its maximum elegance. The Côte Chalonnaise has its champion in Jean-François Delorme, and its pinnacle of wine-making success in the Domaine de la Renarde.

Jean-François Delorme (LEFT) is a trained oenologist who has done much to establish the domain's reputation. He is examining some white Chardonnay vines in the premier cru vineyard at Moulesne.

Before being taken to the chais, the grapes are unloaded from the hottes on the backs of the pickers into the trailer (LEFT). These white Chardonnay grapes are in excellent condition.

These white Chardonnay grapes are being placed into the Vaslin horizontal press (LEFT, INSET). Delorme specializes in long slow pressings for the grapes to avoid any possible bitterness in the wine. If white grapes are pressed harder, some bitter- *ness may come from the stems and pips. The disadvantage of slow pressing is that it lessens the amount of juice obtained.*

The Varot vineyard (LEFT) is one of the great assets of this domain. All 18 hectares belong to the domain. The Chardonnay is being picked. On the Côte Chalonnaise, large parcels of vineyard in one piece are more common than on the Côte d'Or where the plots tend to be small and scattered.

The leaves on this Chardonnay vine in the Varot vineyard (LEFT CENTRE) are just beginning to turn colour. This older vine (LEFT) is a Pinot Noir. Older vines produce ·reduced amounts of very fine wines. The juice from these grapes has more concentration and extract than those from a younger vine.

Jean-François Delorme (LEFT) is tasting a very young wine which is still cloudy. He uses a glass which is almost the same shape as a cognac glass. Such glasses are good for tasting and drinking Burgundy. A glass which narrows towards the top helps conserve the wine's bouquet.

Grape pickers are renowned for the huge appetites produced by the long hours and physically demanding work. The estate caters for this by providing a substantial meal for all the pickers (ABOVE).

The hydrometer (ABOVE) is used for checking the density of a wine in the laboratory. Such checks are carried out throughout fermentation, but especially towards the end to see how much sugar remains.

PAUL JABOULET AINE—HERMITAGE

Hermitage La Chapelle from the house of Paul Jaboulet Aîné has the full-powered depth of one of the world's greatest red wines. The dark, purple colour of the young wine is typical of the Syrah grape when grown on the steep granite slopes of Hermitage. But the tannin and power gradually age to a concentrated complexity of flavours that live long in the mouth and linger on the palate.

The family firm of Paul Jaboulet Aîné, now headed by Gérard Jaboulet is the undoubted king of the northern Rhône, producing wines of majestic proportions from their own estate at Hermitage and Crozes-Hermitage, as well as selecting excellent wines from all parts of the Rhône. There is a division, both geographic and in the types of wine made, between the northern part or Rhône Septentrional of this large viticultural area, and the Rhône Méridional in the south, where Châteauneuf-du-Pape reigns supreme.

In the northern Rhône, south of Lyon and beginning at Vienne, lie the steep slopes which make some of the longest lasting red wines produced in France. This granitic, hard land is part of the Massif Central, with the river Rhône providing the border. But the great hill of Hermitage somehow found itself on the left—or eastern—bank, of the river, in spite of being geologically part of the granite mass. It faces full south, and attracts all the sun possible, while the granite reflects the heat onto the vines. The slope is so steep that the vines are planted in terraces, locally called *chalais*. The top soil is a thin, loose layer of decomposed flint and chalk, which can be dislodged by heavy rain.

The red Syrah grape rules supreme here. Some consider it is named after the Shiraz of Persia, others that it emanated from Sicily. In Australia it is known as the Hermitage, after the great hill in the northern Rhône where it does so well, and the Shiraz in South Africa. In the Rhône, the Syrah is often known as the Sérine or Petite Syrah. The characteristics of this grape variety are its deep, purple colour, which hardly changes for years, the all-enveloping, slightly peppery intense scent, and the structure and body which produce wines eminently suitable for ageing.

Hermitage La Chapelle, the great red wine of Jaboulet, is named after the chapel of St Christopher, which was probably built by the Chevalier de Stérimberg on his return from the medieval Albigensian Crusade in southwest France. The chapel stands on the top of the hill. The vineyard is below it, beautifully exposed, although obviously vulnerable to any high winds. It is for this reason that the bush *gobelet* system of training is used for the Syrah so that the vines are each tied to a stake.

The whole of the Hermitage vineyard *appellation* covers 130 hectares, and Jaboulet own 19 hectares planted in red Syrah, and 5 hectares planted in white Roussanne and Marsanne. These white grape varieties of the northern Rhône are related, with the Marsanne slightly more distinguished. The total annual production of Hermitage (red and white), is normally over 3,000 hectolitres, with the prolific 1973 vintage amounting to 4,108 hectolitres, and 1979, 4,895 hectolitres. The proportion is approximately 80 per cent red wine, 20 per cent white wine.

The Syrah grapes for making Hermitage La Chapelle are usually picked at the beginning of October. The must is fermented in vat at about 19°C or 20°C for between 15 and 18 days. At the end of this period, the grapes are pressed, with the result put into concrete vats to fall bright. At the end of two or sometimes three months, the wine is racked, well protected from the air to avoid all risk of oxidation, and then put into oak barrels with a capacity of 225 litres. There is usually no problem with the malolactic fermentation, which follows the alcoholic fermentation. In June, the wine is again racked and put once more into oak barrels. Sometimes another racking is performed in September, but generally the wine is bottled around October, a year after the harvest. Its alcohol is usually around 12.5 per cent. Hermitage La Chapelle rouge is fined with albumen for about eight days but is never filtered.

Jaboulet's Hermitage Blanc Le Chevalier de Stérimberg is made in the modern idiom, with the emphasis on freshness, rather than in the traditional, cask-aged tradition. The wine is made from an equal mixture of the two white grape varieties—Roussanne and Marsanne. The grapes, once picked, are immediately pressed in Vaslin presses and the juice put into refrigerated stainless steel vats to clarify at a very low temperature; this is necessary in order to prevent the fermentation beginning prematurely. After about 48 hours, the clear juice is put into another cooled vat with the temperature maintained at about 16°C. Thus, fermentation begins very slowly and with the temperature maintained at this level; the transformation of the must into alcohol is slow and gradual, giving the wine maximum aroma and fruit.

When the fermentation is completed, the wine is racked away from the air and placed in a stainless steel vat, together with the carbon dioxide amassed during the fermentation, in order to protect the new wine from all risk of oxidation. Bottling is very early, usually in February, after a light filtering. Thus, the aim is to capture this youthful fruit and freshness and to maintain it. So the white Hermitage, unlike the red, sees no wood ageing. Malolactic fermentation only takes place rarely on the white Hermitage—the last time that this happened was with the 1964 vintage. Jaboulet prefer to keep all the natural acidity of the wine, as they consider that the grape varieties grown here need this to give the wine balance. White Rhône wines used to have a tendency to flatness and flabbiness, due to low acidity and, in some instances, oxidation.

However, the great glory of the northern Rhône, and of the domain of Paul Jaboulet Aîné, is the red Hermitage La Chapelle. Dense, inky, rich, with great *charpente* or structure, these wines need ageing in bottle for their complexity and breed to come through. The Syrah always produces a high amount of tannin, but there is great

Hermitage La Chapelle is the greatest red wine produced by these winemakers; the 1973 vintage was very attractive. The white wine Hermitage Le Chevalier de Sterimberg is made from equal proportions of Roussanne and Marsanne grapes.

The firm is very much a
Jaboulet family affair with
father Louis and sons Jacques
and Gérard (BELOW) in-
volved in every aspect of the
winemaking.

The maitre de chais is in
general charge of the cellar
work under the direction of the
winemaker. Here the Jaboulet
maitre de chais (BELOW) is
standing in front of some of
the old 225 litre barrels used
in the cellars.

The northern Rhône is
dominated by the very steep
granitic slopes of the Massif
Central which rise up from
the river.

depth of fruit to balance this, and the wines maintain their
'fat' and gloss, without drying out. The 1978 is a classic
year, with enormous concentration and extract, but 1979
has the same style, if not perhaps the weight; 1976 was
most successful, while 1973 produced wines of great
attraction. 1972 was really at its most successful in the
whole of France in this area of the northern Rhône, while
the trio 1971, 1970 and 1969 were all top wines. Wines
of the 1960s at La Chapelle have stood the test of time with
honour, culminating in the glorious 1961. When the
pocket does not run to Hermitage, there is always the
Jaboulet estate wine of Crozes-Hermitage Rouge Thala-
bert, some of the very best value in France, even in a year
not generally highly regarded in the northern Rhône such
as 1977. The white Crozes-Hermitage from Jaboulet is
Mule Blanche; it should be drunk young while waiting
for the reds to mature.

PAUL JABOULET
AINE–HERMITAGE

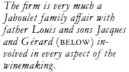

The terraces in the La
Chapelle vineyard on the
Hermitage hill (LEFT) are
called chalais locally. They
help prevent the soil from
being washed down during
rainstorms. On occasion, after
a particularly severe storm,
soil has to be carried back up
the hill from the bottom of the
slope.

After fermentation the vin
de presse or press wine
(LEFT) is pumped into concrete vats to clarify before
being mixed with the free-run
wine.

The bottle barrow (LEFT) is
used for transporting bottles
in the cellar before they are
laid down for ageing. Capsules and labels are only put
on shortly before the wine is
sold.

The St Christopher Chapel
(LEFT) stands on top of the
Hermitage hill. The vines are
trained around a single stake.
This helps the vines withstand
the strong Mistral winds.

CHATEAU DES FINES ROCHES

Château des Fines Roches is a Châteauneuf-du-Pape in the classic mould. The product of the southern Rhône sun and the great, flat stones of the vineyard of Châteauneuf, the wine starts life as vivid, robust and well-structured, but develops subtleties and a beautiful, earthy, rich flavour when mature.

In contrast to some of the properties of the region which were growing vines in the Middle Ages, Fines Roches is relatively new on the scene. The actual medieval-style château was built at the beginning of this century by the Marquis de Baron Celly Jabon as a hunting lodge. There were no vines on the property at the time and development of the vineyard has been a gradual rather than a swift or sudden one.

In 1940, Claude Mousset—grandfather of the present owners—purchased the property, but it was his son, Louis, who was responsible for extending the planted area. He also established a wine-shipping company bearing his name which now has its offices in a separate building behind the château, itself now a deluxe restaurant with a splendid view over the valley.

Château des Fines Roches is a *société civile* belonging to Louis's daughter and two sons. The vineyard, however, is run by Robert Berrot, who is the son-in-law of Louis Mousset. He is assisted with the winemaking by both Guy and Jacques Mousset, the two sons.

The vineyards cover 40 hectares with 10 of these planted within the last few years. Production of Châteauneuf-du-Pape is legally limited to 35 hectolitres per hectare, and at Fines Roches it is generally between 30 and 35 which is relatively small compared to other areas in France. It is not possible to reclassify if there is any excess, and wine produced from vines less than five years old is blended into their Côtes du Rhône.

The system of controlled appellation wines was first applied to Châteauneuf-du-Pape and is now used nationally. In addition to the ruling on low yield, there is a required minimum of 12.5 per cent alcohol, and this must be acquired naturally, because chaptalization, or the addition of sugar or concentrated grape must, is forbidden.

The microclimate of this part of the Rhône Valley is characterized by several particularities. Winters are dry and anticyclonal, followed by a spring that is generally short with torrential rains. Summer is long, hot and dry, succeeded by autumn which is also long, but with heavy rain during the equinox. The entire valley is buffeted by the strong Mistral wind which blows down from the north at high speeds. Even though its force has diminished during the last 20 years, it is still strong enough to knock grapes off the bunches and break vine branches. However, it has the advantage of bringing the good dry weather of early summer and drying the grapes of their moisture in the autumn. Without the Mistral, the area would be subject to chronic fog and greater rainfall. At Fines Roches, it is estimated that the Mistral blows for more than 200 days in the year.

The topsoil of the vineyards is characterized by the famous *galets* or rounded stones often as big as a fist which completely hide the subsoil from view. They were deposited and rounded while part of the riverbed of the Rhône that once flowed through the entire valley. The subsoil is composed of gravel, sand and limestone of a reddish colour. Its structure enables the vineyards to drain quickly, which is one of the factors responsible for the relatively small grape production on the estate.

Three-quarters of the vineyards of Château des Fines Roches are on hillsides exposed to the south and south-east. Along with Château Fortia just to the west and Château de la Nerth to the east, Fines Roches has traditionally been considered one of the top winemakers. The field has broadened to include others, but Fines Roches remains one of the most respected.

No less than 13 grape varieties are permitted for Châteauneuf-du-Pape, but no one uses them all. About 60 per cent of the Fines Roches vineyards are planted with Grenache, this is a fairly common proportion for the commune. Grenache produces good wine in this region, but is very vulnerable to bad weather during flowering—another factor limiting production.

The second most important grape is the Syrah which is planted in a proportion of 15 per cent. This variety is used at Hermitage and Côte Rotie and gives a very pronounced dark red colour and longevity to the wine. The other planted varieties include 10 per cent Mourvèdre, 5 per cent Cinsault with the Clairette and Muscardin taking up the rest. The Carignan, widely planted in the Côtes-du-Rhône, is forbidden fruit in Châteauneuf-du-Pape.

The harvest is finished in three or four weeks by a group of 30 Spaniards and a few local pickers. According to Guy Mousset, it never rains two days in succession at Châteauneuf and, because of the stony vineyard surface, the harvesters can enter the vineyard as soon as it stops. Picking begins with the Syrah and Cinsault which ripen earlier than the others.

After eliminating any unhealthy grapes and twigs, the grapes are transported to the second floor of the *chais* by a stainless steel worm screw and into a truck that circulates on rails above the vats. The grapes can either be dumped directly into the vats or passed through a *fouloir*—an apparatus for gently breaking the skins to release the juice. The grapes are left on the stalks in order to give more tannin to the wine. Because of its particular characteristics, the Syrah grape is vinified separately from the other varieties and blended with the rest when the vinification is completed. It is one of the variables of the *appellation* that some domains go so far as to vinify all their varieties separately, blending them several months later.

As soon as a vat is full, yeasts are added to help start the fermentation. Given the usually warm autumn weather, there is

Château des Fines Roches is one of the most highly respected wines in Châteauneuf-du-Pape. The wines are full-bodied and have power. The wine should be at least five years old when drunk.

rarely any need to heat the must. Indeed it is often necessary to cool the must once fermentation has begun, in order to prevent the temperature from rising too high and too fast.

The must is immediately pumped over the *marc* for an hour to homogenize the juice. On the second day, it is pumped over again for 30 minutes, and a sample taken to determine if the wine needs to be acidified. With grapes such as the Grenache, whose acidity is very low, tartaric acid must be added in order to balance the wine. Legislation carefully limits the quantities which can only be added in small amounts when the wine is pumped over the *marc* every two or three days.

The density and temperature of the must are closely observed. Each fermenting *cuve* has an automatic system of electrovanes regulated by a thermostat set at 30°C for cooling and in addition a system of heating by electrical resistance located beneath the vat if the temperature falls below 15°C.

The first fermentation takes about two weeks and the wine is run off when the density has reached the correct level; the *marc* is removed from the vat and put directly into the Vaslin horizontal press by the worm screw and pressed six times. The juice from the first two pressings is added immediately to the free-run wine and the rest is set apart to be drunk later as ordinary table wine.

Due to the fact that a good proportion of grapes is put into the vat whole and remains intact throughout the alcoholic fermentation, the press wine has a high density and considerable grape sugar. Mixed with the free-run wine in another vat, a second, but minor, alcoholic fermentation ensues lasting five or six days. The malolactic fermentation follows, but can begin immediately after the alcoholic has finished, or as long afterwards as a year later.

When the secondary alcoholic fermentation is complete, the largest solid particles are separated off; an operation that is repeated one or two weeks later. The wine is aged for at least six months in the metal vats used for vinification, then one to three years in oak vats with a capacity of between 50 and 60 hectolitres. Bottling of the wine, therefore, varies from two to four years after the harvest depending on the quality of the vintage, and also on whether or not the malolactic fermentation is complete.

Part of the production is put into bottles embossed with the coat-of-arms of Châteauneuf-du-Pape reserved for members of the local syndicate. However, the wine sold through the large French wine-store chain Nicolas is marketed in a plain bottle with a slightly different label.

The wine of Château de Fines Roches is full-bodied and powerful, registering around 14 per cent alcohol. It is best drunk aged five years at minimum and can last until it is 15 or 20 years old. It is characterized by a solid ruby colour, good structure and backbone, depth, richness and that earthy flavour of Châteauneuf-du-Pape.

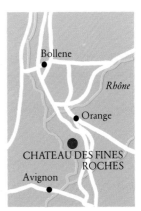

Château des Fines Roches, one of the best situated properties on the plateau of Châteauneuf-du-Pape, has extremely stony soil. The Mistral blows in this area for more than 200 days each year. The summers are long, hot and dry.

The marc (LEFT) *is the solid matter after the free-run wine has been run off the vat. At Fines Roches, the equipment is extremely modern with the latest devices for controlling fermentation temperatures. Such a high degree of technology is unusual for this area.*

This large plot of vineyard is in front of the château at Fines Roches. The château looks medieval but was only built at the beginning of this century. It is now a restaurant. The estate's cellars and offices are still beside the château.

These dense stones (ABOVE) are a feature of the soil of Châteauneuf-du-Pape. At night they reflect the heat absorbed from the sun during the day. This means that the grapes can ripen 24 hours a day.

Meticulous cleanliness is vital at all stages in the wine-making process. Here one of the stainless steel fermentation vats is being hosed down (ABOVE). It is easier to clean stainless steel than wood, which is still used for making some vats.

*H*UGEL ET FILS

The wines of Hugel show the dazzling array of tastes that Alsace can provide. Whether it be refreshing Riesling, or a great late-picked Gewürztraminer or Pinot Gris, the wines always display the true fruit character of each individual grape variety, the most natural expression of what wine can be. The pronounced, fruity bouquet always leads the taster in to a wine of perfect balance and harmony.

To many people, particularly in Britain and America, Hugel is synonymous with Alsace. The Hugels have been wine-growers in Riquewihr since 1639, and it seems generally acknowledged that the firm's strength lies in the family, and the determination of each successive generation to maintain exacting standards, even during times of expansion. In fact, it is partly because of their size that Hugel can make great wines, as the techniques and expertise necessary for taking full advantage of what nature gives in Alsace can be developed more readily in a large enterprise.

When trying to understand the secrets of Hugel's success, it is important to start with the raw material, the grapes, as the house holds the firm view that the quality of its wine is entirely attributable to nature. The way man moulds this raw material should be as unobtrusive as possible. With the sad death in 1980 of the great Jean Hugel, the high principles of the house are now carried on by his three sons, Jean, Georges and André. Jean is a Montpellier-trained oenologist who travels throughout the world on behalf of the firm.

Hugel manage 25 hectares of vineyard, composed of 48 per cent Riesling, 47 per cent Gewürztraminer, 3 per cent Tokay or Pinot Gris, and 2 per cent Muscat. The planted area includes a good part of the Sporen, the south-facing, saucer-shaped hill at Riquewihr, and vines on the other famous slope of Riquewihr, the Schoenenberg. They grow no Pinot Blanc, Sylvaner or Chasselas, and 97 per cent of all Hugel wine sold is made from grapes which come either from their own vineyards or from contract growers in the surrounding area.

Training of the vine is on five wires, usually to a height of about 1.80 metres. This type of training makes working in the vineyard relatively easy, and the worst frost dangers are also avoided, which is important in such a northerly wine-growing region as Alsace. There are between 4,000 and 5,000 vines per hectare. The vineyards are located in the foothills of the Vosges, on east-west spurs off the main north-south range. Like other soils in the western half of the Rhine rift-valley, partly as a result of glacial action, they are extremely mixed in constitution. The vineyards, in general, do not go above a height of 200 to 400 metres.

There is such a variety of soils and situations that the ideal sites for each grape variety can be chosen. This is especially useful with regard to such grapes as the shy, late-ripening Riesling, which must have a particularly favoured portion of land. Tokay (Pinot Gris) and Pinot Blanc are hardier. The Sporen produces the finest Tokay and Gewürztraminer in Riquewihr; the Hugel Sporen holding consists of approximately 60 per cent Gewürztraminer, 25 per cent Tokay, 15 per cent Riesling and Muscat.

Obviously, the Pinot Noir must have shelter from cold and winds if adequate red or rosé wines are to be made. The very best soil is calcareous clay but there is also sandy soil, loamy soil, and various combinations of these.

Harvesting by hand takes place over a period of four weeks, beginning with the Chasselas and Sylvaner grapes and ending with the Riesling and Gewürztraminer. In common with most growers and winemakers in Alsace, Hugel use Vaslin horizontal presses, but with certain modifications, such as the central screw which turns in the opposite direction to the rotating outside cylinder. This avoids pressure problems with grapes like the Sylvaner, that squash easily. The Hugel presses also have non-standard stainless steel fitments (screw, hoops, chains, nuts and bolts) to avoid possible iron contamination and thus the need to fine the must against this. All new presses the firm buys have glass-fibre reinforced plastic cylinders to avoid even slight contact between the must and the enamelled metal frame and wooden cage. Hugel were the first in France to acquire pneumatic Willmes presses, in 1952, but found they gave tannin problems from the pips, so since 1961 only the more adjustable Vaslin horizontal presses have been used.

Bentonite fining against excess proteins is not performed on the wine, but only on the juice as the former procedure tends to remove some substances necessary for the best wine. The must is also centrifuged to clean it before fermentation, as it is especially important in fine white wine that the ferments work on a thoroughly clean base. Fermentation is caused entirely by the natural yeast present on the skins of the grapes. The white wines are fermented at between 20°C and 25°C for between 10 days and six months or, in the case of the 1976 Sélection de Grains Nobles, nine months. Wherever possible—and always in the case of the great wines such as really ripe Rieslings, Pinot Gris and Gewürztraminers—fermentation takes place in wood. The casks are oak, and most of them are over 100 years old. The lesser, lighter wines are fermented in glass-lined concrete tanks.

About 30 per cent of the storage is in wood, but the casks have thick tartrate linings, which prevent much contact between wine and wood, and help precipitate tartrates in the wine. Hugel require a wide mixture of cask sizes, to deal individually not only with the many grape varieties, but also with the often small lots of superb wines, which need topping up at racking with exactly the same wine as that contained in the cask. Therefore, in the Hugel cellars in Riquewihr, cask sizes range from a maximum of 324 hectolitres down to 200 litres. The magnificent Ste Caterine cask dates from 1715 and holds 88 hectolitres.

Malolactic fermentation is desirable in lesser years with high total acidity. It

The 1976 Tokay d'Alsace Vendange Tardive had an extremely high sugar reading of 106° Oechslé. In Alsace, the German system of sugar readings is used.

requires ideal conditions of temperature, pH, and atmosphere in order to foster the action of the bacteria. Such conditions not only demand great skill in manipulation, but can in any case be exceedingly difficult to achieve, especially in new cellars, where stainless-steel or glass-lined concrete *cuves* can discourage even artificially introduced bacteria. The malolactic fermentation needs to be followed daily, to pick the optimum moment to rack the wine off the lees, before too much acidity has been lost. The malolactic fermentation can be accompanied by a rather disagreeable taste, which has to be eradicated by racking in contact with the air. In this way, in a year with bad weather conditions, if the malolactic fermentation is successfully accomplished, the wine remains attractive and supple, without being sweetened to camouflage the malic acidity it contains.

In years of great ripeness, it is desirable to retain the malic acidity so that the higher total acidity balances the sugar. However, in these circumstances it can be very difficult to prevent the malolactic fermentation, because the combination of ripe grapes and a high degree of natural sugar can take a long time to ferment. Malolactic fermentation thus may begin unnoticed, in parallel with slow alcoholic fermentation. But, in a great year, over 80 per cent of the total acidity is often composed of tartaric acid, which is not affected by the malo and thus remains intact. If malolactic fermentation begins, it can usually only be detected by chromatographic analysis, and in most cases the 'taste' of the fermentation is so muted as to pass unseen. Naturally, if some practical way could be found to prevent simultaneous alcoholic and malolactic fermentations, this would be ideal in a great year, when any loss of acidity is undesirable.

All Hugel wines are bottled between April and September following the vintage, with the great Gewürztraminers and Pinot Gris being bottled last. Hugel consider that bottling young retains freshness, while the wine can continue to develop in the bottle. This is why even the most modest wines mature in bottle for six months before being listed. The greater wines are only put on sale after 12 or 18 months, and certain exceptional wines are kept for three years before being made available.

Hugel say that the art in winemaking is not to lose any of the grapes' best qualities during the winemaking. Their practice is conclusively vindicated in the great 1976 vintage, where late-harvested grapes gave extraordinary sugar readings which are always measured in Oechslé degrees in Alsace. Hugel consider that Pinot Gris, for instance, should have over 88° Oechslé to be good, and great ones should be over 100°. In 1976, a Tokay Vendange Tardive measured 106° Oechslé, and the unique Tokay Sélection de Grains Nobles reached an amazing 135° Oechslé. The wine also shows the inimitable taste of noble rot, with which the grapes were totally affected, and thus epitomizes the best in the wines of Alsace.

The current generation of the Hugel family (LEFT) involved in carrying on the family's great winemaking traditions include (from LEFT to RIGHT) the brothers Jean, André and Georges.

A sample is being taken from one of the large wooden foudres in the Hugel cellars (LEFT). The wide range of cask size in the cellars is needed for the many different types and varieties of wine made in each vintage.

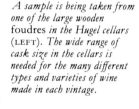

The Hugels' own vineyards are centred on Riquewihr. They include holdings on two of the most famous Alsace vineyard sites—the Sporen and the Schoenenberg.

The entrance to the Hugel cellars is on the main street at Riquewihr, one of the most beautifully maintained villages in Alsace (LEFT).

DOMAINES SCHLUMBERGER

Domaines Schlumberger specialize in wines of great, almost earthy richness, with great fullness in the mouth and long finish. The steep slopes of their own vineyards and the special soil of southern Alsace give the Gewürztraminer in particular a roundness and an individuality that is rarely seen.

Domaines Schlumberger is a family firm, managed by Eric Beydon, and the largest proprietor of vineyards in Alsace. They own 140 hectares of vines and only ever make wines from the produce of their own estate. They have their offices, cellars and vineyards at Guebwiller, at the southern tip of the Alsace wine-growing area. Only Thann is further south. They are not only set a little apart geographically from the centre of the Alsatian wine-growing industry round Riquewihr—their wines are also rather different, with a highly individual taste derived from the extraordinary setting of most of their vineyards and their winemaking methods, which are designed to bring this out.

The great individual vineyard site at Guebwiller is the Kitterlé, a really impressive hill, crowned by trees. The vineyard faces due south, and therefore benefits from every ray of sun, giving the grapes marvellous maturity. The Wanne is another site on a steep hill which faces east-south-east. The soil is poor with some sand, and the vines are grown on narrow terraces. Schlumberger own about 12 hectares on the Kitterlé where working the terraces is difficult. The Gewürztraminer is ideally suited to both these slopes, and the ripeness gives it great body. The vines are grown at altitudes of between 250 metres and 450 metres, with the more ordinary grape varieties tending to be in the lower part. However, this is not followed strictly, because, with dangers such as frost, it is better to have the vineyard slightly mixed so that in one year the whole of one grape variety is never completely lost.

The Schlumberger holdings are divided between the grape varieties in the proportions of 35 per cent Gewürztraminer, 30 per cent Riesling, and the remaining 35 per cent divided between Sylvaner, Pinot Blanc and Chasselas. The root stocks used are the SO_4, the Téléki 5BB and the 3309. They are chosen to adapt to the soil, particularly with regard to its limestone content, and the grape variety which is being grafted on.

The vines on the slopes are treated against mildew and oidium from a helicopter, which greatly eases the task. There is, however, a certain amount of wastage with this method, so the helicopter is abandoned for the more expensive anti-botrytis treatment and when the spraying has to be done with great precision. Vintaging takes a long time with such a big domain, and, even with a big team of pickers, the last grapes are not brought in until five weeks after the start of the harvest. The Gewürztraminer are the last to be picked. So, with this lengthy vintage, there are always some grapes being picked as *vendange tardive* or late harvest grapes.

Selection is a word much used at Schlumberger—selection of the grapes from the individual plots of vineyard, and selection of qualities, so as not to mix the very good

with the merely good. Great care is taken to see that the grapes, and then the wines, keep their respective individuality. The grapes are pressed in Vaslin presses, and the must is clarified by centrifugation. Fermentation is in wood, which is something the domain finds suits their wines, imprinted as they are with marked character and a good deal of richness. The winemakers find that the fermentation temperatures do not go up in wood, and they particularly like the fact that they never have any trouble with starting perfectly natural fermentations using the yeasts from the grapes. It is obvious that in a very old cellar, with wood that has been used for many vintages, there are a great number of yeasts present in the atmosphere, which can only encourage fermentations. In some modern cellars, with nothing but brand new stainless-steel vats, fermentations can be unwilling to start, or can 'stick'—or stop—which is always highly dangerous. This is why proprietors in such new wineries often have recourse to selected yeast strains. They might, in some cases, prefer this, as greater control is possible over fermentation started by yeasts which have been specially chosen; but in Alsace, if there is a 'natural' way, they prefer to take it. Because the cellars and casks have been used for numerous vintages this contributes to the fact that the malolactic fermentation takes place naturally following the alcoholic fermentation. The only exception to this was the 1972 vintage, where it seemed as if the malolactic fermentation would never start. This was a problem found in other parts of France in that year, notably in Burgundy.

There is a very wide range of cask sizes for all the various qualities and types of wine, enabling the domain really to keep apart their finest wines and to preserve each category in as pure a form as possible. Casks range in size from 17,000 litres and 15,000 litres for straightforward, standard wines. All top quality wines are stored in 7,000 litre barrels, allowing about 10,000 bottles of this individual quality. In good years, the domain might have about 20 of these top quality barrels. The cellar is always at a constant, cool temperature, influenced by the water under the Lauch valley where Guebwiller is situated.

Bottling is carried out throughout the year, as and when it is necessary. The only period without any bottling is over the vintage. Schlumberger like to keep an average stock of wine of about $2\frac{1}{2}$ years of age, as, only producing from their own estate, they have to try to protect themselves against a poor vintage or harvest. They always have between two and three million bottles in stock. They like to give their wines between three and six months of rest after bottling before despatch, and so try to plan sales and shipments accordingly. They consider that this is vital for their wines, which take more time than some Alsatian wines to smooth out and show at

The Cuvée Christine Schlumberger is the finest wine made at Schlumberger. It is only made in the best years and has a majestic noble rot character.

their best, as they are relatively richer. Finesse and bouquet improve in bottle. This richness and weight is partly due to the site and soil, but also to the lowish yield that these produce, giving concentration to the wines. Yields here often average about 50 hectolitres to the hectare, which is somewhat less than in other parts of Alsace. Schlumberger consider the 1971 vintage as one of their greatest ever. They finished picking their last grapes on 7 December, and these wines reached 14.2 per cent alcohol and 48 grams residual sugar; 1976 was also a great vintage, but they sadly lost two-thirds of the vintage through violent hail storms.

Domaines Schlumberger have the usual paradoxical mixture of the old and the new that is the mark of so many intelligent winemakers. They have old cellars and old casks, but young staff, centrifuges and they use a helicopter. Their efforts are crowned in certain years by the production of the famous Gewürztraminer Cuvée Christine Schlumberger, renowned for its majestic *pourriture noble* character. This wine has fetched the highest price at auction, so far, for any wine from Alsace. This is a proud feat, but the Domaines Schlumberger's real pride is in the consistent quality, character and personality of their wines.

Domaines Schlumberger are based at Guebwiller in the extreme south of the Alsace wine-growing area. The two main vineyard sites are the Kitterlé and Wanne, both on steep hills.

The technical director of Domaines Schlumberger is Jean-Marie Winter (FAR LEFT). The harvest in the Schlumberger vineyards (LEFT) is always prolonged because of the size of the vineyards and the many different grape varieties planted which all mature at slightly different times.

The Gewürztraminer is always harvested last (FAR LEFT, LEFT)

This section of Schlumberger vineyards (TOP) shows the high training used for vines in Alsace. Schlumberger are the largest proprietor of vineyards in Alsace. These slopes are near Guebwiller where vines have to be planted on terraces because of the steep terrain.

The Gewürztraminer does very well on the steep slopes found in some parts of Alsace (ABOVE). It achieves great ripeness thanks to the excellent exposure to the sun.

GERMANY: INTRODUCTION

Germany is the most northerly of all wine-producing countries. It is not surprising, therefore, that geography and the study of microclimate has assumed a greater importance here than in any other wine-producing country. Here the hours of sunshine, the incidence of frost and rainfall, the protection from cold northerly winds, the altitude and soil moisture, the warming influence of slatey soils and adjacent rivers, are all carefully recorded and studied. The classification of vineyards by microclimate is even more important in these northern latitudes than the study of soil.

Since 1971, German vineyards have been divided into 11 specified or designated regions called in German *bestimmte Anbaugebiete*, dominated by the river valleys of the Ahr, Mosel, Saar, Ruwer, Nahe, Rhine, Main and Neckar.

The Mosel-Saar-Ruwer area is one of the most important and most individual of Germany's wine regions. There are 11,500 hectares under vine, of which 65 per cent are Riesling and 19 per cent Müller-Thurgau. The region stretches from the Luxembourg border along the Mosel and its two tributaries until it flows into the Rhine at Koblenz. The vineyards are planted, often precariously, on the steeply sloping, slatey hills along the river banks. This is the land of classic Riesling wines, and more Riesling is planted here than anywhere else in Germany. The region is divided into four sub regions or *Bereiche*, of which the most important are Saar-Ruwer and Bernkastel. The Saar-Ruwer is one of the two great quality areas of the region, especially noted for producing great wines in the best vintages. There are many outstanding estates, notably—in addition to those featured in this volume—the Staatliche Domäne, Vereinigte Hospitien, Friedrich-Wilhelm-Gymnasium and von Kesselstatt, all of which are in Trier, as well as von Schubert and the Karthäuserhof on the Ruwer.

Bernkastel embraces all the best vineyards of the Mittelmosel, which produces slightly fuller, more earthy Rieslings than the Saar-Ruwer. Here are to be found two of the most famous vineyards in Germany, Bernkasteler Doktor and Wehlener Sonnenuhr. In addition, the domains of Dr Thanisch and von Schorlemer at Bernkastel and of Bergweiller-Prüm at Wehlen are particularly worthy of note.

The Rheingau is the smallest of the great quality wine regions, with only 2,900 hectares of vines, but it makes up in quality for what it lacks in quantity. Like the Mosel, this is Riesling country, and the percentage planted is even higher—75 per cent, with Müller-Thurgau accounting for a further 10 per cent. Geographically, the Rheingau is that small stretch where the Rhine flows from east to west, so that the exposure of the vineyards on the slopes of the Taunus mountains is due south. This is a region of many great estate wines. The major estates include Schloss Reinhartshausen at Erbach and Königin Victoria Berg at Hochheim. With an un-German illogic-ality, there is only one *Bereich*—Johannisberg.

The largest of the German wine growing regions, with nearly 22,000 hectares under vine is Rheinhessen. This large and varied area lies between the Rhine and Nahe rivers, and stretches from Bingen to Worms, via Mainz and Alzay. It is mostly a region of rolling hills producing large quantities of rather ordinary wines made from the Müller-Thurgau (36 per cent) and Silvaner (28 per cent) and destined for sale as Liebfraumilch and Bereich Niersteiner. However, on the Rheinfront between Nackenheim and Oppenheim, some fine and even great wines are made, much of it from the Riesling, which only accounts for 5 per cent of the plantings overall. Some fine wines are also made around Bingen. Apart from the estates featured here, the following are also worthy of special note: in Mainz, the Staatliche Domäne; in Nierstein, Franz Karl Schmitt, Schuch, Strub, von Heyl and Guntrum. The *Bereiche* are Nierstein, Wonnegau and Bingen.

Nahe is a small, but very picturesque region running along the Nahe valley from Bingenbrück south to Bad Kreuznach, then along several tortuous valleys. The finest wines have the raciness and spirit of fine Mosels with a little more body, while many others made from Müller-Thurgau (30 per cent) and Silvaner (27 per cent) are more like good Rheinhessens with rather more elegance and floweriness. The Riesling accounts for 23 per cent and produces most of the outstanding wines. In all 4,400 hectares are under vine. The *Bereich* names are Kreuznach and Schloss Böckelheim.

In English-speaking countries, the Rheinpfalz is often called the Palatinate. With nearly 21,000 hectares under vine the region is only just beaten into second place by Rheinhessen. It is traversed by the Deutsche Weinstrasse which runs for 80 kilometres down to the French border with Alsace, the largest unbroken stretch of vines in Germany. This is the warmest of the German wine regions, protected from cold northerly and westerly winds by the Vosges, the Pfälzer Wald and the Haardt mountains. The greatest wines are produced in the Mittelhaardt between Kallstadt and Neustadt. It is here that most of the region's Riesling (13 per cent) is planted. The mass-production area is in the Oberhaardt to the south, dominated by Müller-Thurgau (24 per cent) and Silvaner (23 per cent). But Morio-Muskat, Scheurebe, Kerner and Ruländer are also important. After Baden and Württemberg, this is the most important region for red wines in Germany, mostly made from the Portugieser. Apart from those featured in this volume, the most important estate is Bassermann-Jordan in Deidesheim. This is an important area for co-operatives, which account for 27 per cent in the region as a whole. In the Mittelhaardt, they produce some very fine wines, and in the Oberhaardt have done much to raise standards. The *Bereich* names are Mittelhaardt Deutsche Weinstrasse and Südliche Weinstrasse

Two general points need to be made when introducing

this small selection of top German wine estates. The most important concerns the new German Wine Law of 1971, which sets the standards and objectives against which German winemakers can measure their success. The second is the special character of the German estate or domain in contrast to what is found in other countries.

The new German Wine Law, which first became operative with the 1971 vintage, defines a scale of qualities which appear on every German wine label, as well as radically reforming the nomenclature of German vineyards. The way in which quality is defined and controlled is certainly the factor which most clearly differentiates German wine practice from those of other European countries. Basically, the measurement of the sugar content of grapes at the time of harvesting determines how the wine will be labelled and therefore how it is to be handled in the cellar. In Germany the Oechsle system is used; this is a measurement of the specific gravity of the grape must (grape juice before fermentation has commenced). For example, 100° Oechsle means that 100 litres of must contains 25 kilograms of sugar.

The following classes of quality are defined by the Wine Law. *Deutscher Tafelwein*, sold with a minimum alcoholic level of 8.5 per cent, is the lowest category. Wines of this quality made on estates would not be sold under the estate label. A wine classified as *Qualitätswein bestimmte Anbaugebiete* (normally abbreviated to *QbA*, or *Qualitätswein*) has, with a minimum of 60° Oechsle, the quality of which is improved by chaptalization, the addition of sugar to must during fermentation to increase alcoholic level. The third category *Qualitätswein mit Prädikat* (normally abbreviated to *QmP*, or *Prädikatswein*) has six sub-categories. All wines must be unchaptalized. *Kabinett* wines have a minimum 73° Oechsle. *Spätlesen* have minimum 85° Oechsle and are made from late-harvested, fully-ripe grapes. *Auslese* wines have a minimum 95° Oechsle from selected fully-ripe grapes. *Beerenauslese* wines have a minimum 125° Oechsle and are made from selected overripe grapes affected with botrytis, called *Edelfäule* in German. A *Trockenbeerenauslese* wine needs a minimum 150° Oechsle and is produced from selected, shrivelled botrytized grapes. *Eiswein* is made from grapes picked and pressed when frozen. An *Eiswein* is also always then classified in one of the *QmP* categories such as *Auslese*. These minimum Oechsle degrees in fact vary slightly according to the district and grape variety, especially in the Mosel where they are lower.

The degrees Oechsle set for each category are minimums. One of the main differences between the best estates and more run-of-the-mill wines is that they frequently set themselves higher standards, so that, for instance, some top *Spätlese* wines could be downgraded to *Kabinett*.

One of the consequences of this system for grading quality is that one can easily compare the quality of one vintage with another. Another is the premium which is placed on sweet wines. There is a tendency to think that a *Beerenauslese* or *Trockenbeerenauslese* must *per se* be the best wine. But, in practice, of course, it is the *Kabinett* and *Spätlese* wines, with their lively acidity and balancing fruit, which are the most suitable wines for drinking with meals. *Auslese* wines are usually best as an aperitif, or sometimes to accompany a first course of smoked fish for example, while the great sweet dessert wines can either be enjoyed at the end of a meal or on their own.

Usually, the high acidity of *QbA*, *Kabinett* and *Spätlese* wines is softened by the addition of *Süssreserve* (unfermented grape juice). The residual sugar in *Auslesen* or the still higher categories of *QmP* comes from residual sugar which is naturally left in the wine when the fermentation stops, whereas *Süssreserve* is added to completely fermented-out dry wines. Recently, a vogue for *Trocken* and *Halbtrocken* wines has caught on in Germany, although the main export markets, the UK and USA, seem less convinced. *Trocken* wines usually contain a few grams of residual sugar which are not detectable on the palate, and are rated as suitable for diabetics. *Halbtrocken* wines contain a few grams more residual sugar but, again, this is not detectable on the palate. Such wines are normally found in the *QbA* and *Kabinett* categories, although some *Spätlesen* and even *Auslesen* have been made. Basically, successful *Trocken* or *Halbtrocken* wines need a good ripe flavour, where the acidity is not too dominating and the fruit can show through. For this reason, the most successful examples tend to come from the top estates.

Another result of the 1971 Wine Law was a dramatic reduction in the number of vineyard names, and the introduction of designations giving the *Bereich* (district or sub-region) and the *Grosslage* (collective site or vineyard name). Most domains use their individual site or vineyard names, but *Grosslage* names are often found on wines like *Trockenbeerenauslese* or *Eiswein* where grapes from a number of different vineyards may have to be used to make even a small quantity of such rarities. Normally, estate wines are sold under the name of the *Gemeinde* or village, for example Rüdesheim, and some estates may have vineyards in three or four of these. The Bischöflichen Weingüter in Trier has no fewer than 11 villages amongst its vineyards. But there are a few special cases where, for reasons of historic usage, the name of a property has been sanctioned without the use of the village or site names. Examples are Schloss Vollrads, Steinberg and Scharzhofberg, where the famous site names are used without reference to the village name.

In general, the German estate compares more closely to a Burgundian domain than to a Bordeaux château, for example, because of the many different wines it normally produces. But, whereas a Burgundy domain will usually produce wines from a number of different *appellations*, its German counterpart will also produce a number of different qualities within each vineyard name, and will also usually cultivate a number of different grape varieties. So, the number of different wines which can be produced in any one vintage, especially in good years when the whole gamut of qualities are produced, is often formidable.

The Riesling is still the class grape variety which is most prized and produces the greatest wines on the Mosel, in the Rheingau, on the Rheinfront in Rheinhessen, and in the best vineyards of the top estates in the Mittelhaardt district of the Rheinpfalz. But the Müller-Thurgau is now more widely planted. Its mild fruity wines produce most of the cheap commercial wines and are ideal for early consumption. For this reason, the proportion planted in the leading estates is small.

Other varieties—traditional and experimental—are also planted. Traditional varieties include the Gewürztraminer which produces very aromatic wines and is mostly grown in Rheinpfalz and Baden; the Silvaner has declined in importance. Mainly grown in Rheinhessen, Rheinpfalz and Franconia, it can, in the right place, produce spicy, full-bodied wines of breed. New varieties are mostly the result of many years of experimentation, much of it at Geisenheim, aimed at producing crosses which will ripen fully in less favoured sites, while retaining some regional characteristics. Ehrenfelser is a Riesling-Silvaner cross which resembles the Riesling and has proved most success-ful in the Rheingau. The Scheurebe is a Riesling-Silvaner cross which produces wines with a marked aromatic character. It is at its best at the *Auslese* and higher categories when it can produce great wines. It is mainly found in Rheinhessen and Rheinpfalz.

One distinctive feature of the German estates is the number which belong to the state. These have their origins in the royal estates of the Prussian state, in some cases expropriated from the church. Since the creation of the Federal Republic, these have passed to, and are now administered by, the various *Länder* or provincial governments.

The German estates today represent all that is best in German viticulture and winemaking. They are the bastions of quality wine production and the guardians of the best traditions in winemaking. At a time when Germany's export markets are flooded with cheap, often branded wines for immediate consumption, they are still trying to produce wines which need time to mature and which show the individual character of the districts from which they come and of the grapes from which they are grown. It is upon such wines that Germany's reputation as a producer of some of the world's greatest white wines is justly based.

Vintage chart

1945 Potentially a fine vintage, but the chaotic conditions of the aftermath of the war meant that even fewer fine wines were made than would have been the case. Some magnificent *Spätlese* and *Auslese* wines, especially on the Rheingau and in the Rheinpfalz.

1947 A fine vintage that did not quite come off. Many wines had too much albumen and producers lacked the facilities to treat them properly.

1949 A really great year in all districts, but especially on the Mosel, Great *Auslese*, *Beeren-* and *Trockenbeerenauslese* wines were made. Ripeness combined with elegant acidity.

1952 A sound year with some top quality wines.

1953 A great vintage with fine acidity if not the weight of the 1949s. Noble rot enabled top quality wines to be made, which have lasted well.

1959 A very dry warm summer produced powerful wines, often lacking acidity and therefore breed. But wines of high Oechsle were made, although there was little noble rot. Those wines with good acidity have lasted well.

1961 A year chiefly remembered for some fine *Eisweine* which marked the beginning of the vogue for these rarities. Most successful on the Mosel but no more than good overall.

1964 A very good year on the Mosel, where the best *Auslesen* have lasted well.

1966 Light, charming wines, especially for the Riesling.

1967 Excessive noble rot made for difficult winemaking conditions, but some very successful wines were made in the Rheinpfalz.

1969 A good year for the Mosel. Some light, elegant *Kabinett*, with some *Spätlesen* were also made in the Rheinpfalz, Rheinhessen and Rheingau.

1970 The largest vintage since 1945. The best wines were to be found in the Rheinpfalz, where some fine *Auslesen* were made.

1971 A very great year, especially on the Mosel and the Rheingau. The Rieslings have outstanding balance and breed, and are still improving. Even simple *Kabinett* and *Spätlese* wines are still magnificent after 10 years, if well cellared.

1972 A poor year, when even *Prädikatsweine* often had a pronounced green acidity.

1973 The largest vintage of the century. Attractive, well-balanced, fruity wines, especially in the *Kabinett* category. A few good *Spätlesen*. Not a year for keeping now.

1974 A small vintage of below average quality. Better than 1972, but much inferior to 1973.

1975 A very good vintage for the Mosel, and for Riesling wines in general. On the Mosel they have great elegance and charm, especially at the *Kabinett* and *Spätlese* levels.

1976 The second great vintage of the decade. A much higher proportion of *Auslese* wines than in 1971, but with less acidity. The top wines should continue to improve for some years.

1977 A very large vintage, largely of *QbA* wines. A rather dull year.

1978 Average size crop overall, but small in the Mosel. Some *Kabinett* wines, but only marginally better than 1977.

1979 Larger vintage than 1978 in which the Riesling did particularly well. Elegant, stylish *Spätlese* wines on the Mosel, with some good *Kabinett* wines in the Rheingau. Much better than 1977 or 1978. Comparable to 1975.

1980 Yield very small in Mosel-Saar–Ruwer and Rheingau, better in Rheinpfalz. Mostly *QbA* and *Kabinett*, with some *Spätlese* in the Rheinpfalz.

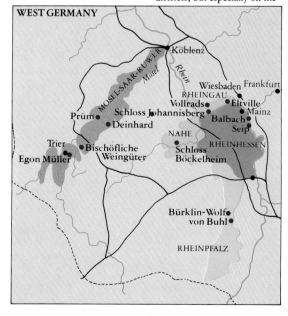

Care in the vineyards, during vinification and the hand of master winemakers ensure the continuing quality of German wines.

BISCHÖFLICHE WEINGÜTER

This greatest of Mosel estates is particularly famed for its fine Saar and Ruwer wines. No estate has done more to preserve the traditional crisp, fresh acidity which should mark a great Riesling wine from this region. For this reason, the Bischöfliche Weingüter wines keep exceptionally well, even at the <u>Kabinett</u> and <u>Spätlese</u> levels.

The church has played a major and distinguished part in European viticulture. Almost at a stroke, the French Revolution and its crusading armies, which soon engulfed most of the rest of Europe, swept this patrimony away. The three estates, now united for administrative purposes under the *Verwaltung der Bischöflichen Weingüter* in Trier, today represent the most distinguished European survival of a tradition which goes back over 1,000 years.

The first of the estates is the Bischöfliches Priesterseminar which is the Episcopal Seminary of Trier where the priests of the diocese receive their training. The vineyard property goes back to 1773 when the Elector (also the Archbishop) Clemens Wenzeslaus presented them to the Seminary. They were confiscated in 1794 but returned in 1809. The second is the Bischöfliches Konvikt, which is the Episcopal Hostel. The major part of this estate originates in bequests from the Elector (also the Archbishop) Philipp Christoph von Sötern (1623-1652). In 1653 these came under the control of the Trier Cathedral Chapel. After its confiscation, it was acquired by a prelate, Johannes Endres, who was director of the Episcopal Hostel from 1860 to 1892. He left the vineyards to this institution on his death with the condition that education must be provided for poor scholars at the school. The third estate, the Hohe Domkirche, encompasses the vineyards belonging to the Cathedral of Trier. Parts of this estate, such as the Scharzhof, were originally monastic property. Other parts, such as Avelsbach, were privately owned. The monastic portion of Scharzhof was bought after confiscation by Egon Müller, one of whose daughters on her death left her portion to the Cathedral.

In 1966, these three estates merged their administration and cellars. The result is the largest and most varied estate on the Mosel, with important holdings on the Saar, Ruwer and Mittelmosel, whose wines represent the Riesling at its most traditionally classic. The estates for long seem to have been fortunate in their administrators, and the wines have maintained an enviable reputation with great consistency over the years. The cellars are most extensive and combine the best of the old and the new. Parts of them are over 400 years old, parts are a recent addition, made so that all the wines of the three estates could be under one roof.

The offices and cellars are in the centre of Trier. The vineyards are in Urzig, Erden, Trittenheim, Dhron and Piesport on the Mittel-Mosel; in Kasel, Eitelsbach and Avelsbach on the Ruwer; and in Canzem, Wiltingen and Ayl on the Saar.

The holdings of the three estates are as follows—it should be noted that each still uses its own label showing its arms and name. Bischöfliches Priesterseminar has holdings at Urzig, Würzgarten 0.39 hec-

tares; at Erden, Treppchen 3.10 hectares; at Trittenheim, Apotheke 1.70 hectares and Altärchen 1.65 hectares; at Dhron, Hofberger 2.42 hectares; at Kasel, Nies'chen 6.53 hectares; at Canzem, Altenberg 9.60 hectares and Scharzberg (for red wine) 0.50 hectares; at Wiltingen, Kupp 5.20 hectares; and at Ayl, Kupp 10.20 hectares. The holdings of the Bischöfliches Konvikt are at Piesport, Goldtröpfchen 2.84 hectares; at Eitelsbach, Marienholz 19.10 hectares; at Kasel, Kehrnagel 4.10 hectares; at Avelsbach, Herrenberg 4.77 hectares; at Ayl, Herrenberger 4.60 hectares (sole owner). The third estate is Hohe Domkirche which has holdings at Avelsbach, Altenberg 9.7 hectares (sole owner); at Wiltingen, Rosenberg 2.20 hectares, and Scharzhofberg 7.88 hectares.

The total vineyard area of the Bischöfliche Weingüter is 95.48 hectares, all on steeply sloping sites with the schist soil typical of the area. Although the domain has some fine vineyards on the Mittelmosel, by far the most important part is on the Saar and Ruwer, and it is these parts that constitute its real glory and claim to fame. On these steep slopes, with their slippery covering of schist so suited to the Riesling and vital for reflecting the day's heat and storing it during the night, favourable expositions are all-important. This estate has a high proportion of the best sites.

Some 95 per cent of the plantings are Riesling, the rest experimental varieties. This is at a time when the proportion of Riesling for the region as a whole has dropped to 65 per cent. The Riesling can only ripen satisfactorily, especially on the Saar and Ruwer, in the most favourable conditions, so this is eloquent testimony to the siting of the domain's vineyards.

The grapes are pressed in the vineyard areas and the must is then brought to the cellars in Trier. About 90 per cent of the storage here, amounting to 1.5 million litres, is in wood, mostly 1,000 litre casks. The philosophy here is to produce very classic Rieslings with the minimum of residual sugar, preserving a very marked, crisp acidity which gives longevity to the wines. This is an exemplary instance of the happy blending of modern cellar techniques with an old tradition. *Süssreserve* is used sparingly, for the *Qualitätswein, Kabinett* and *Spätlese* qualities. The sweetness in the *Auslesen* and better qualities comes from residual sugar naturally retained in the wines after fermentation is completed.

These estates produce wines which carry on the very highest traditions for producing dry, fruity Riesling wines, especially from their Saar and Ruwer vineyards. Because of their acidity, such wines last extremely well, and not only the great *Auslesen*. For example a simple Dom Scharzhofberger 1952 was still fresh with a wonderful, lingering, mature Riesling flavour when 28 years old. It was characteristic of this domain that,

These labels show that the constituent domains in the Bischöfliche Weingüter have retained their own identities since the cellars were amalgamated.

A clear impression of the spacing of the Avelsbach vines can be seen (BELOW). This is done to ensure that each vine receives maximum exposure to sunlight and to the slatey soil.

Vintaging is taking place in the Avelsbach vineyard (ABOVE). The small plastic receptacles ensure that the grapes are not crushed by their own weight. The steepness of the vineyards can also be seen in the background.

when the 1971 wines were first tasted in 1972, the acidity of the *Auslesen* made them seem almost dry and certainly austere, at a time when many wines of this great vintage looked nearly ready to drink. Harmonious fruit acidity is the key to these wines, to their classic composition and the many years of pleasure they give.

At first sight, it seems strange that an estate with so many sites where ripening is often a problem, should now produce many *Trocken* and *Halbtrocken* wines, since these require well-balanced musts. But then, the domain has never had to rely on hefty doses of *Süssreserve* to make their wines palatable. In the 1979 vintage, 35 per cent were *Trocken* and 20 per cent *Halbtrocken* wines.

Since the Second World War the great years here have been 1945, 1949, 1953, 1959, 1971 and 1976. In such years the greatest *Auslesen* have been produced. But delightful and fine wines abound. In 1979 Saar and Ruwer production was decimated by frost. This resulted in a *Kabinett*, *Spätlese* vintage, with 42 per cent *Kabinett* and 25 per cent *Spätlese* and just 4 or 5 per cent *Auslese*. The wines have a lovely flowery, spicy acidity. In 1978 the quantity was only half a normal harvest with very few wines of *Prädikat* quality; but the wines have fullness and character. In 1977 there was normal-sized harvest. Again the wines were mostly *QbA*, with just a few *Kabinett* wines. The wines are high in extract with a typical Riesling character. The famous year 1976 yielded only half a normal harvest, with 45 hectolitres per hectare, but it is one of the century's greatest wines. Some 80 per cent of the wines were *Auslese* or better. These wines have great character, a unique spiciness and a ripe fruit acidity. 1975 was a typical year for fine Mosel Riesling, producing flowery, fruity wines with a certain piquant fruit acidity. The top wines were few in quantity but very fine, with outstanding *Spätlese* and *Auslese* wines.

The main offices of the estate are at Trier, but this large and important estate has major holdings on the Saar and Ruwer, as well as the Mittelmosel.

The village of Kasel (LEFT) is seen overshadowed by the Nieschen hill. The system of terracing gives easy access for work in the vineyards.

DEINHARD—BERNKASTEL

**The jewel of this estate is the Bernkastel Doktor, the most
famous vineyard on the Mosel and one of the great wines of Germany.
The wines of the Doktor have a finesse which sets them apart from other
Bernkastel wines. The true characteristics of this vineyard are best seen
in the great <u>Auslesen</u> and sweet wines of the best years.**

There is a large amount of misapprehension and misrepresentation existing in the trade as to the produce of the Bernkasteler Doctor vineyards, states a quotation from a letter by Deinhard to its customers in 1908. This demonstrates Deinhard's concern that the fame of the vineyard had proved an irresistible attraction to so many wine merchants—an attraction which has persisted to the present day.

Bernkasteler Doctor, spelt with a 'c', rather than a 'k' at that time, was found in every wine list, both in Germany and abroad. Its distribution was so widespread that it would have been impossible for all the wine sold under the vineyard name to have been of genuine origin. The aim, therefore, of the Koblenz-based company of Deinhard, when acquiring a major part of the vineyard in 1900, was not simply to become the owner of a valuable piece of real estate. It was also to secure and control for itself a source of supply of certainly one of the best-known wines on the Mosel, if not of the whole of Germany. By owning the estate, Deinhard could guarantee the quality and authenticity of its Bernkasteler Doktor.

The origin of the name 'Bernkasteler Doktor' dates from the fourteenth century, when the ailing Elector of Trier, Archbishop Bohemund II, revived miraculously—or so legend has it—after accepting a glass of wine from the vineyard on his sick bed. However, the subsequent history of the Doktor is somewhat incomplete. It is known that the site was owned by a Count von der Leyen from 1750 until 1760, and that in 1794, under French rule, the vineyard was declared communal property. During most of the nineteenth century the Doktor was leased to one family, until eventually it became the property of Dr Hugo Thanisch. His family are known to have lived in Bernkastel since 1636 and were already important wine growers in the Bernkastel district in 1800. Thanisch sold part of his holding in the Doktor vineyard in 1899 to the Mayor of Bernkastel and the celebrated sale to Deinhard followed. The price of 100 marks per vine which Deinhard paid, compared to the 60 marks that the Mayor had paid Thanisch the previous year, was the highest amount that had ever been spent on purchasing a vineyard in Germany. The trade was suitably impressed.

According to Common Market regulations an individual vineyard must be of at least 5 hectares. This has meant that the original Doktor site of 1.47 hectares has had to be extended, but the exact size and ownership of the new Doktor is still not finally settled. All that can be said at present is that the owners of the Doktor vineyard before the enlargement—Deinhard, Thanisch and the firm of Lauerburg—will, between them, own by far the biggest portion of the enlarged Doktor vineyard in the very near future.

Dr Guyot, a famous French oenologist,

writing in the last century, described the Mosel valley as a land where 'sweet fruits can all attain refinement and perfection.' To this might unkindly be added the qualification 'but sometimes do not,' for the Mosel-Saar-Ruwer, to give the area its correct title, displays all the symptoms of a region on the edge of the practical vine-growing area of Europe. On average, it enjoys less sun than most of the other 10 vine-growing regions of Germany, so that, in many years, only those vineyards on steep slopes, facing south-south-east to south-south-west, are able to produce well-ripened Riesling grapes.

The Doktor vineyard faces due south, on the steepest of slopes that fall directly into the town of Bernkastel. The Deinhard Doktor lies on a bulge of the hillside that might be said to reach out to the sun. The result is that rays of the sun strike the vineyard first thing in the morning and continue to do so until nightfall. The slate roofs of the town reflect the heat of the day, which rises upwards to warm the vineyard in the evening.

As elsewhere on the Mosel, the slate of the soil also retains the heat. Its texture is somewhat reminiscent of demerara sugar, for it has a similar moist quality even after a dry summer. This, no doubt, helps it to remain in position on the steep hillside, without being eroded too severely by the rain. There is regular analysis of the soil, which is appropriately enriched when necessary with stable manure.

Because of the nature of the terrain, work in the Doktor vineyard is carried out by hand. As a result of this, there are over twice as many hours per hectare spent on vineyard work in the Doktor, as there are, for instance, in the Palatinate, where slopes are less severe and the use of viticultural machinery is possible. Some 45 per cent of West Germany's viticultural area has been reconstructed in the last 35 years in order to reduce costs. This has often meant completely altering the contour of a vineyard slope, whilst new service roads and drainage systems were built. Such a process, however, is not likely to happen on the Doktor, for the vineyard is so valuable as it stands that not even the demand for rationalization can allow one square metre of the soil to give way to any road-widening or similar scheme.

The Deinhard Doktor is planted almost 100 per cent in Riesling vines, of which some 50 per cent remain, ungrafted, on their own root stocks. Phylloxera has not flourished on the Mosel to the same extent as it has in most other vineyard areas of the world, so the planting of grafted vines at Bernkastel is not absolutely necessary, although it is wise as a precaution. The harvest in the Doktor usually starts towards the end of October, but the date varies according to the vintage, and the quality of the wine that it is hoped to produce—*Kabinett*,

*Before the new wine law of 1971, Deinhard
were the only producers to sell wine from the
famous Doktor vineyard unblended with
wine from other vineyards. This 1970 wine
is called Christ Eiswein because it was
harvested at Christmas.*

Spätlese, and so on. About 36 pickers are engaged for the harvest. Most of these come from the surrounding countryside; amongst regular attenders are often found a few students and members of the British army from northern Germany. The grapes are preselected in the vineyard, different qualities of grape being placed in different containers. A further selection may be made at the *Gutshaus*, or estate house, across the Mosel at Kues. This is a solid building, dating from 1906, with a splendid view of the Doktor vineyard. It is also the home of the estate manager, Norbert Kreuzberg. Felix Wegeler is the member of the Deinhard family who has overall charge of the estates in the Rheingau and Palatinate, as well as those on the Mosel. Together both men hold a top level meeting with technical advisors from the main Deinhard house at Koblenz to decide how the harvesting of the grapes is to be conducted. It is a time when good organization is absolutely essential.

The grapes are crushed immediately on arrival at the *Gutshaus* by a Willmes horizontal hydraulic press. Every effort is made to avoid any unnecessary delay at this point in order to retain the freshness so characteristic of Mosel wines. After clarification, the grape juice is taken to the main Deinhard cellar, downstream in Koblenz.

In Koblenz, the wine is made by a combination of modern and traditional methods, with, of course, the technological support that a large cellar can offer, but which would be impractical to maintain on a small estate. The style and quality of the wine is determined by the vintage and the Riesling grape and it is the job of the cellar master to 'guide' the wine so that it reaches its full potential. The characteristic of all Bernkasteler Riesling wines is that they are positive, firm and have a lingering fruity acidity. The difference between a Bernkasteler and another Mosel, such as a Brauneberger, is not unlike that between an Haut Médoc and a St Emilion—the first is firmer than the second.

To define what differentiates a Doktor from other Bernkasteler wines is not easy, but it is related to size and length of flavour. A cool, steady fermentation helps the delicacy of the wine, so it spends a period in oak casks until the turn of the year. This allows the Mosel Riesling style to develop to the maximum. From January, following the vintage, the wine lies in stainless steel vats to retain its freshness until bottling by the cold sterile process in the spring. A Doktor *Kabinett* quality needs at least a year in bottle to show its best, whereas *Spätlesen* are not normally ready until three or four years after the vintage. Doktor *Auslesen*, and wines of even higher quality, can well be left in bottle for 10 years or more.

Since 1971 wines have been made by Deinhard in the Doktor in the following qualities: 1971 *Kabinett, Spätlese, Auslese, Beerenauslese*; 1972 nil; 1973 *Kabinett, Spätlese, Auslese, Beerenauslese Eiswein*; 1974 *Kabinett*; 1975 *Spätlese, Auslese, Auslese Eiswein, Beerenauslese*; 1976 *Auslese*,

Beerenauslese Eiswein (70 litres only), *Trockenbeerenauslese*; 1977; *Kabinett*; 1978 nil; 1979 *Spätlese, Auslese, Beerenauslese*. 1980 nil. The wines which experts feel are currently at a good stage of development for drinking are the 1971 *Kabinett*, 1973 *Spätlese*, 1975 *Spätlese*, and 1977 *Kabinett*. As with all fine Mosel wines, it is the balance of acidity and fruit that gives a Doktor a long life. In some years such as 1980 the quality of the grapes was not up to the required standard for Deinhard Doktor and the wine was used for other purposes.

The German Wine Law establishes the various quality categories of wine, but the levels of weight that are necessary for a must to reach a certain quality standard are only the minimum acceptable. Deinhard never produces a simple *Qualitätswein* from the Doktor. If the wine is good, but not up to Doktor standard, it can legally be sold as Bernkasteler Badstube. This *Grosslage*, or collective site, covers the five best individual sites or *Einzellagen* in Bernkastel area.

The Doktor vineyard in Bernkastel is one of the most famous in Germany. Facing due south, it is on the steep slopes above the town of Bernkastel.

This splendid view from the Doktor vineyard shows the town of Bernkastel and the adjacent bridge (BELOW). The unusual slatey schist of the Doktor vineyard (INSET) helps to create a very fine wine.

WEINGUT EGON MÜLLER-SCHARZHOF

From the famous Scharzhofberg vineyards, Egon Müller produces wines which are the epitome of Saar Riesling. The lively acidity which these wines retain even in the ripest years gives them a truly remarkable vivacity and longevity.

This estate is dominated in every sense by the famous Scharzhofberg, the great hill at Wiltingen. Its most famous wine comes from the vineyard on this hill, and the house itself stands in its shadow. In 1030 the Archbishop of Trier gave the Scharzhof estate to a Benedictine monastery in Trier. It remained church property until the French Revolution when, together with all church property, it was confiscated in 1796. It was put up for auction, and bought on 1 August 1797 by the great-great-grandfather of Egon Müller, the present proprietor. The present estate only represents about one third of that purchase, the rest having been lost through inheritance. One part has even found its way back to the church, and now belongs to the Hohe Domkirche, part of the Bischöflichen Weingüter.

This estate is typical of the small, high quality family estates of the Mosel. It is essentially a small farm, centred on the house, a large Victorian-style mansion, with outbuildings and the barrel cellar under the house itself. The vineyards are literally at the bottom of the garden. The estate is under the personal supervision of the proprietor, Egon Müller, who assumed responsibility in 1945. He now finds interest and support from his son Egon, and in charge in the cellar is *Kellermeister* Frank.

The Müller house with its adjoining vineyards lie in the heart of the Saar district of the Mosel-Saar-Ruwer region. The Scharzhof is a valley which lies to the east of Wiltingen and the Saar river, and is dominated by the long hill of the Scharzhofberg, the formation of which is somewhat reminiscent of the hill at Corton in the Côte de Beaune.

The most important vineyard is Scharzhofberg, which under the German Wine Law of 1971 is a name independent of any village name (like the Steinberg in Rheingau). There are also smallholdings in Wiltinger Braune Kupp, Wiltinger Klosterberg and Wiltinger Braunfels. This last vineyard adjoins the Scharzhofberg.

The Egon Müller vineyards only cover 11 hectares, but make up in quality what they lack in quantity. No less than 63 per cent is on steeply sloping terrain, with the remaining 37 per cent on gently sloping land. The soil is schist, so that looking down from the top of the vineyard is like looking down on a church roof.

Rieslings are planted in 96 per cent of the vineyards, the remaining 4 per cent with experimental varieties. The exceptional exposition of the vineyard of Scharzhofberg enables a degree of ripeness to be obtained which is unusual and outstanding for the Saar, where on average full ripeness is only achieved every four or five years, and in the lesser years much production goes to make *Sekt*. Egon Müller is a great believer in the importance of viticulture, and quotes with approval a French proverb which says that the vine likes the sun, but even more the shadow of its master. The cellar work is traditional and meticulous. There is no fining of the must before fermentation, which is always carried out in oak casks for the *Prädikatsweine*, so that each batch preserves its own character. The domain uses cultivated yeasts for about half of its wines. The general rule is that the must and the young wines are treated as little as possible. If there are no faults in them, they are centrifuged and then filtered.

Süssreserve is used normally only for part of the *Kabinett* and the *Spätlese* wines, not for the *Auslese* and higher qualities. For wines with more than 100° Oechsle, the fermentation is allowed to stop naturally by placing the wines in the coldest part of the cellar, where they are then racked, so retaining sufficient residual sugar. A few *Trocken* wines are made. The domain is convinced that, in order to give their wines a long life, bottling should be done in the spring after the vintage.

The classic Saar Rieslings are much prized by all lovers of Riesling at its most pure and unadorned. Even the *Kabinett* wines need to be kept for several years before reaching their best, to allow the relatively high acidity to mellow. But the most long-lived wines are the *Auslesen*, where sugar and acidity achieve an ideal harmony which shows the fresh, delicate raciness of these wines to perfection. The 1971 *Auslesen* are exceptional and promise to keep for many years yet, while the 1976s are even more remarkable. Even in less good years, such as 1975 and 1979, very fine *Auslesen* were made. Fine *Auslesen* of the 1964 and 1955 vintages are still in excellent condition. Now that the classification *feinste* can no longer be used to distinguish the best casks of *Auslese*, these are now sold with a gold capsule as distinct from the usual white capsule.

The unusual degree of ripeness achieved on the Scharzhofberg vineyard is well illustrated by the fact that this is one of the few vineyards on the Saar where successful *Beerenauslesen* have been made. The 1976 *Beerenauslese* is one of the outstanding wines of this vintage in the whole Mosel, being of almost *Trockenbeerenauslese* weight and concentration. The 1979 *Beerenauslese* sold at the famous Grosser Ring auction in Trier in October 1980 for the remarkable sum of DM262 per bottle, by far the highest price recorded at the sale for any wine of the 1979 vintage.

In 1979 65 per cent of the wine was *Kabinett* quality, 25 per cent *Spätlese* and 10 per cent *Auslese*. There were no wines of the *Qualitätswein bestimmter Anbaugebiete* designation. The *Qualitätsweine* produced by the estate are sold entirely as Wiltinger Scharzberg under the name Le Gallais. In 1976 the yield from the vineyards was 50 hectolitres to the hectare, and in 1975, 80 hectolitres to the hectare. Even in the ripest years Egon Müller's Scharzhofberg wines retain an acidity which gives the wine vivacity.

The Egon Müller label has remained unchanged for many years. It shows the family house with the adjoining Scharzhofberg vineyards. The 1971 and 1976 vintages were the best of the decade.

The Egon Müller estate house is in Wiltingen, a village dominated, as is the estate itself, by the Scharzhofberg.

Year	QbA	Kabinett	Spätlese	Auslese
1971		40%	40%	20%
1973	5%	40%	12%	3%
1975		55%	35%	10%
1976		45%	40%	15%
1977	20%	25%		
1978	45%	20%		
1979		65%	25%	10%

The Egon Müller house (FAR LEFT) is backed by the vineyards of the Scharzhofberg which rise steeply behind. Egon Müller (LEFT) samples a wine in his library. He is using a Treviris glass, the traditional glass of the Mosel. An aerial view shows the Müller house and estate (BELOW). The orderly system of vineyard tracks give access to all parts of the hill.

These young vines (BELOW) have just been wired onto their posts. The posts are very tall in relation to the vineyard workers. This view is of the

Scharzhofberg (BOTTOM) looking down on to the Egon Müller house and farm buildings. It is the end of the season, and the leaves are already changing colour.

In the cellars racking wine from one cask to another is taking place (BELOW). Here the wine is being filtered as well to remove unwanted yeasts once the fermentation has stopped.

WEINGUT JOH JOS PRÜM

The name of this estate is now synonymous with Wehlener Sonnenhuhr vineyard. Under the direction of the last three generations of the Prüm family, the exceptional opulence of these wines in the great vintages have placed them in the forefront of the Mosels.

Bernkasteler Doktor is the most famous vineyard in the Mosel, but Joh. Jos. Prüm has seen to it that today the name of Wehlener Sonnenuhr is at least bracketed with the great Bernkastel vineyard. The Prüms are a very old family of Wehlen winegrowers. The first mention of the name in local records is of Herhardus Hermann Prüm who lived from 1169 to 1229. It was Jodocus Prüm who constructed the famous Wehlen and Zeltinger sundials (*Sonnenuhren*) in 1842 so that vineyard workers could tell the time of day. In those days, they carried no watches.

It is impossible to go very far in or around Wehlen without bumping into a Prüm, and this can cause confusion. Sebastian Alois Prüm (1794-1871) had six sons, and then his heir, Matthias, had seven children, between whom the estate was divided, so it is not surprising that the Prüm family should be such a conspicuous presence in the area. The main estate today takes its name from the eldest son of Matthias, Johann Josef (1873-1944). There are also many other Prüm estates in the area, but Joh. Jos. Prüm is by far the most important. It was especially the work of Johann Josef's son Sebastian Prüm (1902-1969) which placed this property in the forefront of Mittelmosel estates. Now the task of running the vineyard has fallen on Dr Manfred Prüm who ably continues in his father's tradition. The solid family house on the banks of the river commands a magnificent view of the great natural amphitheatre of vineyards formed on the curve of the river by the vineyards of Wehlen and Zeltingen opposite. Under the house are the vaulted cellars, parts of which are very old. In winter, they frequently have several centimetres of water in them, and on occasion much more, so there is no lack of humidity to maintain a cool, constant temperature.

When the 1949 wines hit the German market, and still more the principal export markets in Great Britain and the United States, a new dimension was added to Mosel wines. The great Prüm estate showed it was possible to make wine at the *Auslese*, *Beerenauslese* and even *Trockenbeerenauslese* level which could rival the great Rhinegaus in the best years, for their majestic ripe Riesling fruit. Indeed, the legendary Wehlener Sonnenuhr Trockenbeerenauslese 1949 was a landmark in the estate's progress, selling for a record DM1,500 per bottle in 1974.

The offices and cellars are centred on the Prüm house in Wehlen. The vineyards are in the four adjoining villages of Zeltingen, Wehlen, Graach and Bernkastel, the heart of the Mittelmosel. The vineyards are in Wehlen: Sonnenuhr (the largest single proprietor, with 4.5 hectares), Rosenberg, Klosterberg and Nonnenberg; at Zeltingen: Sonnenuhr; at Graach: Himmelreich and Domprobst; and at Bernkastel: Bratenhöfchen and Lay.

The Prüm vineyards cover only 14

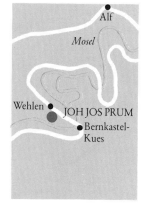

The Prüm estate is based in the village of Wehlen, where some of the vineyards are. The other vineyards are in Zeltingen, Graach and Bernkastel.

The Wehlener Sonnenuhr vineyard is one of the most famous in Germany. The labels show the distinctive sundial from which the vineyard takes its name.

This view from the top of the
Wehlen vineyards across the
Mosel to the village of Wehlen
shows the characteristic
steepness of the vineyards.

Dr Manfred Prüm is seen in
his vineyards (ABOVE) illus-
trating how high the vines are
trained on their individual
wooden stakes.

hectares, which is fairly typical of the small family estates on the Mittelmosel, but little enough for a world-wide demand. About 90 per cent of the vineyards are on steeply sloping sites, and the soil is mainly the schist typical of the finest Mosel vineyards. Some 94 per cent of the plantings are Riesling, in contrast to the 65 per cent for the region as a whole, the remainder being Müller-Thurgau and experimental varieties. It is the combination of the best sites with south and south-west exposures and the Riesling that make possible the exceptional wines produced here.

The philosophy behind the cellar work at Prüm is to try and produce wines which correspond as closely as possible to the quality of the grapes originally picked. To begin with, this is an estate where late harvesting is very much the rule in order to obtain the best possible ripeness. Manfred Prüm cites conscientiousness, care and cleanliness as his watchwords, both in the vineyards and in the cellar.

Once the grapes have been pressed, the main aim is to keep negative influences away from the wine, so as to give

The slatey schist of the Wehlen vineyards (RIGHT) can be seen with a Riesling leaf. The distinctive shape with deep indentations makes the Riesling immediately recognizable to winemakers.

its natural development absolute priority. The cellar-work has not changed fundamentally in the last 25 years, although the estate keeps in touch with modern research and technical improvements, which are used where this advances quality. For instance, methods of filtration have been improved.

For fermentation the winemakers mostly rely on natural yeasts, except in exceptional circumstances when cultivated yeasts are used. *Süssreserve* is only used when the fermentation goes too far, that is when the resulting wine would be too dry. Prüm prefers to leave the natural unfermented sugar in the wine, but when *Süssreserve* is necessary, care is taken to ensure that it is of the finest quality.

In the great years, Prüm's wines have an opulence which is rarely associated with the Mosel, but this is something which these exceptional Mittelmosel sites can produce, if they have the courage to wait for perfect ripeness and the expertise to turn such grapes into wines which correspond to them. In such years, the *Auslesen* have a weight and rich, fruity acidity normally reserved for the Rheingau. But the racy acidity and elegance serve as a reminder of the true Mosel-Riesling origins.

These grapes (RIGHT) are approaching ripeness. In the top left-hand corner some of the grapes have become discolored with Edelfäule *or noble rot.*

This famous sundial gives Wehlener Sonnenuhr its name (RIGHT). Because the division of the Mosel vineyards between many owners, it is customary for a grower to indicate his ownership with a sign at the end of a row of vines (INSET).

SCHLOSSBÖCKELHEIM

This important estate represents all that is best in Nahe wines produced from the Riesling grape. A notable feature are the many wines made at the <u>Kabinett</u> and <u>Spätlese</u> levels, although the estate is justly famed for its famous <u>Auslese</u> and <u>Trockenbeerenauslese</u>, which are wines of great complexity and character.

This domain has the unusual distinction of being the youngest estate producing top quality wine in Germany, and the most recent of the domains owned by the state. Its history only began in 1902 as a scheme by the Prussian state to encourage agriculture in the region. One of the most fascinating documents preserved at the offices of the domain are the panoramic photographs which show the Niederhausen site before work began on it, and then after the first part of the vineyard had been planted. Originally the site was a series of bare rocky crags interspersed with scrub, a less promising place for growing anything would be hard to imagine. Convict labour was used to accomplish the back-breaking work. The result has been a triumph, and there can be few examples of the transformation of a wilderness into something so productive of fine quality since the monks in the Middle Ages transformed other wild places into great vineyards. At Niederhausen, the land was actually cleared in 1902, the first vines were planted in 1903, and the first wines were produced in 1907.

Since the Second World War, the estate has been fortunate in its administrators. The result has been an unrivalled reputation for producing wines of the highest quality. For 30 years Dr Gödecke was the presiding genius, and was responsible for introducing the system of cold fermentation used here. Then, on the retirement of Dr Gödecke, Dr Raquet from Bürklin-Wolf took over. When he decided to return to Bürklin-Wolf, Dr Werner Hofäcker succeeded him in 1977, and today carries on the good work.

The offices and cellars are at Niederhausen-Schloss-böckelheim, some 15 kilometres south-west of Bad Kreuznach, set amidst the largest portion of vineyards in a breathtaking position. The administrator's house is next to the offices, with the cellars beneath. The other parts of the estate are at Münster-Sarmsheim and Altenbamberg in the Alsenztal.

The vineyard names are as follows—at Schloss-böckelheim are Kupfergrube and Felsenberg; at Niederhausen the vineyards are Steinberg, Hermannsberg, Hermannshöhle and Kerz; while at Münster they are Kapellenberg, Dautenpflänzer, Pittersberg and Honigberg. At Dorsheim are the Burgberg and Goldloch vineyards, at Altenbamberg is the Rotenberg vineyard. At Traisen is the Bastei vineyard, and at Bad Münster Ebernburg is Schlossberg. Of these, the finest sites are generally considered to be Kupfergrube, Hermannsberg and Hermannshöhle.

In all, there are 47 hectares under vine, divided between Niederhausen-Schloss-böckelheim with 25 hectares, Münster with 12 hectares, and Altenbamberg with 10 hectares. The vineyards are all on steeply sloping sites, with gradients varying between 25° and 60°. At Niederhausen-

Schlossböckelheim, porphyry is the most important element in the soil, and results in wines of great individuality. In the Münster area, slatey soils predominate, giving more flowery wines, while at Altenbamberg sandstone is predominant.

Although in the Nahe as a whole the Riesling accounts for only 23 per cent of the plantings, here it accounts for 85 per cent, with Silvaner, Müller-Thurgau and experimental varieties making up the balance. With its exceptionally steep vineyards resembling those of the Mosel more than any other region, it is hardly surprising that the Riesling does so well here. Because of the steepness of most of the vineyards, tractors have to be winched up. Parts of the vineyard are equipped with an irrigation system. This serves a dual purpose. In summer, the steepest parts of the vineyard can be adversely affected by drought, so this prevents damage in dry, hot spells. In spring, the irrigation system can be used to spray water on the vines for frost prevention.

The outstanding feature of the cellar work here is their system of cool fermentation, which has been used now for 20 years with excellent results. The fermentation is carried out at between 8°C and 10°C, which makes for a very slow, long fermentation. This seems to bring out the fruity and piquant character of the Riesling grape. At the same time, the winemakers set out to retain a touch of acidity in the wines so that they become more fruity with age. In the main, wooden casks are still used, and there are a wide range of sizes, so that all wines may be individually treated.

Another special feature at Schlossböckelheim is that there is no use of *Süssreserve*. Instead, residual sugar is retained by cooling the wines at the appropriate moment, before fermentation is completed. Dr Hofäcker expresses the philosophy of the estate as permitting as little interference as possible during the winemaking process, with virtually no use of preservatives. Many experts, he says, are astonished with the results, believing that wines will not last without such things. But the wines produced by the domain are the most eloquent possible testimony to the success of such methods.

This domain shows what the Riesling can achieve on such very favoured sites, with wines allowed to speak for themselves, with the minimum of residual sweetness. Even a very humble year such as 1972 produced some delicious bottles. Thus, a Schlossböckelheimer Kupfergrube Riesling Kabinett was surprisingly harmonious and youthful, with a very clean, crisp attractive flavour when eight years old. The green acidity still evident on the nose had mellowed most pleasantly. In another small year, 1977, the best wine of the vintage was a Münsterer Dautenpflänzer Riesling Kabinett. It showed great ripeness

NAHE

1976er
Schloßböckelheimer
Kupfergrube
Riesling Trockenbeerenauslese
QUALITÄTSWEIN MIT PRÄDIKAT

Amtliche Prüfungsnummer 1 750 055/17/77

Verwaltung der Staatlichen Weinbaudomänen
Niederhausen-Schloßböckelheim
Erzeuger-Abfüllung

Like the other state domains, this estate uses the stylized eagle motif with traditional Gothic lettering. Some outstanding wines were produced here in 1976.

and breed on the nose, producing a very pronounced and beautiful bouquet. The flavour was typically fruity and Riesling, with ripeness, delicacy and finesse.

It is nice to find an estate where so many really fine wines are made at the *Kabinett* and *Spätlese* level, and not just at the *Auslese* level and above. Of course, in great years such as 1976 and 1971, many marvellous *Auslesen* were made, also *Beeren-* and *Trockenbeerenauslesen*. But even here, there are examples of *Auslesen* where it is the weight and complexity of the wines, rather than their sweetness, which are their outstanding characteristics. At the *Kabinett* and *Spätlese* levels the wines show their Mosel characteristics most clearly in a lively, racy, fruity acidity, while at the *Auslese* level, the Rheingau affinities become more evident

in a concentrated range of flavours with weight and body.

It is not surprising on such sites that the average yields are normally around 50 or 55 hectolitres per hectare. In the prolific 1979 vintage, 65 hectolitres per hectare were achieved. The year 1979 produced very stylish, elegant, fruity wines. The finest and most characteristic examples were *Kabinett* and *Spätlese* wines, with very little *Auslese*. A similar year was 1975, where the best wines were *Spätlesen*, but mention must be made of a marvellous *Beerenauslese Eiswein* made in the Hermannsberg. In 1973, the wines were again mostly *Kabinett* and *Spätlese*, and once more these were very elegant, charming wines. In 1976, only *Auslesen* qualities and better were harvested, but a number of these were actually sold as *Spätlesen*.

The Nahe is a small river which flows into the Rhine at Bingen. The estate has 25 hectares of vineyards at Niederhausen and Schlossböckelheim. The finest of the vineyards are Kupfergrube, Hermannsberg and Hermannshöhle.

In this view of the Niederhausen-Schlossböckelheim estate (LEFT), the vineyard of Kupfergrube lies to the left, and the Hermannsberg on the precipice to the right above the domain buildings. The picture gives a clear idea of the remarkable gradients of the vineyards.

These Riesling grapes are from the Hermannsberg vineyard (BELOW). A few of the berries have already changed from a bright green to a slightly yellowish tone, denoting the beginning of ripeness.

A view of the Kupfergrube vineyard (BELOW) shows the terracing of the vines. An unusual feature of this vineyard is that the vines are planted vertically. rather than horizontally.

Dr Hofäcker appraises one of his wines (RIGHT). The glass being used is a variant on the classic hock glass.

This portion of the Hermannsberg vineyard is under netting (RIGHT). This is done when the producer is expecting Edelfäule with the possibility of producing Auslesen, the higher grades of Prädikat wines in order to protect the grapes from birds.

Pruning of the young vines is taking place in the Kupfergrube (RIGHT). In the foreground is the overhead piping of the water sprinkling system. This is used either as protection against spring frosts or for irrigation during droughts.

STAATSDOMÄNE ELTVILLE

This estate forms one of the largest collections of vineyards in the whole of Germany and is a model estate which combines new methods with the best of old traditions. Particularly famous are the <u>Eiswein</u>, <u>Beeren</u>- and <u>Trockenbeerenauslese</u>, but the attractive, fruity dry wines now being made also deserve attention as an ideal accompaniment to food.

This remarkable group of estates belonging to the state is today the largest wine producer in Germany. Most of them were monastic foundations, some dating back to the twelfth century. When they were confiscated during the Napoleonic period in 1803, they were given to the Duke of Nassau. In 1866, they became part of the royal Prussian domains. Today, the six estates on the Rheingau and one on the Bergstrasse belong to the regional government of Hesse and come under the jurisdiction of the Ministry of Agriculture and Environment in Wiesbaden. The domain's most remarkable historical legacy is the Kloster Eberbach, founded as a Cistercian monastery near Hattenheim in 1135. It is one of the most impressive and complete medieval monasteries in Europe, with a splendid array of well-preserved buildings, including a magnificent church, superb cellars, including the famous Cabinet cellar from which the name *Kabinett* was born in 1730, and a collection of old wine presses housed in the former lay-brothers' refectory. Two kilometres from the monastery is the great Steinberg, a completely walled vineyard of 32 hectares, the first vineyard created by the monks. Today, the domain uses Kloster Eberbach for its famous wine tastings, wine fair and auctions. State and federal wine competitions are judged here, and the German Wine Academy for English-speaking students is also held here. There is a restaurant at Kloster Eberbach, and the monastery attracts thousands of visitors every year.

Since 1966 Dr Hans Ambrosi has been the head of this great enterprise, and, under his guidance, much progress has been made, so that today the reputation of the domain's wines stands high, and his flair for organization and innovation has ensured a much more fruitful use of the splendid facilities at Kloster Eberbach.

The administrative offices and central cellars of the domain are on the northern outskirts of Eltville. There are separate estate buildings for vinification and the day-to-day running of the vineyards at Assmannshausen, Rüdesheim, Hattenheim, Kloster Eberbach (for Steinberg), Rauenthal, and Schloss Hochheim. The vineyards they encompass include some of the finest in the Rheingau.

The Rheingau vineyards of the domain cover 158 hectares (189 hectares if the Bergstrasse vineyards are included). This makes it the largest wine estate in Germany. Some four hectares of these are devoted to a vine nursery and for rearing rootstocks. Nearly two-thirds of the vineyards are on gently sloping land, with about one-third steeply sloping. Only a tiny proportion is on the flat.

The slatey soil at Assmannshausen is remarkable as the only place in the world where the Pinot Noir (known here as Spätburgunder) is grown on slate. It is also the only estate in Germany devoted exclusively to the production of red grapes. The Riesling reigns supreme in all the other vineyards of the Rheingau belonging to the domain. Actually, it accounts for nearly 75 per cent of all plantings, with 10.5 per cent Pinot Noir (Assmannshausen), 8.4 per cent Müller-Thurgau and the rest new and experimental varieties.

The highest Oechsle degrees are obtained at Rüdesheim, with its steep slopes and southerly exposure. These give the most generous, fruity wines with more body than other Rheingaus. These vineyards do well in wet years, but can have trouble in hot dry ones like 1959. The Steinberg is the vineyard which is highest and furthest removed from the climatic influence of the Rhine, hence its very elegant, racy wines, needing the warmest years to bring out their best. The Rauenthaler vineyards, with the Baiken, an island of red slate in loam and loess, are only slightly less distant from the Rhine. They are also renowned for the breed of their wines which are delightfully perfumed. Hattenheim and Hochheim have deeper, warmer soils, and benefit from the Rhine influence, so they are more full-bodied. The Hochheim wines are especially distinctive, with a broadness of flavour amounting to a *goût de terroir*. Great importance is attached to vineyard work, and the most modern methods are used. As a state enterprise, Dr Ambrosi believes it is the duty of the domain to try out new methods and to experiment.

In spite of the size of the domain, great trouble is taken to produce a large number of individual wines. To begin with, grapes come in from the vineyards in containers holding no more than 600 kilos of grapes—at a domain where an average crop is 1,000,000 litres and the total storage capacity is 3,000,000 litres. Every 600 kilo container of grapes is recorded by computer on arrival at the reception area, as to grape variety, weather when picked, time of picking, weight of grapes and Oechsle degree. All this information is stored on computer and can be referred to at any time during the wine's storage and treatment, and until it is bottled. This provides extremely rigorous control.

Under Hans Ambrosi, there has been a move away from wooden casks to stainless steel for white wines. Two-thirds of the domain's storage is now in stainless steel and only one-third in wood. Today's consumers, Ambrosi believes, look for freshness, liveliness and fruit in their white wines, so contact with the air must be reduced to a minimum. Thus, batteries of small stainless steel tanks have been brought in to cope with the great variety of wines. Bottling is done early, in the spring after the vintage. There is an enormous bottle store at Eltville, capable of keeping 1,000,000 bottles in ideal controlled conditions. The red Assmannshausen wines are aged only in wooden casks for one year, and then bottled. They

This Eiswein won prizes in 1972 at both the Hesse regional and national levels. Eltville's labels all still show the Prussian eagle.

are then given a further year's bottle age before being sold.

The most outstanding point about this domain is the sheer variety of what it has to offer in terms of styles of wines, vintages and qualities. One of their great specialities, the red Assmannshausen, is certainly appreciated more in Germany than outside it, and these almost rosé-like Pinot Noirs, with their fresh acidity, pale colour and often with residual sugar, are certainly unlike wines made from the Pinot Noir anywhere else in the world. In fact, the domain cannot meet the demand for these wines, and they are sold at double the price of other white wines which are of similar quality.

In recent years, two developments of note can be singled out. The first is the move towards *Trocken* and *Halbtrocken* wines. But the variety of these wines offered can be misleading; in percentage terms they still account for a very small part of the wines made. In 1980, it was interesting to note that of eight 1979 *Trocken* wines on the list, six came from the most favoured sites in Rüdesheim, Hochheim and Hattenheim, where the best ripeness is achieved—the

domain certainly realizes that if *Trocken* wines are not to be meagre, the base material must be good.

The other development has been the production of *Eiswein*. The domain is now highly organized to take advantage of the conditions when they arise, and has had great success in producing these highly prized rarities. Such is the price they command that the expense and trouble that goes into producing them are apparently amply repaid.

All labels carry the Prussian eagle, a design that has remained unchanged now for many years. Wines auctioned at Kloster Eberbach have a special label showing the monastery with the words *Ersteigert im Kloster Eberbach* (auctioned at Kloster Eberbach).

The longevity of these wines is especially renowned, thanks to the remarkable collection stored at Kloster Eberbach, some of which is auctioned from time to time, while other wines from the collection are opened for special occasions. After the famous 1921s, some fine examples of 1949 and 1953 are particularly memorable.

STAATSDÖMANE
ELTVILLE   Eltville

Rhein

The domain's offices and main cellars are in Eltville and the vineyards are among the finest in the Rheingau.

This picture shows wine tasting in the Cabinet cellar of the Kloster Eberbach in 1847 (FAR LEFT). It was the Duke of Nassau who first took to using this cellar for keeping the best quality wines. The term Kabinett, *was first used for the 1811 Steinberger.*
These workers (LEFT) are picking grapes for the Eiswein. Their efforts helped create the famous 1970/1971 Heilige Dreikönigs-Eiswein Auslese.

Year	QbA	Kabinett	Spätlese	Auslese/TBA	Yield (hectolitres/ hectare)
1971	0.2%	10%	77%	12%	49
1972	76%	20%	3%	1%	58
1973	58%	38%	3%	0.8%	92
1974	73%	26%	1%		59
1975	18%	60%	18%	4%	82
1976	0.6%	14%	38%	47%	59
1977	83%	16%	0.2%		69
1978	79%	18%	3%		56
1979	27%	53%	18%	1%	88
1980	67%	29%	3%	0.2%	30

This early Steinberg label features the Prussian eagle (BELOW). A special label was created for the 1911 Steinberg Trockenbeerenauslese (BELOW CENTRE). This was an exceptional year, and Trockenbeerenauslese wines were even more rare than they are today.

These two labels for the 1895 Steinberg (BELOW RIGHT AND BOTTOM) show clearly the difference between Kabinett and ordinary varieties.

This medieval gateway leads into the Steinberg vineyard (RIGHT), the original vineyard of the Kloster Eberbach. The 32 hectare Steinberg vineyard is completely walled.

This medallion (RIGHT BOTTOM) appears on all Staatsdomäne wines sold at the Kloster Eberbach auction. The great chapel of the monastery is prominently featured on the medallion.

Dr Hans Ambrosi (RIGHT), director of Kloster Eberbach, is sitting in his office. Dr Ambrosi has been in charge here since 1966 and the estate's reputation has increased considerably since then.

The Kloster Eberbach chapel (ABOVE RIGHT) is the oldest part of the monastery, as can be seen by the Romanesque windows. The topsoil in the vineyard at Rüdesheim (ABOVE FAR RIGHT) is extremely stony.

An identification tag (INSET LEFT) shows the precise origins of a Riesling vine. The dark tinge of the grapes shows they are still not fully ripe. A bunch of grapes is being affected by Edelfäule (INSET. RIGHT). Some of the grapes are heavily affected and

beginning to shrivel, while others are still a yellowish-green and as yet unaffected by the 'noble rot'.

Riesling
Klon 9? Rg
26 Gm

ERSTEIGERT IM · KLOSTER EBERBACH

SCHLOSS JOHANNISBERG

The most famous name in German viticulture is that of Schloss Johannisberg, the first estate in Germany where botrytized grapes were harvested. The wines today retain their deserved reputation for producing classic Rheingau Rieslings of exceptional breed. The great wines of Schloss Johannisberg have a honeyed quality which places them in the forefront of the best German wines.

The names of Schloss Johannisberg and of its owners, the Metternich family, are perhaps the most celebrated names in German viticulture and with wine drinkers throughout the world. As usual in such cases, the reason is a mixture of excellence over a long period of time allied to some historical good fortune.

The history of the estate is long and interesting. Legend records that the Emperor Charlemagne noticed from his palace at Ingelheim, on the opposite bank of the Rhine, that the snow melted early on this hillside, and decreed therefore that vines be planted here. It is known that his son, Ludwig, recorded a yield of 6 *Fuder* (6,000 litres) of wine in 817 AD. Some two centuries later, about 1100, the first monastery in the Rheingau was founded here by Benedictine monks from Mainz. The church was consecrated in 1130 to St John the Baptist, and from this, the hill, monastery and village became known as Johannisberg. During the Reformation the monastery was dissolved and during the Thirty Years War it was seized as a tax pledge by the imperial tax collector.

A new era began when the property was acquired by the Prince-Abbot of Fulda in 1716. Much rebuilding in the baroque style now took place. While the old cellars dating back to the foundation of the first monastery were retained extensive new ones were built. The vineyards were also restored, as many as 294,000 vines being planted during 1719 and 1720. The 38,500 Riesling vines planted marked the beginning of a new era for wine growing in the Rheingau. Today, the term Johannisberg Riesling is used as a synonym for this kind of vine in the United States and some other countries.

In 1775 an accident led to the discovery of the beneficial effects of *botrytis cinerea*, or *Edelfäule* (noble rot). The courier, carrying permission to begin the picking from Fulda, arrived late. The grapes were by then overripe and beginning to rot, to the dismay of the monks. But the results of this first *Spätlese*, or late-harvested wine, took everyone by surprise. On 10 April 1776 the administrator recorded 'I have never tasted such a wine before.' So began the harvesting of *Spätlese*, *Auslese*, *Beeren-* and *Trockenbeerenauslese* wines. The event is commemorated by an equestrian statue of the '*Spätlese* messenger.'

In 1802 both Fulda and Johannisberg were confiscated under Napoleon, and the estate changed hands several times, until finally, at the Congress of Vienna in 1815, the Emperor Francis I presented it to his great Chancellor Fürst Metternich-Winneburg, one of the most redoubtable opponents of Napoleon and one of the chief architects of final victory. The present owner, Fürst Paul Alfons von Metternich-Winneburg, is a great-grandson of the famous Chancellor. In 1942, the *Schloss*, church and farm buildings were severely damaged in an air-raid.

The reconstruction was not completed until 1965. A particular point of interest is that the 50°N line of latitude runs through the property. A marker indicates its exact position.

For a period in the 1950s and 1960s the wines fell below the very high level of quality which they have held for so long, but the 1971s and 1976s have done much to redress the balance, and there is no doubt that the property is again very well run, with Josef Staab as administrator, and Erwin Boos as *Kellermeister*.

Schloss Johannisberg is today easily reached from Wiesbaden. It lies between Winkel and Geisenheim. As with nearby Schloss Vollrads, special provision was made for Schloss Johannisberg in the new Wine Law of 1971, so that the wines can be sold under a single name.

The vineyard covers 35 hectares in one piece which lies immediately below the *Schloss*, with 25 per cent on steeply sloping, 50 per cent gently sloping, and 25 per cent on flat land. The soil is predominantly quartzite and schist, with loess and loess-loam. With its due south exposure facing the Rhine, the vineyard has an especially favourable microclimate, even by Rheingau standards. The observation point on the terrace in front of the *Schloss* is at exactly 180 metres, an ideal height, and the massive buildings of the *Schloss* and the old monastic church, with its park behind it, together with the sheer steepness of the upper part of the hill, affords important additional shelter for the vines. The only problem is that the vines can suffer in very dry years from a lack of moisture. True to its traditions, Schloss Johannisberg today has as much as 99 per cent Riesling planted in its vineyard. The remainder covers experimental varieties. It is their proud boast that since 1716 only Riesling has been planted here.

Harvesting is carried out with great care. Each part of the vineyard is gone through several times, so that very ripe grapes, and those affected by *Edelfäule*, are collected separately. Pneumatic presses have been in use for some time. These apply only a very light pressure of between 2 and 6 atmospheres, compared with 200 to 240 atmospheres for the old hydraulic presses. This produces cleaner must after only a short time, thus reducing the possibility of oxidation. After fermentation, the wines are stored entirely in wooden casks until bottling. Experience has

shown that this produces the best results for the Riesling. At Schloss Johannisberg there is the additional reason that they want the wine to be exposed to the unique influence of their cellars, partly dating from the rebuilding of 1716 and partly from the original twelfth century foundation. This great vaulted cellar is covered with a special fungus *clodosporium cellare*, and is responsible for preserving an ideal environment which gives what used to be called a 'sealing wax' tone to the wines.

Schloss Johannisberg is one of the oldest established vineyards in Germany. Its Auslese wines are especially distinctive.

The Schloss Johannisberg
(BELOW) holds a commanding
position on the hill of
Johannisberg with the vineyards falling away sharply in
front. Harvesting is taking
place in the middle of the
vineyard.

Quartz is commonly found in
the soil at Johannisberg
(BOTTOM LEFT). On a
distinctly shaped Riesling
leaf, the last vestiges of green
are seen around the veins. The
cellar at Johannisberg
(BOTTOM RIGHT), was built
in 1716.

Fürst Paul Alfons von
Metternich-Winneburg
(BELOW), is the present
owner of Schloss Johannisberg
The famous vaulted cellar can
be seen in the background.

Schloss Johannisberg is
between Geisenheim and
Winkel on the Rhine. The
soil in the 35 hectare vineyard
is mainly schist and quartzite
with some loess and loess-
loam.

Since all wines are sold under the name of Schloss Johannisberg, a label and capsule colour code has been evolved. There are two labels, one bearing the Metternich coat of arms, the other showing the *Schloss* and its vineyard. This picture label was formerly reserved for the Cabinet wines, a selection of the finest casks which began here in 1779. The code had to be modified as a result of the new Wine Law of 1971. Since 1830 all labels have carried the signature of the estate administrator.

There can be no doubting the keeping qualities of these wines. The estate has one of the most remarkable collections of old wines to be found anywhere, kept in the *Bibliotheca subterranea*. The oldest example dates from 1748. The estate bottling of the best wines dates back to 1775, and examples of all successful vintages back to 1842 are preserved. When a number of the oldest wines were opened in 1976, even the oldest were still found to be drinkable while some of the great nineteenth century vintages were still great wines, which are still appealing even to today's tastes.

Particularly great wines include the wonderful 1945 Rosalack Spätlese and Auslese. The latter had a wonderfully honeyed nose and lingering flavour which was the quintessence of ripe Rheingau Riesling at its greatest. The 1953s had great elegance without the same depth. Wonderful wines have been made in 1971 and 1976, both years producing classic *Beerenauslesen*. Until 1970, no chaptalized wine was sold under the Schloss Johannisberg label.

Today, very good *Trocken* and *Halbtrocken* wines are being made. The 1979 vintage produced very typical and pleasing examples.

Naturally, one sees Schloss Johannisberg at its most distinctive at the *Auslese* level and above. A tasting of three different *Auslesen* of the 1976 vintage made the point very well. Each was very fine in its own right, with great individuality. The spiciness of the Riesling here is most pronounced, and to this was added a rather smoky overtone, and a beautifully developed botrytis character on the richer wines which was very harmonious with perfectly balanced acidity and sweetness.

To give an idea of the comparative quality of recent vintages, 1979 produced mostly *QbA* and *Kabinett* with only 30,000 bottles of *Spätlese*. The highest Oechsle was 110°, and the overall yield 80 hectolitres per hectare. The year 1976 produced all qualities up to *Trockenbeerenauslese*, while in 1975 there were only *QbA* wines, 1971 up to *Beerenauslese*, but with nearly one gram more acidity than the 1976s of the same quality level. The yields in 1971 and 1976 were of the order of 60 or 70 hectolitres per hectare. The last *Eiswein* was made in 1969. Other *Eiswein* vintages were 1966 and 1965.

A *Sekt* is sold under the name Fürst von Metternich, made from Riesling wines, but very little of this comes from the Schloss Johannisberg vineyard itself, and only in very poor years when a wine is not suitable for sale even as a *QbA* under the yellow seal.

Harvesting (BELOW) *is in progress at Johannisberg. The distance between the top and bottom of the rows is approximately 260 metres. In these Riesling grapes,* (INSET LEFT), *a number of berries have acquired a yellow-gold tinge denoting their* near-perfect ripeness. *The grapes are being unloaded during the harvesting of the vineyard* (INSET CENTRE). *A great deal of trouble is taken to prevent the grapes from oxidizing. These grapes have just been harvested* (INSET RIGHT). *A fair number* suffer from Edelfäule *(noble rot). The resulting wine might be used for a wine of* Auslese *quality.*

Bunches of grapes are being cut during harvesting at Johannisberg (INSET LEFT). After being taken to the winery, the grapes are pumped into one of the horizontal presses (INSET CENTRE LEFT). This monument (INSET CENTRE RIGHT) near the terrace of Johannisberg marks the precise degree of latitude which runs through the estate. The equestrian statue (INSET RIGHT) at Johannisberg commemorates the first Spätlese and Auslese wines created in 1775.

SCHLOSS VOLLRADS

The Matuschka-Greiffenclau family have grown vines for over 800 years. Today, Vollrads is renowned for the verve of its dry Riesling wines and the exceptional longevity and harmony of its great sweet wines.

No estate in the Rheingau more typifies its traditions of long-established aristocratic ownership than Schloss Vollrads. By 1211 the Knights of Greiffenclau were selling wines to the Monastery of St Victor in Mainz. At that time, the family lived in the famous 'Grey House' in Winkel, reputed to be the oldest stone house in Germany still preserved. So this same family have certainly made wine from these vineyards for a period of nearly 800 years.

The most dominant feature of Vollrads today, its moated tower, was built prior to 1330, when the family moved its residence there. The magnificent castle itself was begun in 1684. The last male Greiffenclau died in 1860, but his niece and heir Sophie had married Graf Hugo Matuschka in 1846, and their names and arms were united to carry on the family tradition. The modern Vollrads owes its fame to the grandson of this alliance, Graf Richard Matuschka-Greiffenclau. He died soon after his eightieth birthday celebrations in 1975, having been for many years President of the German Winegrowers' Association. Now, his son, Erwein, presides at Vollrads.

Schloss Vollrads is situated 2 kilometres north of Winkel. The road to the property from Winkel is clearly indicated by green and yellow signs. The castle, which is at the point where the vineyards meet the woods of the Taunus foothills, is approached through its vineyards.

Like Schloss Johannisberg, special provision was made in the new German Wine Law of 1971 to enable the wines of Vollrads to be known solely by this name, without mention of the name of the commune—Winkel—or the names of the different vineyards.

The vineyards of Schloss Vollrads cover 47 hectares and are mostly near the castle, although there are some plots in Mittelheim and Hattenheim. Of the vineyards, 20 per cent are on steeply sloping, 60 per cent on gently sloping, and 20 per cent on flat terrain. The slopes are south facing, and the soil is deep, calcareous, sandy loess or loess, retains moisture well, and is ideal for the Riesling vine, which accounts for 96 per cent of those varieties planted. The microclimate is particularly favourable because the Taunus mountains give protection against cold northerly winds, while the Rhine accumulates heat and acts as a reflector. The additional moisture in the soil here enables the Rieslings at Vollrads to produce wines which are lighter in body than most Rheingau Rieslings, but which have that special fruit acidity, so prized in German wines. Schloss Vollrads produces wines of marked elegance, needing time in bottle to develop their characteristics fully.

Kellermeister Senft is a firm believer in treating the must rather than the wine, whenever possible. So, if the acid needs to be reduced, it is better to treat the must with calcium or double salts rather than have to deacidify the wine. Equally, bentonite fining to remove excess protein takes place before the fermentation. The fermentation mostly takes place in tank. Today, approximately 40 per cent of the wine is matured in wood and the rest in tank. Like most Rheingau estates, Vollrads believes that wood develops just the right harmony between fruit and acidity in their steely Rieslings. There is no malolactic fermentation, because the characteristic taste of the malic acid is a quality prized in German wines. Bottling usually takes place between the March and June after the vintage.

Because the wines here are sold under a single name, with no individual vineyard names as on most German estates, Vollrads has always had its own special system of designating different qualities by using different colours and markings on its capsules.

For the *Qualitätsweine*, the green-silver capsule is reserved for *Trocken* and *Halbtrocken* wines. The *Trocken* has in addition a special necklabel. It should be noted that the best quality *Qualitätswein* is used for these dry wines. Thus, the base wine for the green-silver is the same as for the green-gold capsule. But the addition of the *Süssreserve* makes the ordinary green label, and particularly the green-gold, more harmonious and easy to drink when first bottled. The *Trocken* wines need more time in bottle to mellow. The blue-silver capsule on the *Kabinett* wines is nowadays reserved for the *Trocken* and *Halbtrocken* wines of Kabinett quality. The pink-silver on the *Spätlese* wines is for *Trocken* and *Halbtrocken* wines. The *Spätlese* wines here retain the lean delicacy and relative dryness which characterize wines of the Vollrads.

There are only two qualities of *Auslese*—white and white-gold. The quantity of *Auslese* wines produced at Vollrads is relatively small compared to some estates. On the other hand, only wines with real *Auslese* characteristics, not just the right Oechsle readings, are sold under the *Auslese* label. On *Beerenauslese* and *Trockenbeerenauslese* wines there is a gold capsule with moated tower necklabel. These precious rarities are, of course, only produced in minute quantities in the most favoured years. The last occasion was 1976. At one time the different qualities within each category were only indicated on the capsule; however, today these quality differences are clearly marked on the label as well to avoid any possibility of confusion among consumers—particularly when the wine has been exported.

A very traditional wine such as Vollrads requires some explanation for the wine drinker who is only familiar with the popular commercial blends of German wines. To begin with, the average yield for the whole of Germany is now over 100 hectolitres per hectare. At Vollrads it is about 55 hectolitres per hectare. This means that the resulting wines have more extract and also more acidity. They therefore

As all Vollrads wines are sold as Schloss Vollrads, there is a colour coding for labels and capsules, pink for Spätlese, and white or white-gold for Auslese. 1976 was an outstanding vintage.

Graf Erwein Matuschka-Greiffenclau (BELOW) *is the present head of the family which has presided over Vollrads for nearly 800 years.*

This room in the Schloss (BELOW CENTRE) *is set aside for special tastings. The fourteenth century tower* (BELOW) *is the oldest surviving part of the Vollrads Schloss.*

need more time in bottle to develop their characteristics.

A bottle opened in the first one or two years of its life will probably be disappointing. Schloss Vollrads itself recommends a wait of two or three years for even the *Qualitätswein* and *Kabinett* wines before drinking; and they should remain at their peak for a further four or six years. In the case of vintages with high acidity, such as the 1972, even longer is required. For *Spätlese* wines, this may take between 6 and 10 years, and for *Auslesen* 10 or 15 years, while *Beerenauslesen* and *Trockenbeerenauslesen* can keep for between 25 and 30 years, and even longer if cork and storage conditions are good. These are not instant wines, and just because they happen to be white, this does not mean that they do not require the same careful maturation as do the great red wines.

The development of the *Trocken* styles reflects the interest in Germany in recent years in producing drier wines, and in wines as an accompaniment to food. The present proprietor, Graf Erwein, is especially interested in this, and believes that Vollrads wines, with their very distinctive, delicate fruit-acidity, are especially suitable for this treatment. It must be said, though, that most lovers of German wines in England and the United States find the *Trocken* styles rather austere for their taste, and still prefer wines with some *Süssreserve*, especially when, as at Vollrads, this is sparingly used to bring out the fruit and increase the harmony, rather than to sweeten the wines in a crude way.

Schloss Vollrads is about two kilometres north of the village of Winkel. Most of the 47 hectares of vineyard are near the castle, but there are also some plots in Mittelheim and Hattenheim.

Harvesting is taking place in a piece of the vineyard joining the Schloss (ABOVE). *The estate has 47 hectares of vines near the castle. The slopes are mainly south facing.*

ANTON BALBACH ERBEN

In a region where ordinary wines predominate, the Balbach estate, dedicated to the Riesling wines, is one of the great standard-makers of quality. It is particularly famous for its long-lived <u>Beeren-</u> and <u>Trockenbeerenauslese</u> wines produced in great years.

Since 1654, fourteen generations of the Balbach family have been winegrowers in Nierstein. The daughter of the last male Balbach married Friedel Bohn, who was a schoolteacher. When his father-in-law died unexpectedly, he suddenly found himself in charge of this famous Rheinhessen estate, and had to learn his new trade from scratch. The very high standard of the Balbach wines in the last 20 years is an eloquent testimony to his success. Now the Bohns are assisted by their Geisenheim-trained son. The Balbach family have one special claim to fame in Nierstein; it was they who created the great Pettenthal vineyard from a slope that was still wooded in the nineteenth century.

This estate boasts no historic buildings or cellars, just a comfortable house built in the last century. Its interest is in what it produces. There are larger estates on the Rheinfront, but none has such a high proportion of Riesling vines, and there is nowhere better to see what the Riesling can achieve in Rheinhessen than here.

The house and cellars are only separated from the Rhine by the main road to Mainz. The house is on the road with the cellars behind separated by a courtyard. The garden faces the vineyards of the Rheinfront stretching out towards Nackenheim to the north. All the vineyards are in Nierstein and 80 per cent of them are on the Rheinfront. The vineyard names in Nierstein are Pettenthal, Hipping, Ölberg, Kranzberg, Rosenberg, Klostergarten and Bildstock. The *Grosslage* names are Rehbach, Auflangen and Spiegelberg.

Although there are only 18 hectares of vineyards, 80 per cent are on the Rheinfront. The Riesling now accounts for 78 per cent of the vines planted. Gewürztraminer and Huxelrebe are now being pulled up, and the other varieties planted are Müller-Thurgau and Kerner. Some 30 per cent of the vineyards are on steeply sloping, 20 per cent on gently sloping, and 50 per cent on flat land. The soil is the famous red schist of the Rheinfront.

The Riesling can only achieve ripeness in the best vineyards, and this is what Balbach's vineyards can offer. The red slate stores up the warmth of the sun during the day and then releases it at night. This is ideal for the Riesling, which is a delicate variety and does best when there is not too much temperature variation between night and day. The proximity of the Rhine is another factor which assists in maintaining an even, warm temperature.

Great care is taken to press the grapes slowly to ensure that no tannin is extracted from the stalks. Then the must is centrifuged to clean it before fermentation begins. In this cellar, wooden casks have now given way to stainless steel and fibreglass, but the old carved cask-heads are preserved to decorate the cellar walls. Herr Bohn believes that cleanliness in the cellar is vitally important; 'clean and good wine can only come out of a clean cask,' he states. Since wood is much more difficult to maintain in such a state than stainless steel or fibreglass, this policy is logical enough. The only surprise, perhaps, is that so many estates still manage to persevere with wood. It is interesting to note that in this cellar the Müller-Thurgau is used to prepare the *Süssreserve* which is utilized with all varieties of grape.

Perhaps the most striking feature of Balbach wines is the high proportion of *Auslese* and higher-grade wines produced. It is famous for its great *Beerenauslese* and *Trockenbeerenauslese* wines produced in the great years. One interesting fact about such wines is that they are both lighter in colour and keep this colour better than do similar Riesling wines from the Rheingau. Even one of the legendary 1921 *Trockenbeerenauslesen* tasted in the mid 1960s was still marvellously preserved and not at all dark in colour. Interestingly, it had been recorked several times. Since the Second World War, the greatest years here for *Beerenauslese* have been 1949, 1953 and 1976, when

Balbach are famous for their Beeren- *and* Trockenbeerenauslese *wines. Some outstanding examples were produced in 1976.*

ideal conditions for noble rot or *Edelfäule* were to be found. In 1976, for instance, 11,500 bottles of *Beerenauslese* and 1,800 of *Trockenbeerenauslese* were produced.

When comparing vintages here it is interesting to note that the 1971 vintage produced wines with about 1 gram per litre more acidity for the *Auslesen* than in 1976. But there was not the same degree of *Edelfäule*. The difference in yield between top years and good average years is illustrated by the fact that 90 hectolitres per hectare were made in 1979, and 70 hectolitres per hectare in 1976. In 1979, there was 15 per cent *QbA*, 5 per cent *Kabinett*, 80 per cent *Spätlese*, while 1976 produced no *QbA* or *Kabinett*, only a small amount of *Spätlese*, a few *Auslesen*, and then a lot of *Beerenauslese* and a comparatively large quantity of *Trockenbeerenauslese*.

The interest in dry wines is also now followed by the estate, which produces some very attractive examples. Typical is a Klostergarten Müller-Thurgau Spätlese Halbtrocken 1979, which has 11 per cent alcohol and is really very fruity and delicious.

Friedel Bohn, the present administrator of the estate, (BELOW) holds a picture of Bürgermeister *Anton* Balbach *from whom the estate takes its name.*

All 18 hectares of the Balbach vineyards are in Nierstein. Over three-quarters of them are on the Rheinfront.

The Rheinfront at Nierstein is famous for its schisty, red soil (LEFT).

These vines are on the Rheinfront at Nierstein (LEFT). The red hue of the soil is typical for this area.

This view (ABOVE) shows the Pettenthal vineyard looking along the Rhine to Nierstein.

HEINRICH SEIP KURFÜRSTENHOF

This is a veritable Aladdin's cave for new grape varieties, there is no other estate where it is possible to see and compare so many types at the same time. All are produced with the exceptional flair of Heinrich Seip.

The fame of the Kurfürstenhof is the result of a very old estate coming into the hands of an old family of Nierstein vintners as recently as 1950. The origin of the Kurfürstenhof itself is said to go back to Roman times. It later became a Franconian royal residence and came into the possession of the Electors of the Palatinate in 1375. In 1817, it became private property and was bought in 1950 by the father of the present proprietor, Heinrich Seip. The Seip family itself have been wine-growers in Nierstein since the fourteenth century. The present label incorporates the arms of the Elector Karl Philip who rebuilt the Kurfürstenhof in 1722, with those of the Seip family.

It is the ability and personality of Heinrich Seip which distinguishes this estate from others in the area. Apparently, the people of Nierstein say 'The Kurfürstenhof is a national monument; the owner has yet to become one.' The outstanding features are the work that has been done with new grape varieties and the relatively large number of specialities produced, especially *Eiswein*. Of special interest are the cellar, with its fine vaulting, the old press-house where tastings for up to 100 people can be held, and the small tasting room.

The old house with its cellars is in the centre of Nierstein, near the Martinskirche. The vineyards are in Nierstein, Oppenheim, Dienheim and Nackenheim.

In Nierstein the vineyard names are Paterberg, Bildstock, Kirchplatte, Findling, Rosenberg, Klostergarten, Pettenthal, Hipping, Kranzberg, Ölberg, Heiligenbaum, Orbel, Schloss Schwabsburg, and Goldene Luft (sole proprietor). In Oppenheim is the Schloss vineyard. In Dienheim the vineyards are Tafelstein, Kreuz and Falkenberg, and in Nackenheim is the Engelsberg vineyard. With many small holdings in Nierstein, the *Grosslage* names can be important here for many of the specialities. They are Niersteiner Gutes Domtal, Spiegelberg, Rehbach, and Auflangen.

The vineyards of the estate cover 35 hectares. Of these, 70 per cent are on flat and gently sloping land, the rest on steeply sloping land. The soils are 50 per cent red schist and sandstone, 20 per cent limestone, 10 per cent marl, 20 per cent loess and loess-loam. The Nierstein vineyards are some of the best sited in the district, with a southern exposure and close to the Rhine. This is what is known as the Rheinfront, where all the finest wines of Rhinehessen come from. This, together with the famous red earth, red schist and sandstone, ensures exceptional ripening conditions, especially for the Riesling.

The list of grape varieties planted reads like a catalogue of German grape types. The mixture is 20 per cent Riesling, 20 per cent Silvaner, 20 per cent Müller Thurgau, 25 per cent of Scheurebe, Kerner, Gewürztraminer, Weissburgunder, Ruländer, Huxelrebe and Veltliner, and a further 15

per cent of experimental varieties, such as Bacchus, Ortega, Optima, and Jubiläum. This last variety is of special interest, because Seip is the official producer in Germany for this new variety, and has charge of 30 trials in Rheinhessen. This new variety produces 30° Oechsle more than the Riesling, but with a slightly smaller yield. Even in poor years this enables it to produce wines of *Spätlese* quality, and in most years it can produce *Auslese*, *Beerenauslese* and *Trockenbeerenauslese* wines. The variety was recognized in Austria in 1961. Seip is very enthusiastic about this vine and what it produces, which resembles Ruländer, and believes it stands an excellent chance of being officially authorized in Germany.

Seip believes that the correct use of fertilizers is most important. For his particular soils, he especially needs to add potassium and phosphate. Another important reason for the quality of the wines here is late harvesting. Before the new Wine Law of 1971, this estate was famed for its specialities such as Nikolauswein, Sylvesterwein, Christwein, Heiliger Dreikönigswein. Such picturesque names are no longer permitted, but *Beeren-, Trockenbeerenauslese* and *Eiswein* are still produced. To produce such wines, especially *Eiswein*, it is necessary to net the vines to protect the grapes from birds, and usually the estate has to wait until December for sufficiently low temperatures to harvest the grapes for such wines.

Heinrich Seip places great emphasis on reductive wines, that is, on protecting the wines from air at every stage of cellar-work. He believes that the most important attribute of German wines is their bouquet, and the distinctive bouquet of each grape variety must be safeguarded and brought out, so that the Riesling really smells of Riesling, and so on. He also believes that wooden barrels are the best containers for wine at their stage of bulk ageing. Therefore, the Kurfürstenhof cellars contain mostly wooden casks, although tanks are used for handling quantities of more than 10,000 litres.

There is probably no private estate in Germany where it is possible to learn more about the characteristics of German grape varieties, both old and new. The comparative tastings of the different varieties, especially some of the experimental varieties, are really fascinating. This is not an estate to look for dry wines, because all the specialities are sweet or semi-sweet wines. But the style of the wines is extremely elegant and delicate, with a moderate residual sweetness to bring out their character. At times the new varieties are skilfully blended with the traditional ones to enhance the bouquet, and this results in some most interesting wines. Wines like the *Eiswein* Christwein of 1962 or the 1964 are unforgettable. Such wines can keep their freshness for many years. Even bottles of 1959 *Auslese* are still wonderfully fresh.

In general, 1978 produced wines with low sugar contents, but a new grape variety like the Jubiläumsrebe managed to produce an Auslese.

When a large proportion of the grapes in a vineyard are late harvested, it is important to keep off birds. On this estate flares·are fired from the top of the tower to scare them (BELOW) and nets are put over wires to keep them off

The estate's vineyards are in the villages of Nierstein, Oppenheim, Dienheim and Nackenheim.

Year	QbA	Kabinett	Spätlese	Auslese	Beerenauslese	TBA	Eiswein	Yield (hecto-litres/hectare)
1971	3.7%		71%	24%	1%	0.1%		55
1973	40%	54%	5%	1%				82
1976		25%	22%	41%	10%	2%		50
1978	42%	45%	10%	1%			0.3%	41
1979	5%	50%	38%	6%		0.3%	0.1%	62

The Goldene Luft vineyard in Nierstein (BELOW) is solely owned by Seip. The village of Nierstein can be seen behind the Rehbach vineyard on the Rheinfront (CENTRE).

The vineyards' proximity to the river (BELOW) is an important factor in the microclimate. Flares for scaring off the birds are fired from pistols (CENTRE).

Winzermeister *Heinrich Seip* (RIGHT) *is holding a bottle of Niersteiner Klostergarten. The Klostergarten vineyard is one of the best in Nierstein.*

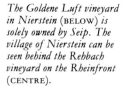

These Riesling grapes (RIGHT) *are growing in the Spiegelberg vineyard. They show the beginnings of botrytis.*

This colourfully painted cask head illustrates a humorous German maxim (BELOW).

In the ancient cellars of Kurfürstenhof there are several sizes of cask (ABOVE). *The small ones are most important for making the small quantities of late harvest wines.*

This painted cask head (RIGHT) *shows the Seip family crest. The Kurfürstenhof estate came into the Seip family only in 1950.*

WEINGUT DR BÜRKLIN-WOLF

This great Rheinpfalz estate produces a range of wines which show what the finest sites in this region can do when planted with the Riesling vine. Because of the exceptionally favourable climate, fine <u>Kabinett</u> and <u>Spätlese</u> wines are produced with great consistency, while the great <u>Beeren</u>- and <u>Trockenbeerenauslesen</u> are much sought after in the great years.

Although the families of Bürklin and Wolf both have old roots in Wachenheim, the importance of this great domain only dates from the beginning of the nineteenth century. The name of Bürklin first appears in the land registry of Wachenheim in 1597. It shows that Bernard Bürklin, mayor of Wachenheim, owned vineyards, most of which now form part of the present-day estate. In 1759 Johann Peter Wolf, a member of an old family of wine-growers, was appointed Master of Wine Cellars by the Elector Karl Theodor. His son, Johann Ludwig Wolf, who lived from 1777 to 1846, was responsible for developing the family estate into one of the most important wine producers in Germany. Both he and his son were mayors of Wachenheim. In 1875 Johann Ludwig's granddaughter married Dr Albert Bürklin, who was a Privy Counsellor, Vice-President of the German parliament, the *Reichstag*, and an honorary citizen of Wachenheim. Under him, the estates of the two families became merged to form what is today the largest wine-producing estate in private ownership in Germany.

The Bürklin-Wolf estate as it exists today owes much to the devotion of Dr Albert Bürklin, who managed it from 1924 until his death in 1979. He was a founder member and Vice-President of the influential German Viticultural Association.

The extensive medieval vaulted cellars of the Kolb Hof, which are over 400 years old, together with their modern extensions, form the cellars of the domain. This is named after the dukes of Kolb von Wartenberg, and the ruins of their castle which dominate Wachenheim also belong to the Bürklin-Wolf domain. Also of note is the beautiful tasting room, and the fine Unteren Hof with its park overlooking the vineyards which is still the Bürklin home.

Since the death of Dr Albert Bürklin, his widow, Jutta Bürklin, has taken charge. She is fortunate to have Dr Raquet, back after a spell at the Staatlichen Weinbaudomänen at Niederhausen-Schlossböckelheim, as manager, together with the *Kellermeister* or cellar-master Herr Knorr. He is the fourth generation of his family to hold this position.

The cellars and offices are in Wachenheim. The vineyards are in the four adjoining villages of Wachenheim, Forst, Deidesheim and Ruppertsberg, the very heart of the Mittelhaardt, the finest part of the Rheinpfalz or Palatinate.

In Wachenheim the best vineyards are Rechbächel, Goldbächel, Gerümpel, Böhlig and Mandelgarten; in Forst, Kirchenstück (the most prized vineyard in the Rheinpfalz), Ungeheuer, Jesuitengarten and Pechstein. In Deidesheim the main vineyards are Hohenmorgen, Langenmorgen, Leinhöhle and Kalkofen; and in Ruppertsberg, Hoheburg, Gaisböhl and Reiterpfad.

The vineyards of Bürklin-Wolf cover 103

hectares on gently sloping or flat terrain. The soils are gravel, loamy sand, clay, marl, loess, and loess-loam. The Riesling does particularly well where there is red sandstone with a loess-loam covering. The subsoil is rich and vigorous, with a high proportion of sand on the surface, so that the durability of the subsoil combined with the capacity of the surface to warm up quickly helps the grapes to ripen very well. The general climate in this part of Rheinpfalz is excellent for grape growing because the vineyards are sheltered from cold winds and excessive rain by the Haardt mountains to the west. The vegetation of the area includes figs, almonds and sweet chestnuts. Exotic shrubs and trees also grow here.

The vineyards are planted with 75 per cent Riesling, and the remainder with Müller-Thurgau, Gewürztraminer, Scheurebe and several new grape varieties such as Ehrenfelser and Kerner. It is the policy of the domain to sell their wines as 100 per cent varietals. This means that a wine labelled as Riesling will contain only grapes from this variety of grape.

There is great emphasis here on viticulture, and the most up-to-date methods are used. The soil is analyzed every 4 years to ensure that it receives the correct amount of the right fertilizers, mostly manure and compost.

The harvesting is meticulously carried out, so that each of the domain's many vineyards—nine in Wachenheim, five in Forst, four in Deidesheim and four in Ruppertsberg—is harvested according to its position and the types of vine, to ensure the highest quality. The winemakers reckon to have the must in the vat within three hours of the grape being picked.

Of the total storage capacity of 1,300,000 litres, about 400,000 litres are in wooden casks, which is considered to be very important in developing the character of the Riesling wines. The *Süssreserve* is kept at −7°C, and separate ones are kept for each vineyard. Generally speaking, this domain combines the best of the traditional and the new in cellar techniques, with the continuity being provided by *Kellermeister* Knorr.

The Wine Law of 1971 lays down minimum Oechsle levels for each grape of quality. However, Bürklin-Wolf are even more stringent with their standards. For example, for *Kabinett* wines the legal minimum is 73° Oechsle while the Bürklin-Wolf standard is 78° Oechsle. For *Spätlesen* the comparative figures are 85° and 88° Oechsle, and for the *Auslesen* they are 92° and over 100° Oechsle.

This is very important for quality. It is often said that many *Kabinett* wines today would be better chaptalized, and that many *Auslesen* do not show the real *Auslese* character. Here at Bürklin-Wolf, because of the higher Oechsle levels which they insist upon, the result is *Kabinett* wines of real character and remarkable *Auslese* wines.

The Wachenheimer Gerümpel Eiswein was the first of this type of wine to be made at the estate.

There are two outstanding points of interest about Bürklin-Wolf wines—the sparing use of *Süssreserve*; and the way in which the individual character of each of the four villages, and indeed of individual vineyards, is brought out and developed.

Some *Trocken* and *Halbtrocken* wines are now sold, amounting in the vintages of 1978 and 1979 to approximately 30 per cent of the wines made in those years. The lower acidities registered, even for the Riesling, in the favoured conditions of the Rheinpfalz make these wines particularly successful here, with more fruit and charm than similar wines from further north. But even the wines to which *Süssreserve* has been added are elegant and well-balanced, fruity rather than sweet at the *Qualitätswein bestimmter Anbaugebiete* and *Kabinett* levels. Even the *Spätlesen* are distinguished by depth of flavour and fruit rather than richness. Higher up the scale, the estate is justly famed for its magnificent *Beerenauslesen* and *Trockenbeerenauslesen*, to which was added for the first time in 1978

an *Eiswein*, the Forster Mariengarten Scheurebe Beerenauslese Eiswein. This wine had an Oechsle level of 140, 12 grams per litre acidity and 190 grams residual sugar. It was picked on 6 December at a temperature of −8°C. The Scheurebe always shows at its most elegant at high sugar levels, when it loses its slightly 'catty' nose and takes on breed and opulence.

In 1979 the proportion of the vintage was as follows. There were 30 per cent *QbA* wines, 60 per cent *Kabinett*, 9.8 per cent *Spätlese* and 0.2 per cent *Eiswein Auslese*. The grapes for the *Eiswein* were picked on 10 January 1980 and resulted in 500 litres of wine. The average yield per hectare was 95 hectolitres. This was a large vintage with no frost and good ripening conditions.

These proportions vary considerably from year to year. In 1977, for instance, the yield was 85 hectolitres per hectare, 78 per cent of the wine was *QbA*, 17.8 per cent *Kabinett* and 3.8 per cent *Spätlese*, there were no *Auslese* wines. Recent great years were 1971 and 1976.

The Bürklin-Wolf vineyards are in the villages of Wachenheim, Forst, Deidesheim and Ruppertsberg. The cellars and offices are in Wachenheim.

Georg Raquet (TOP LEFT), the general manager of Bürklin-Wolf is inspecting a wine for clarity. The large wooden casks in the background are traditional. After the grapes arrive at the winery, they are destalked and then pumped into the horizontal press (FAR LEFT). The town and church of Wachenheim (CENTRE) are set among vineyards. Wachenheim lies in the best of the Rheinpfalz vineyards.

This machine (LEFT) is used for trimming the tops of the vines.

Year	QbA	Kabinett	Spätlese	Auslese	Trockenbeer-enauslese	Eiswein	Comments
1971		30%	40%	25%	5%		Long-lasting wines because of high acidity level.
1976		10%	20%	65%	5%		Average yield 75 hectolitres/hectare.
1977	78%	17.8%	3.8%				Modest quality; yield 85 hectolitres/hectare.
1978	59%	36%	5%			0.3%	Average yield 90 hectolitres/hectare. Higher acidity than 1979.
1979	30%	60%	9.8%			0.2%	Large vintage, good ripening conditions. 95 hectolitres/hectare.

WEINGUT REICHSRAT VON BUHL

This is an estate where the drier wines in the <u>Kabinett</u> and <u>Spätlese</u> categories are in their way as deserving of praise as its great sweet wines for which it is justly renowned. Here the Riesling grape produces wines of great flavour which are full-bodied while retaining the finesse for which this grape is famous.

Although the Buhl name only appears for the first time as an estate owner in the Rheinpfalz in 1849, the origins of this domain are much older. The domain itself developed from the unification, through inheritance, of two properties belonging to two old wine-making families, Jordan and Schellhorn-Wallbillich. They were important vineyard owners in Deidesheim and Forst by the beginning of the eighteenth century. The Jordan family was French in origin and can be traced back to 1575. The last Jordan son founded the Bassermann-Jordan line, while his sister, Barbara, married Franz Peter Buhl, who came from Baden. The Buhls were a merchant family who also traded in wine.

In the 1830s, Franz Peter Buhl inherited vineyards in Forst and Deidesheim from his wife's uncle, but the property was still known as Weingut Peter Heinrich Jordan Erben. Then, when his father-in-law died in 1848, his property was divided into three parts, and at this stage Buhl renamed his portion of the inheritance Weingut F P Buhl. The Buhls had three sons. The youngest, Franz Armand, married Juliane Schellhorn-Wallbillich of Forst, and it was their son, Franz Eberhardt Buhl, who united the extensive holdings of both families in 1909 to form the Buhl estate as we know it today.

Franz Eberhardt who died in 1921 was also a member of the provincial parliament and, later, of the upper house of the Imperial parliament. His title was *Reichsrat*, the equivalent of senator, and this title was incorporated into the name of the vineyard. More importantly though, it was under his inspiration and direction that the *Deutscher Weinbauverband* (German Winegrowers' Association) was founded. As its president, he was active in promoting research in viticulture, vine propagation and the protection of quality wines.

After his death, the estate was managed by his widow, and in 1952 passed by inheritance to the family of Freiherr (Baron) von und zu Guttenberg. Since 1972, the owner has been Freiherr Georg Enoch von und zu Guttenberg, the well-known conductor.

The cellars here are particularly fine, with their vaulting and long rows of wooden casks. Part of the cellars date back to 1789, while some of the estate buildings are of fifteenth century origin. Opposite the offices, the domain has its own hotel, where a wide range of the domain's wines can be tasted accompanied by traditional Palatinate specialities.

There was a time when the reputation of the von Buhl wines fell a little below the very highest level which they have always held. There can be no doubt that today the management of the domain and its wines are of the very highest standards. Helmut Häussermann has been manager since 1977 and his winemaker is Klaus Briegel. Both are young, highly qualified, and dedicated to quality. They are typical of a new generation of enthusiastic German winemakers whose technical accomplishment reminds one rather of Californian winemakers. But they have the additional and inestimable advantage of inheriting a vigorous tradition of winemaking, and combine this with the latest available technical expertise with much success.

The offices and cellars, which extend for 2.5 kilometres, are in Deidesheim, but viticultural equipment is stored at the estate in Forst. The vineyards are situated in the villages of Forst, Deidesheim Ruppertsberg, Wachenheim, Königsbach and Friedelsheim.

The most important vineyards are in Deidesheim. They are called Herrgottsacker, Leinhöhle, Kieselberg and Maushöhle. The vineyards in Forst are Kirchenstück, Jesuitengarten, Ungeheuer, and Pechstein; in Ruppertsberg are Linsenbusch, Reiterpfad, and Hoheburg; in Wachenheim is Luginsland; in Königsbach is Idig; and in Friedelsheim is Gerümpel.

The Buhl vineyards cover 100 hectares of mostly flat terrain. On the basaltic slopes outside the village of Forst, the Riesling ripens more completely than anywhere else in Germany. So, from the vineyards of Kirchenstück, Jesuitengarten and Ungeheuer, Buhl produce some of their greatest late-harvested wines, combining full-bodied richness with great breed. The Deidesheim vineyards produce wines of great finesse. From Ruppertsberger Reiterpfad and Forster Pechstein, typically spicy Gewürztraminers are produced, while Ruländer of fine quality is produced from Königsbacher Idig. The soils are those typical of this best area of the Mittelhaardt — gravel, loamy sand, clay, marl, loess, and loess-loam.

Some 80 per cent of the vineyards are planted with Riesling, which does well in this region. Next in importance is Müller-Thurgau, then Ruländer, Scheurebe and Ehrenfelser. One of the features of this domain is the high proportion of old vines, and the emphasis on totally Riesling wines. When conditions are favourable, over-ripe and raisinated grapes are late-picked to produce the great *Beeren-* and *Trockenbeerenauslese* wines. At vintage time, up to 150 pickers are used, but experiments with mechanical harvesting are being made.

Pneumatic Willmes presses are used, and now a new model that permits a completely closed pressing, excluding all air, is being introduced. Natural yeasts from their own vineyards are used, not selected ones from elsewhere. Häussermann believes this is important in maintaining the individuality of each vineyard. Wooden casks are used for maturing the Riesling wines.

This estate is famous for its Riesling wines. The age of the vines gives von Buhl Rieslings an intensity of flavour and an individuality which are hard to beat. In the

The von Buhl label is very traditional and has remained unaltered for many years. 1976 produced some outstanding Trockenbeerenauslesen.

The terrain of the Ungeheuer vineyard at Forst (BELOW) is relatively flat. The Haardt mountains rise up behind the vineyard and protect the region from westerly winds.

great years, wonderful *Beeren-* and *Trockenbeerenauslese* wines are made which last and develop for many decades. But even modest *Kabinett* and *Spätlese* wines of vintages such as 1969 or 1970 were still in marvellous condition when 10 years old. As elsewhere, *Trocken* and *Halbtrocken* wines have been made in recent years, but these do not account for more than 20 per cent of what is produced. Apart from its great Rieslings, the estate produces some especially fine Gewürztraminers, and the Scheurebes are also fine, rather understated at the *Spätlese* level, but producing marvellous *Beerenauslesen*, with a tremendous concentration of sweetness, yet beautifully fresh and uncloying. It is at this level that the normally rather pungent Scheurebe gives of its best.

In 1979 von Buhl produced 33.5 per cent *Qualitätswein bestimmter Anbaugebiete*, 34.5 per cent *Kabinett*, 27 per cent *Spätlese*, 5 per cent *Auslese* and 0.07 per cent *Beerenauslese*. This last amounted to 370 litres. The yield was 78 hectolitres to the hectare. In 1978 the figures were 54 per cent *QbA* wines, 36 per cent *Kabinett*, 8 per cent *Spätlese* and 2 per cent *Auslese*, with a yield of 65 hectolitres to the hectare.

Von Buhl is a major estate in the Rheinpfalz. The vineyards are in the villages of Forst, Deidesheim, Ruppertsberg, Wachenheim, Königsbach and Friedelsheim.

This 400 year old sculpture (LEFT) commemorates the purchase of the Kirchenstück vineyard in Forst by the Spindler family. Even at that time, this vineyard was the most prized in the village.

Year	QbA	Kabinett	Spätlese	Auslese	Beerenauslese	Yield (hectolitres/hectare)
1978	54%	36%	8%	2%		65
1979	33.5%	34.5%	27%	5%	0.07% (370 litres)	78

The general manager, Helmut Häussermann is standing in the Jesuitengarten vineyard (TOP). This is one of the most famous vineyards in Forst. Like many growers, von Buhl have distinctive signs (ABOVE) to mark those parts of the vineyard which they own.

Ehrenfelser (ABOVE) is an important new grape variety with Riesling characteristics. Von Buhl grow a small proportion of this type.

ITALY: INTRODUCTION

Italy is the largest wine-growing country in the world, usually producing a little more than France, and, in a prolific vintage such as 1979, exceeding 80 million hectolitres. It is a matchless source of supply for good, medium-priced wines, with a few great wines and a great deal of wine sold to other countries to 'improve' their own produce. The vine is grown in all of Italy's 19 regions. In view of the great length of Italy, with its mountains, long coasts, and river valleys, it is easy to see that the variety of wines is almost endless. Red, white and rosé wines are made everywhere, some of them sparkling and some of them fortified.

Italians have always appreciated the wines of their own land. But in the early 1960s they realized that they would have to put some order into the production and marketing of these wines if they wished to make a true impact on export markets, allowing the wines to be sold in their own right and not as anonymous components of a future blend. So, in 1963 the *denominazione di origine controllata* or *DOC* was born, and provided the platform on which the remarkable success of Italian wines abroad has been based. In the United States alone, over 50 per cent of wines imported are now Italian.

About 200 wines have been accorded the *DOC*, which was roughly based on the French *appellation contrôlée* laws. Thus, place of origin, grape variety or varieties, yield,

alcohol level, and viticultural and vinification processes are controlled. In some ways, the law is tighter than the French law, specifying for example, the percentage of the permitted grape varieties in a *DOC*, and, in some cases, the material in which wine should be aged and the length of time it should spend in wood barrels. Another difference from France is that chaptalization is not allowed in Italy. Therefore, if the alcohol content of a wine has to be raised, this is achieved by the addition of concentrated must, which can often come from outside the area of production of the *DOC* wine. All Italian wines claiming *DOC* have to be tasted. There are some excellent wines in Italy that are not *DOC*, largely due to some winemakers preferring to make wine from grapes, often French varieties, not permitted for their particular *DOC*. There exists also a *denominazione di origine controllata e garantita* (*DOCG*), which was first awarded to Brunello di Montalcino in 1980.

The Veneto and Emilia-Romagna in the north, and Sicily and Puglia in the south, are the four regions of Italy which produce the most wine. The Veneto produces the most *DOC* wine, but Tuscany produces only a little less, and by far the greatest proportion of *DOC* wine in relation to total output is made by Trentino-Alto Adige. The north-east of Italy—the Veneto, Friuli-Venezia-Giulia and Trentino-Alto Adige—are greatly influenced by French and German grape varieties. The red Cabernets, Merlot

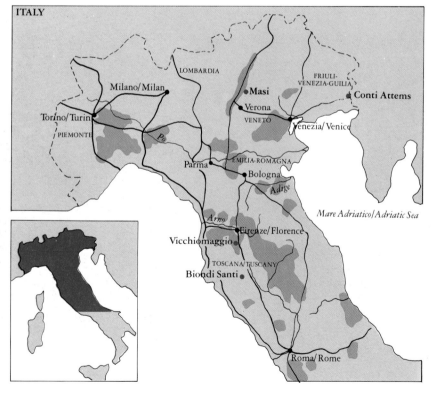

ITALY

| Vintage chart | 1980 This looks very good. |

Tuscany

1967 A very good year for both Chianti Classico and Brunello di Montalcino.

1968 Good to very good.

1969 Very good.

1970 Good.

1971 Very good indeed.

1972 Wines mostly rather astringent, lacking fruit.

1973 Good, with occasional very good wines, but should be chosen with care.

1974 Some nice, solid wines. Some rather dull.

1975 Very good indeed.

1976 Poor because there was much rot.

1977 Very good.

1978 Very good indeed.

1979 Good.

Veneto

1971 Very good year—only Amarones would be good drinking now.

1972 Poor year.

1973 Fairly good, but most wines would be tiring now.

1974 Fairly good.

1975 Medium quality vintage.

1976 Good.

1977 Good.

1978 Some hail reduced crop. Good wines.

1979 Large crop of very nice Valpolicella wines.

Friuli-Venezia-Giulia— Collio

1970 Good quality.

1971 Good whites, reds with deep colour, suitable for ageing.

1972 Below average quantity

and Pinot Noir, and the white Riesling (both Rhine Riesling and Italico Riesling), Pinot Gris, Gewürztraminer, Pinot Blanc, Silvaner and Sauvignon are proof of this. There are, of course, important local varieties as well, such as the Corvina, Molinara, Rondinella and Garganega of the Veneto, the Marzemino and the Teroldego of Trentino-Alto Adige, and the Ribolla and Verduzzo of Friuli-Venezia-Giulia. On the whole, these north-east wines rely on youthful, fragrant fruit for their charm and attraction, with either no or minimum wood ageing, and the grape variety characteristics being brought out.

It is a very different picture in the north-west, in Piedmont, where some of the most suitable red wines for ageing are made. The Nebbiolo grape is responsible for the finest Barolo and Barbaresco, Gattinara and Carema, with the Barbera making wine of the same name. These tannic wines are often aged in wood for years before sale, acquiring a concentration and strong flavour that is adored by some, found difficult to understand by others. Piedmont is also the region for the best sparkling wines made from the Muscat, or Moscato, grape—lovely, luscious wines, with a tempting fruity sweetness to them. Emilia-Romagna makes a great deal of quite modest but very drinkable wine. Its greatest export is Lambrusco, the slightly frothy red wine that enlivens parties and converts many people to wine drinking.

Tuscany is the region of Chianti, where the Sangiovese grape reigns supreme—it is in fact the most widely grown red grape in the whole of Italy, particularly in the centre and south of the country. In Tuscany it makes wines both for delicious young drinking and for ageing. Umbria is famous for white Orvieto, traditionally a little sweet but now made more often dry, and the Marches, on the Adriatic side of Italy, are well-known for the delicious dry white Verdicchio, made from the grape of the same name. The ubiquitous white Trebbiano grape appears almost everywhere in central and southern Italy. Important in both Orvieto and Frascati, the wine made near Rome, it is often blended with grape varieties that are as old as Italy's history, some of them of Greek origin. Both the red and white wines of southern Italy used to be described as heavy, high-alcohol wines. The reds were pungent and powerful, the whites often oxidized. But although red wines high in alcohol and colour are still made for blending with lighter wines both at home and abroad, there are now much lighter red and rosé wines made in Puglia, Calabria and Sicily, while the whites are lighter and fresher than of old. Both represent wonderful value for the wine drinker. The same could be said of the wines of Sardinia. The island has always made large quantities of blending wine and red and white dessert wines, but now it is an excellent source of delicious, light wines.

Many Italian estates are still family based. The grapes are often harvested by the estate workers and their families or other local people.

with top quality. The whites have style and body, especially the Pinot.

1973 Large vintage, but rainy. However, good whites with finesse. Reds are medium-quality, not really for ageing.

1974 Low quantity. Excellent whites, harmonious, quick maturing. The reds are deep coloured and have good alcohol.

1975 Average quantity, low acidity, therefore not keeping wines. Tocai very characteristic and full of body. Some wines almost too aromatic. Pinot Bianco and Grigio have some finesse.

1976 Average quantity. Acidities fell noticeably after the malolactic fermentation. The reds were suitable for ageing if they were harvested early.

1977 Average quantity. Above average quality. Very scented white wines, with Pinot, Sauvignon and Traminer especially good.

1978 Low production. Excellent quality, with early maturing white wines possessing balance. Good alcohol levels. Deep coloured red wines, with body and scent.

1979 Abundant quantity with very high quality. The white wines are medium-bodied, very characteristic and perfumed, especially those wines of an aromatic nature—such as Malvasia, Riesling Italico, Sauvignon and Traminer. Excellent year for the Cabernet Franc. Merlot has body and character with a well projected bouquet.

1980 Quantity varied with the grape variety, with Pinot Bianco having only a modest production and therefore a higher than average quality. Small production for the Cabernet Franc and less than average for Merlot. Wines are not deep coloured, but have harmony and balance. The whites lack real body, but have a nice, clean scent and are fruity.

BIONDI SANTI

Biondi Santi make the Brunello di Montalcino that established the wine as great. Produced from severely selected grapes and aged in wood, it is a wine that has almost legendary capacity to mature with interest, and its intensity of bouquet and diversity of flavours open out in the glass as it is being drunk.

In Biondi Santi's Brunello di Montalcino are linked Italy's most famous wine producer and its most famous wine. The wine is perhaps an acquired taste, and certainly needs careful serving to show at its best, but the finest examples reach new dimensions of taste. On the whole, the taste of top Brunello is appreciated in Italy more than anywhere else, partly because limited production has meant limited distribution. Although there are massive new plantings of Brunello in the area, it is doubtful whether these are aiming for the very special status achieved by the greatest, traditional producers. An indication both of the longevity of these great red wines and the extent to which they are appreciated in Italy can be given by the fact that at Sotheby's 1978 wine auction in Florence, US $385 was paid for a bottle of 1917 Brunello di Montalcino from Biondi Santi, and US $250 for a 1945 Riserva from the same house.

Brunello di Montalcino is produced in the commune of Montalcino, 24 kilometres south of Siena in Tuscany. The countryside is hilly and quite beautiful. The vineyards are planted on the slopes, with very good exposition to the sun, at heights up to 600 metres but on average, in the case of Biondi Santi, at 525 metres. It would be quite accurate to say that the Biondi Santi family made this wine famous. At the beginning of the 1800s, Clemente Santi, the great-great-great-grandfather of the present head of the family, was already exporting wine to France and to Great Britain. In 1867, at the Universal Wine Exhibition in Paris, he received awards for making wine of high quality. But the real fame of Brunello di Montalcino stems from the grandfather of the current owner, Ferruccio Biondi Santi, an oenologist and viticulturist, who did an immense amount of work in vine selection with the Brunello at the end of the last century. This was necessary because Montalcino was originally known for its white wine and not its red. The estate still has four bottles from this first production—in 1888—of Brunello di Montalcino as it is known today; and in April 1970, the last bottle opened was in very good condition.

The Brunello grape of Montalcino is the Sangiovese Grosso. It buds during the first part of April and flowers around mid May. The grapes gradually turn colour towards the latter part of August, becoming ripe at any time from the end of September to the first two weeks in October. The skins of the berries are quite thick and are of a black-purple-blue hue, resulting in deep-coloured wine, which browns to deep brick with age.

Biondi Santi accredit much of the quality of their wine to careful grape selection. Only the best, ripest and most healthy grapes are used for the estate wine; in some years, only 30 per cent of the total production is used to produce the Brunello. In years, such as 1960, 1962, 1965, 1972 and 1976, when the quality is not judged high enough for making Brunello, the production is sold in bulk as red table wine. This happens because the weather can sometimes be quite cruel in these hills, particularly affecting the ripeness of the grapes. The age of the vines is also taken into account in the grape selection process. For making Brunello, the vineyards have to be over 10 years old at Biondi Santi, although they only have to be three years old under *DOC* law. For the Brunello Riserva, the top wine most suitable for ageing, the vineyards have to be over 30 years old. There is in effect a three-tier system, with Brunello Riserva only being made after an exceptional year from magnificently ripe grapes from old vines, while the wine from those vines of between 10 and 30 years of age becomes the Brunello Annata. In years thought of as no more than good, only the Brunello Annata is produced. There is another condition attached to the production of these wines, but which is imposed by the *DOC* authorities, not the estate owner. Brunello di Montalcino cannot be sold to the consumer before being aged in wood for at least four years. With five years of age, it is allowed to use the term *riserva*.

Dr Biondi Santi and his son run the estate with a high degree of personal supervision. Fermentation is in closed 80 hectolitre, glass-lined barrels at between 22°C and 28°C. There are *remontages* twice a day for half an hour. At the end of January, the wine is racked into barrels of Slovenian oak. At first, the wine is kept in young oak, and as it ages, it is racked into older oak barrels, some of which were bought by the grandfather of the present owner at the beginning of the century. There is a great range of barrel size, varying from eight hectolitres to 80 hectolitres, which allows flexibility with the different types of wine. Obviously, with the large-sized barrels, the wine-wood contact is not the same as in a small Bordeaux-size *barrique* of 225 litres, and thus makes it possible to age wines for as long as five years in wood. The malolactic fermentation follows naturally after the alcoholic fermentation. During the second and third year in wood, the wines are fined with fish glue. Before bottling, the wines receive a light filtering through wide Seitz filters.

Dr Biondi Santi and his son are both qualified oenologists. They say their wine is made with 'passion and tradition.' They consider the fact that they only vinify their own grapes is very important to the wine's ultimate quality. Quality is obtained from close attention at all stages. Even before the grapes are picked, between 15 and 20 days before the harvest, the leaves are stripped from around the bunches, giving the grapes more light and therefore enabling them to ripen better. The vines at Biondi Santi are not asked to produce as much as the law allows. By law, a winemaker can produce 100 *quintali* of grapes per hectare, which is about 70 hectolitres of

The Brunello di Montalcino is the most renowned wine produced by the Biondi Santi estate. It is made from the Sangiovese Grosso grape and ages very well.

At Biondi Santi a small hand press (BELOW) *is used in preference to a mechanical one. This is so that the grapes are only pressed very lightly* (BELOW CENTRE).

wine per hectare. At Biondi Santi, they never produce more than 50 *quintali*, and in 1979, a very good year for the area, 15 hectares produced about 400 *quintali* of wine, far less than the quantity permitted by law. The land and soil where vines are to be planted are chosen with great care, and Dr Biondi Santi attaches great importance to the vines being trained evenly along the wires so that the bunches can attain maximum ripeness. A hand press is used, rather than a mechanical one, in order to give the grapes only a very soft pressing indeed.

Technical qualifications and tradition have often made good partners, and Biondi Santi is a famous example of this happy blend. The wines are warm and generous, deep and full, with a tannic background and a long finish. With decanting a good time before drinking (the actual time varies according to personal taste, the character of the year and the age of the wine), the nose can open out to a bouquet of huge dimension and interest.

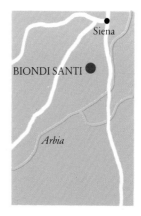

The estate is based in the area of Montalcino which is to the south of Siena in Tuscany.

The age of the vines in the Biondi Santi vineyards is very important as the estate imposes stringent controls on vine age. For example, vines have to be over 30 years old (LEFT) *before the estate will use them for their top wine, Brunello Riserva.*

Biondi Santi place much emphasis on the quality of the grapes in their wines, only using the very best for the estate wine. These newly picked Sangiovese Grosso grapes (LEFT) *are in excellent condition.*

Dr Biondi Santi (TOP LEFT) *runs the estate with the help of his son. The skins of the Sangiovese Grosso grape* (TOP RIGHT) *contribute to the deep colour of the Brunello di Montalcino wine for which Biondi Santi is famed. The estate* (ABOVE) *surrounds the modest Biondi Santi family house.*

CASTELLO VICCHIOMAGGIO

**Castello Vicchiomaggio is a Chianti Classico that is deep
flavoured and fruity, with the capacity to age. It has the body and
velvety texture associated with the wines of the Greve area of the
Classico zone, and the careful treatment and use of the <u>governo</u> system
ensure that it has vigour as it ages.**

The focal point of the Tenuta di Vicchiomaggio, the estate of Vicchiomaggio, is an imposing castle, high on a hill above the village of Greve. Greve is the centre of the Chianti Classico area, the original heartland of the Chianti-producing region, between Florence and Siena. The wines of the Greve district are regarded as the longest lived of all Chianti Classico.

Vicchiomaggio has belonged to the Matta family since 1965, and the present proprietor and manager, John Matta, lives at the Castello. He is a qualified oenologist, trained at the famous oenological school at Alba, Piedmont, which specializes in the vinification of red wine. John Matta is aided by his estate manager, Pier-Paolo Brandani, a specialist in viticulture. The fact that John Matta was brought up in England has undoubtedly contributed to the success of Castello Vicchiomaggio in the British market, but Vicchiomaggio is also exported to Germany, Switzerland and the United States.

Vicchiomaggio consists of 25 hectares of specialized vineyard and only three hectares of the old-style mixed vineyard. The domain itself covers 150 hectares, including 50 hectares of forest and wood and 20 hectares of olives. The rest is given over to other sorts of agriculture, mainly wheat. Chianti has always had this mixture of specialized vineyards, where only vines are planted, and some mixed crop vineyards, or *promiscuo* planting, where the vines are combined with other agricultural products, such as olives or cereals. This was the old way which enabled a grower to be self-sufficient, and many people thought that the additional crops enriched the earth and helped the vines. However, this pattern is gradually being replaced by exclusive vine cultivation. The plots are obviously easier to work and the change is perhaps inevitable in large, successful vineyards run on commercial lines by trained winemakers.

The classic Chianti grapes are used at Vicchiomaggio: about 75 per cent Sangiovese, 10 per cent red Canaiolo and 15 per cent white Malvasia and Trebbiano varieties. The vineyards lie at an altitude of about 350 metres. This can lead to the danger of spring frosts, and it certainly becomes a factor at harvesting which, in common with all Chianti Classico, does not usually begin until mid October. The permitted Chianti Classico yield is 45 to 48 hectolitres to the hectare.

Without doubt a *riserva* wine, the very best Chianti Classico and the most suitable for ageing, is born in the vineyards. The grapes come from the best vines, usually the oldest, giving the most concentration. The pruning is harsher for the grapes that will give a *riserva* wine, in order to ensure maximum quality, body and colour.

Vinification takes place in the cellars underneath the castle. There is a new cellar exclusively for the fermentation, with both stainless steel and fibreglass vats of a total capacity of up to 2,000 hectolitres. The fermentation is as prolonged as possible, lasting from two to three weeks, and the temperature is kept low by keeping doors and windows wide open. This slow fermentation undoubtedly contributes to the bouquet and body of the final wine, which makes Vicchiomaggio one of the most rewarding of the top Chianti Classicos when aged. There is a Garnier horizontal press, which ensures that the grape skins are not given too much pressure, although no press wine is added to the estate wine.

The unique aspect of the Chianti Classico winemaking process is the *governo* system. This is an extra manoeuvre in the vinification system, and one that has been abandoned by many Chianti Classico producers because it is too costly in labour. However, a personally-run estate like Vicchiomaggio is still convinced of its benefits and considers that the extra work involved is more than compensated for by the effect on quality.

The *governo* system involves the addition of the juice of about 5 per cent of dried grapes which, when fermenting, is added to the main body of the wine in December to start a second fermentation. This makes the end product richer and smoother, and the young wine can be very lively, with a touch of freshness on the palate, through the extra carbon dioxide created by this second fermentation. The *governo* fermentation is, again, very slow and can last up to two months at temperatures of between 15°C and 18°C. It is immediately followed—or joined—by malolactic fermentation. The cellar is heated to help the completion of this process.

When the malolactic is finished and the wine completely stable, the wine goes to the part of the cellar set aside for ageing. Here traditional Chianti 50 to 85 hectolitre casks are used. When new wood is necessary, it is first employed as a fermenting container for two years, to take away the rather harsh new wood taste. John Matta considers that the qualities of Chianti Classico, with its particular blend of grapes such as the delicate Malvasia and Canaiolo, are masked by the flavour of new oak. In the first year, racking takes place at least four times, and up to six times if the vintage is not so good.

By law, *riserva* wines need three years' development before being sold. Not all of these need to be in wood. However, the old Chianti Classico traditions often ensured that the wine remained in cask for four or five years. Vicchiomaggio takes the view that bottle ageing before sale is just as important and the non *riserva* wines are now usually bottled after 18 months in the Chianti casks. The fact that Chianti Classico is one of the few red wines in the world to contain a proportion of white grapes (Côte Rôtie is another) means that it is not a wine of

*1971 was a superb vintage. The
Vicchiomaggio castle is shown on the label.*

The castle high on a hill above Greve (LEFT) was first described in a document of 957 AD. During the Renaissance Leonardo da Vinci mentioned it in his writings. Some of the wine cellars are in the dungeons of the castle.

immense body requiring massive ageing in wood. In fact, too lengthy keeping in wood clearly dries out the wine more than is desirable.

Equipment is regarded as very important during every stage of vinification. There are piston pumps for the racking, which do not allow air to reach the wine during this process. There is also a stainless steel bottling machine which ensures rapid and hygenic bottling, and the vats, floors and equipment are kept clean with the aid of a high pressure washing unit.

Bottling time for the non *riserva* wines is flexible. Usually bottling takes place after 12 or 18 months, but a notable exception was the light 1976 vintage, when John Matta decided that seven or eight months was quite adequate, and the rest of the ageing should be in bottle. The wine was not substantial enough for lengthy wood ageing, and the delicacy was preserved by shortening this part of the process.

The ideal Chianti Classico would have about 12.5 or 13 per cent of alcohol and between 5.8 and 6.2 grams per litre acidity. In Italy, acidity is expressed in terms of tartaric acid. Chaptalization is forbidden in Italy, and adjustments to the alcohol level are made by adding concentrated must. John Matta stresses that this should come from good grapes. At Vicchiomaggio it comes only from Vicchiomaggio grapes, concentrated using specialized equipment at another good Chianti Classico estate, Villa Cafaggio.

One of the great qualities of Vicchiomaggio is the richness it preserves as it ages in bottle, and the great bouquet it develops. John Matta, unusually for the region, is a great supporter of decanting, and the effect this has on his wine, especially on vintages of a decade old or more, is quite startling. Very good vintages for Vicchiomaggio include 1969, 1971 (which is superb), 1973, 1975, 1977, 1978 and 1980.

John Matta, owner of Vicchiomaggio (FAR LEFT) trained in oenology at Alba. His father bought the estate. John Matta has improved the quality of the wine and gained it an international reputation. The inner courtyard of the castle (LEFT) has interesting arched balconies.

The grape pickers set off to start the vintage (ABOVE). As well as vineyards, Vicchiomaggio has a large farm growing olives and wheat, as well as 50 hectares of woods.

Vicchiomaggio at Greve is in the centre of the Chianti Classico area, a hilly zone between Florence and Siena. Chianti Classico wines from Greve are known for their body and are good for ageing.

This area at Vicchiomaggio (LEFT) is an example of a specialized vineyard. A specialized vineyard is devoted solely to vines rather than, as was the custom in much of Italy, mixing grape growing with the cultivation of other crops such as olives.

These Sangiovese grapes (LEFT) have just been harvested. This variety makes up 75 per cent of Vicchiomaggio's planting. Three red and two white varieties are used for Chianti Classico.

After being harvested, the grapes are taken to the cellars (LEFT).

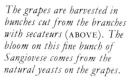

In the cellars, John Matta is experimenting with small Bordeaux-type barrels for wood ageing (LEFT). They hold 225 litres as opposed to the normal size for Chianti Classico which is between 50 and 85 hectolitres. The glass stoppers are expensive but help stop oxidation.

The grapes are harvested in bunches cut from the branches with secateurs (ABOVE). The bloom on this fine bunch of Sangiovese comes from the natural yeasts on the grapes.

CONTI ATTEMS

The wines from the estate of Conti Attems enchant by their projection of fruit and tempting bouquet. The whites all show their grape variety individuality and enticing aromatic flavours, and the reds are glossy examples of Cabernet and Merlot, intended to be enjoyed in zestful youth rather than aged splendour.

Count Douglas Attems presides over an estate producing impeccable red and white wines in the Collio district of the Friuli-Venezia-Giulia. Vines cover under 30,000 hectares in Friuli, producing about 1,250,000 hectolitres a year. Unfortunately, the area is prone to earthquakes, the last serious one taking place in 1976. In the far north-eastern corner of Italy, between the Veneto and Yugoslavia, with Austria to the north, the area was called Julia by the Romans. Collio, or Collio Goriziano, is a small *DOC* (*Denominazione di Origine Controllata*) district in hilly country west of Gorizia, on the border with Yugoslavia. It is considered the best *DOC* area in Friuli-Venezia-Giulia.

Count Attems comes from a very old family, which probably originated in Germany but came to Gorizia in about 1000 AD. Being so near a national frontier, the family and estates suffered many vicissitudes through the centuries, especially during the First World War when houses, vineyards and farms were destroyed. Before the First World War, the great estates of the then Count Attems supplied wines to the court in Vienna, but the Austro-Hungarian Empire is no more and the estates have grown smaller and become a compact commercial enterprise. It has, however, lost nothing of the quality in the transition.

There was one beneficial effect of this destruction in the early part of the century—when the vineyards came to be built up again, only noble grape varieties were replanted. The vines grown show that this area is at a national crossroads. There are vines of local origin, as well as French, German and Austrian origin, and if grape variety is added to the denomination of Collio, this always means that the wine is made completely of that grape variety. White wines predominate in these hills, although there are also delicious reds.

The Pinot Grigio probably produces the white wines that have won the most fame outside the region, especially on export markets. Count Attems makes a glorious example, fresh and enticing, full of flavour, with none of the neutral character of some too 'technical' wines. The Pinot Grigio is the same grape as Pinot Gris of Alsace and the Ruländer of Germany. It has a slight pinkish hue to the skin giving the wine an aromatic tinge. This colour in the wine should not be confused with oxidation, which would give a totally different, and less pleasant bouquet, and, although the skins should not be in too much contact with the juice if an unpleasant, dark colour is to be avoided, a little contact gives the wine its character.

The local Tocai grape gives another white wine of well-defined personality. This variety has nothing to do with Hungarian Tokay, but gives a spicy, golden-yellow wine which can take a few years of

bottle age to develop fully. Count Attems also produces dry white Sauvignon, with a slightly pungent taste, and an elegant, understated, somewhat flowery Riesling Italico.

The reds are wines of verve and a certain 'sappiness' which concentrate on grape flavour rather than woody maturity. Count Attems makes a superb Cabernet Franc, and a fruity, soft-styled Merlot. Both wines drink well throughout their life. They are bottled with a little carbon dioxide gas in them, and they can be decanted to let the few tiny bubbles escape. It is considered that this carbon dioxide helps retain freshness. The Cabernet Franc develops a little more slowly than the Merlot.

The climate is rather mild in the Collio area around Gorizia, but has quite heavy rainfall. Luckily, the hills protect the vines from the cold strong wind from the north-east called the *bora*, which blows very strongly at Trieste on the Adriatic. The Collio only receives a fraction of this wind, and in fact it can help blow away fog and rain clouds. However, hail can be a problem in the vineyards, with the hills susceptible to summer storms. As Count Attems only makes wine from his own grapes on the estate, he realizes that it would be a tragedy if he lost a good part of his crop, so he has invested heavily in nets to protect against hail. A percentage of his vineyards have this net covering, which is expensive both in terms of investment and labour. He prefers this to insurance, which he says gives you the money but not the wine. The soil is basically marl and sandstone, and the effect of the hill sites is to produce relatively low yields, unlike the flat, fertile plains.

Count Attems considers that the prime danger to his wines is oxidation, because they are not high in acidity, which can provide some protection against the influence of the air. The low acidity comes from the mild climate and soil, and everything is done to handle the wines with great care and personal attention. The white wines are kept in concrete vats and not in wood, to minimize contact with the air. Bottling is done meticulously, when the wines are young, by the wives of local labourers, with the emphasis on careful preparation and good corks. Generally, the white wines are bottled in the spring following the vintage, and the red wines in late summer or autumn. Everything, including the vintage, is personally supervised, with a cellarman and one main assistant.

The result is wines of immense charm, even seduction. Count Attems considers both his white and red wines are at their best when between one and three or four years old. 1979 was a year with both superb quality and quantity—a happy occurrence. Apart from running his own estate, Count Attems helped to found 15 years ago the Collio consortium of growers which is dedicated to high standards in quality, control, research and advice.

The Pinot Grigio wines from this estate have attracted much international attention. The wine is full of flavour with a slight aromatic tinge.

Count Douglas Attems
(BELOW) *owns the property
which bears his name and has
done much to bring the wines
of the Collio region of Italy to
the attention of the world.*

These barrels in the cellar
(BELOW) *are used for ageing.*

*The Conti Attems estate is
in the Collio district of the
Friuli-Venezia-Giulia. The
estate is very close to the
Yugoslavian border.*

*The rolling hills of the Collio
area* (ABOVE) *are ideal for
wine growing. It is an old
established wine growing area.
This small* DOC *is con-
sidered the best area in Friuli-
Venezia-Giulia because of its
terrain and slopes.*

A small tractor is best for tilling the land (BELOW). The nets above these Tocai vines protect them from hail. The white Sauvignon grapes (BELOW CENTRE) are trained high. They produce a dry white wine which is crisp and fragrant.

Although the red wines from the Collio region are not meant for long ageing, bottles like these of Merlot (BELOW) need a year or two to soften.

The river Isonzo (RIGHT) is on the border of Collio with the Grave del Friuli area.

These rows of Tocai (RIGHT) are planted horizontally to the slope. It is more usual to plant the vines vertically to the slope, but this configuration gives better exposure to the sun on this site.

The capsules for the bottles are put on by hand at the estate of Conti Attems (RIGHT). The estate remains a small family concern.

These Merlot vines are planted vertically to the slope (RIGHT). This is the conventional method. However, the main concern with all vine planting is to ensure maximum exposure to the sun. The nets are put over the vines if hail seems imminent.

AZIENDA AGRICOLA MASI

Masi produce Valpolicella and Soave of note, but the full, rich red wine of Campo Fiorin has a glorious velvety texture and richness of flavour that demands attention. The great Recioto wines are there to be sipped and savoured, strong, many-dimensional and intriguingly complex.

The Veneto is one of the four most important wine regions in Italy from the point of view of production, the others being Emilia-Romagna, Puglia and Sicily, and one of the regions best represented on export markets. Its 'liquid assets' include Soave, Valpolicella and Bardolino, and the revered Recioto Amarone wines. The family firm of Masi is headed by Dr Sandro Boscaini, and has its cellars and offices at S Ambrogio di Valpolicella, right in the heart of the area. The Boscaini family has run this wine estate since the 1700s, combining all that is good in the formidable traditional winemaking methods of the Veneto with modern techniques where they are better suited to obtaining high quality in the finished wines.

Masi wines are the produce both of their own vineyards and of grapes which are bought in. With Soave and Bardolino, the grapes are entirely bought in from small wine farms or *aziende agricole* which are under the supervision of the Boscaini family. This means that the firm exercises control over such aspects of viticultural activity as pruning and picking, thereby ensuring that they receive high quality grapes. Where Valpolicella is concerned, Masi own 80 per cent of their requirements, with the remaining 20 per cent coming from supervised grape growing farms.

The Veneto, with its gently rolling hills, and relatively mild climate with rare frost in winter and warmth rather than scorching heat in summer, is an ideal region for growing the vine. It is for this reason that the vine covers about 124,000 hectares as principal crop, and a further 72,000 hectares as a secondary crop, producing about 10 million hectolitres a year with only about 1.5 million hectolitres *denominazione di origine controllata* wine. Naturally, over such a large area, there is a great variety of soil, usually marn or marl mixed with calcareous matter, but near Lake Garda there is some sand, which influences the production of the light red Bardolino. Vines are trained high, sometimes in the pergola system.

The white wines of Masi aim for fragrance and lightness, rather than heavy-flavoured weight. The Soave of Masi is made predominantly from the Garganega grape, with some Trebbiano. It is *classico*, which means that the grapes come from the defined, original area of Soave, and *superiore*, denoting that the wine has a minimum alcohol content of 11.5 per cent when sold to the consumer, which is not before the 1 July following the vintage. The Soave of Masi is not wood-aged, but is kept in concrete and stainless steel vats for six months before being bottled in the late spring after the vintage. The Masianco white table wine, which is not *DOC*, for which Masi is well known, is made in the same, freshness-orientated manner, but some Riesling and Durello are added.

The Valpolicella *DOC* production area is to the north of Verona, with the Bardolino zone on Lake Garda in the west and the Soave zone in the east. The vines are on gently sloping hills about 200 metres high, with the best grapes grown at altitudes of between 300 metres and 600 metres. These higher sites are used for the production of the Recioto Amarone della Valpolicella, which is the highly individual reserve wine of the area. There is a good deal of calcareous soil in these hills. Unlike the ubiquitous Trebbiano of Soave, to be found all over Italy and even in France under other names, the grape varieties that make Valpolicella are unique to the region. The main variety is the Corvina Veronese, giving finesse, then the Rondinella which gives strength, the Molinara which gives a charming, fruity suppleness, and the Negrara which contributes colour and body. Again, Masi make Valpolicella Classico Superiore. Here, *classico* means that the wine must be produced in the communes of Negrar, Marano, Fumane, S Ambrogio and S Pietro in Cariano, the oldest vine-growing part of the area, and *superiore* denotes that the wine has a minimum alcohol content of 12 per cent when sold to the consumer and is only sold after at least one year from 1 January after the harvest.

The Valpolicella at Masi is kept in Slovenian oak barrels of between 80 to 100 hectolitres, so the influence of wood on the wine is slight, an altogether desirable process since Valpolicella should rely on its perfumed, fruity taste rather than oak-flavoured pungent flavour. The great Recioto della Valpolicella is made from grapes planted on the mid to high slopes of the hills. The grapes are picked at the beginning of October and are then left to dry on racks for about three months until the end of December or the beginning of January. After pressing, these *passito* grapes then ferment in the cold temperature of January for 30 days or more. The yield is very small, normally a maximum of between 20 and 25 litres of wine for every quintal of fresh grapes. This produces a wine of marked character and concentration, often reaching 15.5 or 16 per cent alcohol, which is suitable for ageing. This wine begins its life as sweet, and if it is bottled after two or three years in 25 or 30 hectolitre Slovenian oak barrels, it can be sold as Recioto della Valpolicella Amabile. If the slight, slow fermentation is prolonged until spring or autumn, until the wine is completely dry, and the wine is aged for 4 or 5 years, it can then be bottled and sold as Recioto della Valpolicella Amarone.

Only very small amounts of these Recioto wines are made, and Masi are proud of these high-quality individual lots of wine which are kept quite separate one from another. There are 3,800 bottles of the 1975 Recioto Amarone Campolongo di Torbe, with an alcohol level of 16 per cent; 2,500 bottles of the Recioto Amarone Mazzano

Masi's Campo Fiorin wine has a high reputation throughout the world. The grapes come from the Campo Fiorin vineyard and the wine is made in a very traditional way.

To make the Recioto della Valpolicella, the grapes are picked at the beginning of October. They are then left to dry on racks (BELOW) until late December or early January.

1974 with 16 per cent alcohol and drier than the 1975 wine, and 3,700 bottles of the Recioto Amabile Mezzanella 1975, which had two years in cask and one year in concrete vat before being bottled. These are all red Recioto della Valpolicella wines, but Masi also make a Recioto Bianco Campociesa which is not *DOC*. Here, the method is slightly different. The Garganega, Trebbiano and Durello grapes are destalked, and 60 per cent of the skins are preserved for the fermentation, which lasts for 20 or 30 days. The wine is aged for about a year in glass demijohns, which are kept on the roofs in winter to help the precipation of tartrates. Then the wine is put into 6 hectolitre barrels for two years before bottling. The result is a most unusual wine, which does not appeal to everyone but wins its proportion of admirers.

The Masi wine which has perhaps attracted the most admiration around the world is the superb, non-*DOC* red wine, Campo Fiorin. It is produced using methods which are very traditional in Valpolicella but which have now largely fallen into disuse. The wine begins by taking the normal Valpolicella grapes from a particular site called Campo Fiorin. As usual, they are picked in October and the wine is made. The wine is then put into vats containing the skins or lees of the Recioto; this induces a secondary alcoholic fermentation which gives the final result a marked scent and flavour, a good deal of glycerine and probably 1.5 to 2 per cent more alcohol, the wine being sold with an alcohol level of around 14 per cent. It is thus a wine somewhere between a Valpolicella Classico Superiore and an Amarone in character–rich and fascinating.

The Valpolicella DOC production area lies to the north of the city of Verona.

Grapes from the Campo Fiorin vineyard (LEFT) are used to make the non-DOC red wine which has brought the Masi estate to the attention of the world.

Rondinella is one of the grape varieties used for making Valpolicella. For picking the vines which are trained high (ABOVE), baskets are suspended on wires. The main grape variety used here in the Valpolicella is Corvina Veronese. Molinara and

Negrara are also used. The vineyards in the Veneto region are mostly planted on rolling hills. However, some terracing is still needed (ABOVE). The region produces about 1.5 million hectolitres of DOC wine per year.

AUSTRALIA: INTRODUCTION

In terms of international wine production, Australia occupies a position in the charts, a drop or two above Brazil and a bottle or two ahead of South Africa, making about one twentieth the amount of France or Italy; in world terms this is small. About half of the 3.5 million hectolitres of wine produced a year in Australia is distilled into *eau de vie* for fortifying wines or for industrial alcohol.

Australia occupies a land mass rather smaller in size than the United States, some 15° on either side of the Tropic of Capricorn. Vines grow in every state of Australia, the areas themselves being in some cases as far apart as Paris is from Moscow. This distance produces an abundance of possible soil, weather and vine combinations which have yet to be explored. Few of the vineyard areas exist because of dictates of, for example, soil or rainfall; for the most part the face of Australian viticulture has tended to result more from socio-economic factors.

Areas were established by farmers long before viticulturalists. The first settlers from Britain lived to the north of Sydney and fortunately grew grapes in the fertile soils of what was to become the Hunter Valley. Likewise, in South Australia, the Barossa Valley came to shelter the vine along with German settlers.

If the selection of these first vineyard sites seems random, the choice of grape variety was restricted to availability and demands such as ease of growing and yield. For those reasons, the Rhône variety, Syrah, which is called Hermitage or Shiraz in Australia, became the work-horse of the Australian wine industry. Likewise, viticultural

practice rested as much on hearsay as it did on tradition.

Despite all this, by the mid nineteenth century Australia had built up a thriving export trade and established vineyards to the north and south of Adelaide, in Victoria and New South Wales. To withstand the long trip to Europe, the wines were fortified until it was realized that the climate alone could produce the necessary sugar and alcohol. This had the effect of developing a tradition of winemaking where 'big was beautiful' which gave Australia internationally the reputation for hearty 'Burgundies', wines as unique as the indigenous marsupial fauna.

Phylloxera struck Australia at the same time as the economic depression of the last century. The disease devastated the vineyards of Victoria, scarred South Australia but spared New South Wales. At the end of the First World War, the Australian Government instituted a policy of settling repatriated soldiers on the land which led to the development of grape-growing areas of the Murray River in South Australia, the Murrumbidgee River in both South Australia and New South Wales, and Merbein in Victoria. The settlement of these areas—not solely for vineyard cultivation—was really due to the abundant water supply.

The depression of the 1930s saw local consumption fall by 60 per cent and a disastrous decline in exports. The government intervened with subsidies which caused many of the vineyards to be pulled up in favour of dairy farming. Victoria lost thousands of hectares of vines around Arrarat and Stawell, all the vineyards in the Yarra Valley and most in central Victoria. In South Australia, the Barossa firms

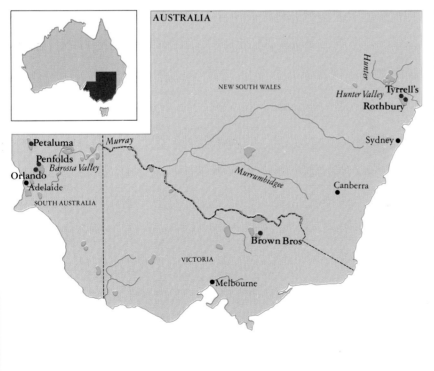

AUSTRALIA

survived by concentrating on fortified and sweet wines which led to the importance of the production of those styles in that area today.

Up until this time, wine had for the most part been sold in bulk or bottled under customers' labels as required. Apart from a vague mention of the state from which the wine came, the region, winemaker or variety were only rarely mentioned. The designation 'Claret' was more important than Coonawarra, as was 'Hermitage' in comparison to Hunter Valley. These market conditions contributed to the emergence of the generic titles under which so much Australian wine is still classified. The Second World War had the effect of increasing home consumption, as well as allowing further repatriation of soldiers, this time near Loxton and Renmark in South Australia, and Robinvale in Victoria—and once again, the vineyard sites were chosen because of potential yield and convenience.

With wine shortages after the war, the big wine companies took the opportunity to sell under their own labels. What started as a foot-hold became a strangle-hold viewed from the position of the small vineyard and the development of regional styles. With these large companies that still dominate Australian wine came the development of multi-area blending and the formulation of 'cellar-styles'.

Large companies, such as Penfold's, Lindeman's and Hardy's, had access to vineyards, in some instances, thousands of kilometres apart. In wine terms, they recognized the importance of flavour balance and harmony over regional character. It was not only common but considered quite good sense to blend the wines of these regions—not

only did the companies buy wine from growers in different areas, but they swapped wine among themselves.

The 1950s saw the beginning of a wine renaissance in Australia. Stainless steel made its appearance and the large companies began to increase their vineyard holdings. By the mid 1960s a wine boom of the sort which hit California began. At no other time in Australia's history has there been such interest and public enthusiasm for wine.

For the first time the small winemaker gained importance and generic labels were replaced by varietal ones. Not only was there a revival of established areas, but great experimentation and planting was begun to find new flavours. In white wines, Chardonnay, Traminer and, to a lesser degree, Chenin and Sauvignon Blanc have become established, along with the traditional Rhine Riesling and Sémillon. For reds, Cabernet Sauvignon was introduced to areas formerly dominated by Shiraz. As exciting as that enthusiasm which ran into the 1970s was, in quality terms it now seems to have done little more than establish the old vine-growing areas where selectivity and tradition of winemaking produced unique regional flavours.

The real development in Australian wine is only just beginning to take place. The cool climates of southern and higher areas such as in Tasmania, the Margaret River in the West and, a century after phylloxera devastation, the Yarra Valley in Victoria have just begun to produce wines which demand the attention of the traditionalist. However, experimentation is still enormous and the production of the new areas is miniscule. It may be that the real age of Australian wine is yet to come.

—generally above average. Reds are sound, regarded as good, with some very good wines produced.

1979 Very hot January resulted in wines regarded as better in their flavours than the 1978s although the acids tended to be lower. Eden Valley, less affected, produced the best wines of the vintage.

1980 Very cool in January resulting in well-balanced wines with excellent acids. Both red and white were regarded as excellent.

Milawa

1970 Reasonable pre-vintage conditions; wettish vintage. Reds light, firm wines which filled out with bottle age. The whites are elegant.

1971 Hot and dry before and during vintage, which produced red wines with full styles tending towards a

'porty' character, and whites which were big and broad, lacking elegance.

1972 Virtually identical with the 1971 vintage.

1973 A very dry year with low yields. Reds—full, but not jammy; whites—full bodied.

1974 Very wet year. Both red and white were light.

1975 An even wetter year than 1974. Red and white lighter and thinner than the 1974s.

1976 Good rainfall produced very good yields. Reds—dark in colour, attractive and fruity; whites—fruity and flavoursome.

1977 The 'storm' year at Milawa, almost everything was lost. What wine was made was straight-forward.

1978 Long dry vintage. The red wines medium bodied, elegant with attractive fruit.

Whites flavoursome and showing finesse.

1979 Dry year, reds and whites similar to 1978s.

1980 Good vintage with sound fruit and very good yields. Both reds and whites are rich in flavour and not broad or jammy.

Hunter Valley

1970 Good year. A dry vintage produced full-flavoured reds, some very good styles, but with a tendency towards bigness. Whites rich in character and reasonably soft in acid but generally regarded as excellent though not long lived.

1971 Wet—a disastrous year.

1972 Sound vintage, good straightforward reds with depth of rich soft fruit but showing their age. Whites

better—soft and rich with reasonable acid.

1973 Quite a warm vintage with scattered rainfall. Big reds showing pronounced regional character. The whites were big and rich which filled out quickly in the bottle, but are still drinking well.

1974 Generally wet but cool causing some variability in reds because of late rain. Red lighter styles, developed quickly as they lacked tannin to have any staying power; whites much better, intensely flavoured, some outstanding.

1975 Generally a dry year with some rain late in vintage. Reds —very good, balanced with considerable depth of flavour. Whites, initially underrated, have developed into elegant well-flavoured wines.

1976 A fair bit of rain during vintage which affected reds more than whites, with a warm growing season. Reds

were soft and round with great depth of flavour, though developing quickly now. Whites full and rich.

1977 Dry spring, hot summer leading to some rain during vintage. Reds were strong and well-balanced. Whites a little light and lacking intensity.

1978 Dry spring but some rain fell during the early vintage; yields below average. Reds showed some variation in styles, broad to firm but not outstanding. Whites, medium bodied and filled out quickly.

1979 A dry spring and mostly dry vintage. The reds were big tending towards the firmer styles. Whites big with a tendency towards broadness.

1980 Perfect spring growing season followed right through to vintage above average yields. Reds were balanced, elegant and soft. Full-flavoured but elegant whites.

CHE ROTHBURY ESTATE

The Rothbury Estate quickly established itself as one of the finest producers of wine in the Hunter Valley. It is above all renowned for the superb dry white wines made from the Sémillon grape - rich and powerful. They have a great capacity to age in bottle. The reds show promise for the future, and include Hermitage (Shiraz), Cabernet Sauvignon and Pinot Noir.

Although any form of *appellation contrôlée* has taken many years to come into being in Australia, there is now some progress in that direction. Small, insular regions such as Mudgee in New South Wales and the Margaret River in the south-west of Western Australia, have found it easier to proceed with some form of regulation, but the movement is not widespread as yet. However, the New South Wales Pure Food Act stipulates that any named varietal shall contain 80 per cent of that variety, and there are other names which have come into common, and apparently self-regulated, use.

One such name is 'Estate', taken to mean that the produce of the company must come from its own, local vineyards. The Rothbury Estate in the lower Hunter Valley district of New South Wales, the oldest existing vineyard area in Australia, is one of the largest of these, having approximately 350 hectares under vine. The major varieties planted are Hermitage (Shiraz), Cabernet Sauvignon, Sémillon and Chardonnay. From these some distinctive wines are made under the various vineyard names, and these may be single varietals or blends of two varieties, depending on the year.

However, all the varieties are made separately from each vineyard block, before being put together by the tasting committee. Each year the most distinctive of the white and red parcels are kept to one side, and, if they mature successfully, are then sold as 'Individual Paddock' wines. Among the reds there have been some fine examples of both the locally named Hermitage and also Cabernet Sauvignon. Some excellent Pinot Noir examples are beginning to appear, and it could well be that this variety will one day achieve 'Individual Paddock' status. But it is the success of the white wines which has been responsible for the prestige of the Rothbury operation, and of these none is better than the best produced from the Sémillon variety.

As most devotees of the grape are aware, this variety is not regarded as one of the great classics. Though very important as a basis for dry wines in Bordeaux and as a component of the wonderful sweet wines of Sauternes and Barsac, and grown freely in other countries, it has rarely reached the heights of Riesling or Chardonnay. However, the Hunter Valley can be said to produce quite the best examples of the variety in the world, in top years producing a wine which, with bottle maturation, provides a uniquely flavoured non-European style. In the 1970s, with their Individual Paddock whites made from the Sémillon variety, the Rothbury Estate appeared to produce a consistent range of great wine.

There are some interesting reasons for this. The microclimate of the lower Hunter Valley favours that absolute common factor of all the major quality wine areas of the world—a slow ripening cycle, during which fine flavour and sugar gradually build in the grape, while acid levels fall even more gradually.

Theoretically, the area is too near the Equator to achieve this cycle, but a unique feature, a cloud cover from the Hunter basin which appears almost daily throughout summer, provides the means by which this is achieved. In very hot years, the reds prosper, the acidity factor being very stable, with the sugar content rarely rising above 12° Baumé. But these years tend to produce heavy, full whites of deep flavour which may lack finesse.

Wetter vintages may arise because the area is subject to the tail-end of cyclonic patterns in January and February. These may produce rot which results in whites being picked quickly and not fully mature. The ideal situation for the white appears to be one of spring rain, followed by a cool, dry summer with only inter ittent showers. Such years were 1972, 1974, 1976, 1978. 1973 was a year which produced whites which matured early. A good red year was 1975, and the whites were 'sleepers'—big, alcoholic wines which developed surprisingly well. The year 1979 was dry and produced both white and reds of high quality. But the even years and the 1979 wine showed that exceptional whites can be made from the Sémillon grape.

These wines have a duality. When young, say in the first year or two of their bottle life, they are fresh, lively, clean, well balanced, and lightly fruity—altogether a most enjoyable refreshing drink. Then they suffer an intermediate phase as they start to mature, being neither fresh and lively, nor developed. After a period of bottle age which may vary from three to six years, they begin to show their true quality, and a further period of development or maturation may last for up to 20 years. It is possible to enjoy old Hunter whites, of great flavour and vitality, which are over 30 years old, while the 1972 Rothbury, for example, showed no signs of reaching its peak after eight years in the bottle.

The colour of the matured wine is a deep, rich golden green. The latter colour is an essential part of the whole and, if it is not present, the wine invariably lacks refinement. The nose is estery with a volatile lift which has nothing to do with acetic acid. 'Burnt toast', 'vanilla pods', 'honeydew' are terms used to describe the bouquet, which deepens and intensifies with further age. This element carries onto the tongue, and the flavour swells in the mouth. One of the features of these wines is their rich, mouth-filling flavour, the best ones never showing a sign of flabbiness or coarseness. This fullness is emphasized by softness and roundness, and the finish is long and sustained.

There are many different soils in the Hunter with seven major ones, and the best Individual Paddock whites come from the worst of them, a mixture of loam and sand on a clay base. This land looked so bad the local viticultural officer recommended

Rothbury's 'Individual Paddock' wines are made from grapes grown in selected parcels of vines. Rothbury makes excellent white wines, especially from the Sémillon variety.

Rothbury made its first vineyard plantings at Pokolbin in 1968. The estate has a reputation for producing wines of very high quality. The main varieties grown are Hermitage (Shiraz), Cabernet Sauvignon, Sémillon, and Chardonnay. The estate produces both single varietals and blends of two varieties, depending on the year. Behind the vineyards (BELOW) is the Brokenback Range of mountains.

One of the assistant winemakers at Rothbury is examining a glass of Sémillon for clarity (LEFT). This white variety established Rothbury as a maker of fine wines. Rothbury's Sémillon ages well and is superb after about five years.

The young white wines in these stainless steel tanks (LEFT) are being film cooled during fermentation.

not planting it, and, even when mature, the vines are poorly wooded and lack vigour. The yield is light, approximately 3.5 or 4 tonnes per hectare, and rarely rising to 4.5 tonnes. The vineyards are close to the winery, and the grapes can be delivered to the crusher within 30 minutes of being picked. An Amos mill crushes and de-stems the grapes, which are then drained in enclosed Buchot-Guyer presses filled with carbon dioxide. The juice is chilled and settled overnight, a process called 'debourbaging'. A surprising amount of solids is removed in this way, the clean must being racked off in the morning. This is centrifuged for further removal of solids and inoculated with cultured yeast, strain 729 (*saccharomyces cereviseae*). Fermentation takes place in enclosed stainless steel vessels under a controlled temperature of between 12°C and 15°C, and lasts about three weeks. Sulphur treatment is kept to a minimum. Cold stabilization takes place about three months after fermentation and considerable tartrate is thrown out of suspension. The final chemical analysis of the 1979 Individual Paddock serves as an example of proportions considered ideal—5.4 grams per litre tartaric acid, a 3.2 pH level, a reducing sugar of 1.4 grams per litre, alcohol 11.65 per cent by volume, and 138 parts per million total sulphur.

During the 1970s the winemaker was Gerry Sissingh, a Dutchman from a wine family who was trained as an oenologist in Australia by Lindemans, which may explain some of the similarities between the respective Hunter whites produced by Lindemans and Rothbury. From 1980, the winemaking team was under the charge of Murray Tyrrell, one of the founders of the Rothbury Estate, famed for his own Hunter wines which appear under the Tyrrell label.

The Great Cask Hall (LEFT) is used for tastings and other functions. The winery at Rothbury has won architectural awards and has very modern equipment.

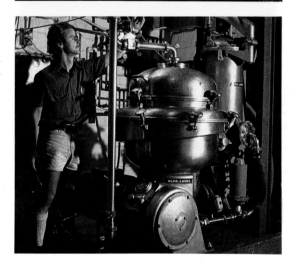

The Sémillon juice is cooled and then the solids are removed by high speed centrifugation (ABOVE). This method results in the wine retaining full varietal and regional characteristics.

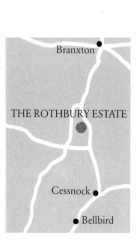

Branxton

THE ROTHBURY ESTATE

Cessnock

Bellbird

The Rothbury estate is in the Hunter Valley, one of Australia's main winemaking areas.

TYRRELL'S VINEYARDS

Tyrrell's is a mighty name in the Hunter Valley. Their Chardonnay is on a level with the best the world can produce from this grape variety, capable of ageing beautifully. The Pinot Noir has astonished by its sheer class, and Sémillon, Sauvignon Blanc, Traminer and Cabernet are other wines to look for.

One of the most famous of all Australian wine districts, and the oldest still in existence, is the lower Hunter Valley, centred on the shires of Rothbury and Pokolbin. Of the old, traditional family companies not taken over or sold, only Drayton's, Elliott's and Tyrrell's remain. The last is certainly the most famous, having developed remarkable winemaking abilities coupled with a genius for the promotion of an image of rugged simplicity. The dirt floors and corrugated iron roofs of the winery remain. 'We're a winery not a refinery,' they say to those who inquire about this.

Murray Tyrrell is at present in charge of the vineyard. Some 20 years ago Murray was desperate to sell his wines in bottle instead of bulk and, by his own confession, 'was still trying to make drinkable soft white wines.'

In 1970, he became devoted to the top red and white styles of Burgundy, and decided to develop a programme of Chardonnay planting, for he was convinced the variety would do extremely well in the district. There were already some clones of Chardonnay available in the valley, which were not appreciated by the growers, and new clones were being imported.

Murray obtained selections, and various plantings were made over the years. In 1973 he presented the first white made from that famous grape and it was hailed throughout Australia. Rarely has any wine achieved such instant recognition. The 1973 wine won various prizes, including many gold medals and several special trophies for Best Dry White, Best Young Wine, Best Table Wine of Show.

Since those days the style of the wine has developed, and particularly good wines were made in 1976 and 1979. Other sources of material also became available from local growers or people whose properties were managed by the Tyrrells, and there are now different layers of the variety available at different prices.

The top wine, the one which is selected to be shown, remains among the outstanding examples of Chardonnay available in Australia, even though there are now many different companies producing the variety from many different areas. The Hunter Valley still wins the majority of prizes awarded to examples of the variety and, of those, Tyrrell's wins more than most.

The top wines at Tyrrell's are grown on red podozolic soils which have a mixture of limestone, clay and shale underneath. It is a difficult variety to grow, because, like Pinot Noir, it has a thin skin and is very susceptible to oidium and downy mildew. Rigorous spraying programmes have to be maintained to guard against these diseases. In addition, the vines are framed on vertical trellises and the fruit so spaced that maximum draught from the wind is obtained, thus further reducing the risk of the two diseases. However, the vintage in the Hunter is generally accompanied by rain, which is a fringe effect of the cyclonic patterns which harass parts of Queensland 1,000 miles to the north. There is often a touch of mould present in the grapes, and Murray Tyrrell prefers to pick grapes in this condition, for it acts as a balance to what is sometimes called a 'sunburnt' character. The sugar level is between 12° and 13.5° Baumé, the latter preferred, as long as balancing total acidity is between 7.2 to 7.5 grams per litre. This may sometimes be lower in occasional hot, drought years.

The must is settled overnight, then racked off heavy solids into old French puncheons—casks with a capacity of 100 gallons—for initial fermentation. This takes between eight and 10 days at 15°C until only 3° to 4° Baumé remains in the wine.

Final fermentation of this remaining sugar may take another three or four weeks and this takes place in new French oak puncheons in a cool part of the cellar. The wine remains in these new casks for approximately three months, before being racked off into one year old French oak casks for five or six months. Then it is back to new oak for a final two or three months so that the wine picks up new, strong oak flavours. At this stage there appears to be too much oak, though the Chardonnay fruit is strong. Sometimes an element of residual sugar, no more than four grams per litre, remains to swell further and augment the fruit, adding to the wine's character.

A characteristic of all the best Hunter whites is that they grow and mature in the bottle. The fruit softens and broadens, and a most generous, rich, mouth-filling flavour evolves. Of course, the oak extraction factor remains constant, and the fruit grows to surround it completely. When the fruit and oak are totally integrated, neither flavour is dominant and the whole impression is one of extreme richness. However, it is fair to state that when the wine is young, the oak dominance is often disconcerting, and sometimes the wine is marked down at shows by judges who are not completely familiar with the development potential of the area which transmits to the style. Though the balancing acidity in the wine can be quite high—up to 7 grams per litre—it appears to be lower because of the volume of flavour apparent.

When the wine is mature, which varies between four to 10 years of bottle age depending on the vintage, the nose is huge, with an extremely rich, straightforward bouquet which is the result of the developed fruit and oak integration. This carries onto the palate. On the tongue the flavour fills out. Although the style may lack refinement or extreme finesse, the superbly satisfying flavour more than compensates for this.

The style of the wine is more that of Puligny and Chassagne than of Chablis, more like the richer recent Napa wines than the lighter Sonomas. However, it remains

1979 was a good vintage for Tyrrell's Chardonnay (TOP), one of Australia's premier dry white wines. Although best known for Chardonnay Tyrrell's also produces Pinot Noir (ABOVE).

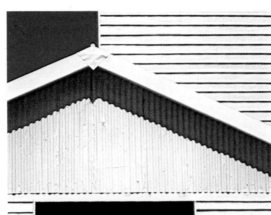

Murray Tyrrell (BELOW) *is one of the pioneers of fine wine-making in the Hunter Valley, particularly of wines from noble grape varieties such as the Chardonnay and Pinot Noir.*

a particular style of the Hunter, and indeed a particular style of Murray Tyrrell.

However, a certain flatness sometimes occurs on the back of the palate which might be overcome by the addition of, say, from 10 to 20 per cent of the more estery Sémillon of the area. Increasingly, the Chardonnay variety in the district appears to achieve maturity quicker than the Sémillon.

The wine was bottled in a light green Moselle-style bottle, the one generally used for the white Tyrrell's wines. However, in the early 1980s Murray decided that a new presentation was required for his top Chardonnay. This is sure to make the same impact as did the advent of the Tyrrell's Chardonnay in the 1970s, which continues to have enormous effect in Australia.

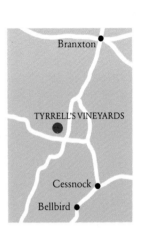

The lower Hunter Valley is one of the most renowned of Australia's winemaking areas. Although Tyrrell buy in wine from other growers, Tyrrell has 4 hectares planted with Chardonnay and 2.8 hectares planted with Pinot Noir. The Tyrrell family have owned vineyards (FAR LEFT) *at Pokolbin since 1858.*

The cap is pushed down into the fermenting must (LEFT), *so that contact is maintained between the juice and skins to help produce a good colour in the wine.*

Large barrels (LEFT) *are used for initial fermentation of the Chardonnay. This variety is traditionally fermented in oak. Aged oak is used as new oak would give the wine too strong a flavour.*

The skins on most of the Pinot Noir grapes are broken before fermentation. Some whole grapes with unbroken skins are added (ABOVE). *The weight of the whole grapes on the top induces intercellular fermentation in the grapes as the juice runs* *from the grapes at the bottom. During this process of partial carbonic maceration, carbon dioxide is released. Full carbonic maceration tends to produce light, fruity wines, but after partial carbonic fermentation the wine retains more body.*

ORLANDO

Orlando make an impressive array of wines, the result of some really technical winemaking. They were among the first to see the potential of late-picking on the Rhine Riesling, and the results have been some amazing Auslese wines, with the finesse and fruit of this distinguished grape variety combined with great, rich flavour on the palate.

The six acre Steingarten vineyard (BELOW) was established in the early 1960s. It is at an altitude of over 500 metres on a mountain behind Rowland Flat. The soil is so rocky that stone hammers had to be used to break up the soil.

Orlando have long been to the forefront in the development of the white table wine industry. They were the first to use imported pressure fermentation tanks in 1953, and wines made then are still holding up remarkably well. They were early with their late-picked styles, and their Green Ribbon Riesling has always been among the best of that type.

Their experiments with a sweeter style, the Auslese, started in the early 1960s, the first being the 1964. It was produced from Rhine Riesling grown on the floor of the Barossa Valley near Rowlands Flat. The soil is clay loam over limestone and clay. Sometimes the grapes are left on the vine up to 12 weeks after vintage. They become raisin like rather than botrytis infected, with sugar levels sometimes rising to 17.5° Baumé. However, the sugar is balanced by a high concentration of acid due to dehydration although an Indian summer is needed for this to occur.

After 1964, no Auslese appeared until 1971. Since then, there have been several vintages– in 1972, 1974, 1976, 1977, 1978 and 1979. It is possible that increased knowledge of the style as well as development of yeast cultures, juice clarification and 'cold' fermentation techniques, enable the makers to handle the material to greater effect.

The wine is made to finish quite sweet, but with full Rhine Riesling character and a fresh acidity. It is bottled fairly early, and benefits greatly from some bottle age—the 1974 drinks remarkably well in 1980, for example. At the time, the nose was very full and voluminous with a luscious, ripe character to it. This carried through onto the tongue, and again the characteristic of many top Australian wines—a spreading, swelling mouth-filling flavour—dominates. The flavour stays in the mouth, and with no 'dip' at the end of the palate. The finish is clean but does not cut off sharply.

Orlando has an excellent reputation for its German-style white wines, especially the sweeter Auslese quality. As in Germany itself, 1976 was a good year.

The winemaker at Orlando is Gunter Prass (ABOVE). He is seen in the cellars. The Barossa Valley was an area of early German settlement.

These fermentation tanks (BELOW) are used for cold fermentation of white wines. The tanks are jacketed to achieve this type of fermentation. Cold fermentation of white wine is one of the ways in which Orlando has helped establish Australia as a producer of quality white wines. Fermentation tanks for red juice (BELOW) have no jackets.

Orlando uses much modern technical equipment in all stages of their winemaking. Here the white wine is being bottled (BELOW).

  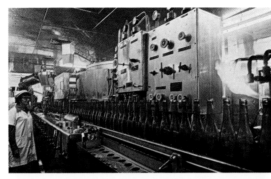

This rugged scenery (LEFT) is typical of the Barossa Valley. The area has a natural water supply. The climate of the area is unreliable but its relatively cool temperatures are conducive to good acid retention which helps produce wines with both finesse and delicacy.

197

PENFOLD WINES

Penfold Wines make wines of almost every colour and type, but their world class contribution is Grange Hermitage, an amazing, mysteriously produced red wine of the utmost interest. The intensity and depth of this wine, its range of flavours and dimensions, and its majestic capacity to age defy even the most literary of wine tasters.

The famous Australian dry red wine, Penfolds Grange Hermitage, was created in the early 1950s by Max Schubert, aided by his manager, Jeffrey Penfold-Hyland, a direct descendant of Dr Rawson Penfold, the Englishman who founded the great company of Penfolds in 1844. Dr Penfold had a cottage and vineyard in the foothills of Adelaide in South Australia which he called Grange. The cottage remains as a museum. Although the vineyards are now surrounded by houses, it is unlikely that they will be lost in the foreseeable future because, apart from any other consideration, they have given their name to the most celebrated of all Australian red wines.

This was not always so. When the style first emerged, it was subjected to ridicule, insult and degradation. Very few wine experts supported the evolution of the style. Letters were written to the company board, and unfavourable publicity appeared. The reason was the total departure from any other style. The major feature of Grange was the use of small new wood. This practice, widely known throughout France, is mandatory in the great châteaux of Bordeaux. Yet its use was almost unknown in Australia. Schubert reasoned that he could add another dimension of flavour to his wines with the use of new hogsheads. He found that the fuller, more pungent American oak was better suited to the full red wines he made. In fact, he wanted to make a bigger wine, yet one which did not contain too much tannin, in order to carry the characteristic full oak flavour.

From the outset, he wanted his wines to be made so that they would last and improve for 20 years or more. The early Granges were—and are—wonderful wines. Though rarities, they may still be enjoyed with the greatest pleasure. The celebrated 1955 wine still emerges in museum classes at shows, and almost always wins yet another top prize. For the next five years there was a lull. Management had taken notice of the almost incessant criticism, and forbade the yearly purchase of new oak. Fortunately, at the end of the decade, there was a change of board direction, and since 1960, Grange Hermitage has been the almost undisputed leading red wine in Australia. It commands the highest release price each year and at times is even issued on allocation or quota.

Although Max Schubert issued carefully phrased statements about the making of the style, most serious students of winemaking processes are fairly sure there is more to it, and that the whole picture has not yet emerged. It is known that most of the new material comes from the Grange vineyard itself and the Kalimna vineyard situated on the northeastern perimeter of the Barossa Valley. However, it appears there may also be other sources, perhaps the Paradise suburb of Adelaide or the company's vineyards at Coonawarra to the south. Although the

wine was named Hermitage, it is known that Cabernet Sauvignon was used 'in small proportions as a balancing factor.' The details remain uncertain.

There is also some mystery about the winemaking techniques. It is known that the sugar levels of the grapes used are between 11.5° and 12.5° Baumé and that the total acidity is between 6.5 and 7.5 grams per litre, prior to the beginning of processing.

Skin contact is important during fermentation, which lasts up to 14 days before the latter part of the fermentation takes place, off skins, in new oak hogsheads. A graph is made of the reducing sugar level and length of fermentation, a line being drawn between the two. The sugar must reduce at a constant rate, never departing from the straight line between original sugar level and end of fermentation. This is achieved by temperature control, and is said to be one of the secrets of the style. However, the spectacularly black, purple and red colour of Grange would seem to indicate that the pressings material is also used. Yet there is none of the tannin, bitterness or hardness associated with such material. It is possible that use is made of pre-pressings material. The normal first drainings are taken from the completed fermentation, then, prior to the skins being pressed, there is an overnight settling, a sort of gentle percolation. In the morning this may result in as much as 15 centimetres of dark wine. This is run off. Obviously, a lot of other red wines are being made to allow that small amount of special, dark, yet unpressed, soft wine to come from each cask. If some of this is added to the Hermitage, it could explain why the wine has an element of acetic acid volatility. However, this remains largely a matter of conjecture.

Once made, the wine has an incredibly deep, black-red colour. It is matured in new oak for 18 months, and then bottled, being held three to five years before release. Of all the Granges, the wines of 1952, 1953, 1955, 1960, 1962, 1963, 1966, 1967, 1968, 1970, 1971 and 1972 were outstanding. Some suffered from a noticeable degree of acetic acid volatility, but this fault, so pounced upon by Australian wine judges, seemed to matter less, such was the power of the fruit and oak flavours.

Of all the Granges, the 1962 is the most famous. It is said that it contains a fair amount of Cabernet Sauvignon and

that much of the material came from Coonawarra. However, it won 30 championship trophies, 123 gold medals and 68 silver medals in major Australian wine shows, an outstanding record which is probably unlikely ever to be equalled.

Although the information available about the winemaking process is sketchy and much is conjecture, the fact remains that Max Schubert has evolved a unique style of wine, one quite different from anything else in the world and, particularly, one which

The 1955 Penfolds Grange Hermitage is still a celebrated wine. Since the early 1960s Grange Hermitage has undisputedly been the leading red wine made in Australia.

Max Schubert (BELOW), is
the chief winemaker at
Penfolds. He created the
famous Penfolds Grange
Hermitage in the early 1950s.
Behind him are some of the
small barrels made from
American oak which are
important for this wine.

The headquarters of
Penfold are near Adelaide in
South Australia. Many of the
company's vineyards are in the
Barossa Valley at Kalimna,
Nuriootpa and Eden Valley.

was completely different from anything else made in
Australia. There have been many copyists since, yet none
have equalled Grange quality. Whatever the blend of
grapes, wherever the origin of the material, however it
was and is made, Grange is unique, a tribute to the vision
of one man.

Max Schubert (BELOW), is
the chief winemaker at
Penfolds. He created the
famous Penfolds Grange
Hermitage in the early 1950s.
Behind him are some of the
small barrels made from
American oak which are
important for this wine.

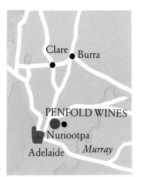

Grange Hermitage receives
18 months' maturation in
small wood before being
bottled, and over three years'
bottle maturation before being
sold. The bottles are kept in
wooden bins in the bottling
warehouse (LEFT). Grange
Hermitage can be drunk when
it is first sold, but reaches its
peak after about another
10 years.

Penfolds Winery is one of the
largest vineyard owners and
winemakers in Australia.
These large stainless steel
tanks (CENTRE) are at
Tanunda in the Barossa
Valley. Fermentation takes
place in 6,000 gallon stainless
steel tanks (LEFT).

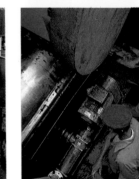

White grapes are put into the
top of the crusher (CENTRE).
After crushing, the juice
comes out of the bottom part
of the machine (LEFT).
Penfolds produces a wide
variety of wines and wine
styles.

The famous Grange vine-
yard (TOP) is very close to the
city of Adelaide. The
Kalimna vineyard (ABOVE) is
at Nuriootpa in the Barossa
Valley.

THE PETALUMA WINERY

Petaluma is a comparative newcomer to the Australian wine scene, and its remarkable, flowery Rhine Riesling can at first startle by its novel taste, then delight by its finesse, length and aromatic qualities. The Petaluma Chardonnay has style and discretion, needing bottle age to fulfil its character potential.

The Petaluma Rhine Riesling vineyard is at Clare in the northern part of South Australia. The town of Clare is 145 kilometres north of Adelaide at a latitude of 34°S and at an altitude of 450 metres. The Clare valley has long been renowned for the quality of its fruit, and wine grapes, because of the flavour conferred by cool nights during the ripening period.

The Petaluma vineyard is the highest in the area, at over 500 metres above sea level. It is on the side of a steep, barren hill which faces south-west, so that the sun does not creep onto the lowest vines until 10.00 am, even in the lowest night temperatures in South Australia. This combination of aspect and low night temperatures results in the cool ripening conditions for the Rhine Riesling grapes of the Petaluma vineyard.

Less than two centimetres of rain falls on the vineyard on average during the growing season, and the vines rely heavily on soil storage of winter rain to help them through the dry summer. Because of the shallow skeletal nature of the shale-based soils of the Petaluma vineyard, the only way sufficient soil moisture storage could be provided for each vine was to plant the vines with extremely wide spacing—3.5 metres between rows, 2.5 metres between plants, giving only 800 vines per hectare compared to the Australian norm of 1,500 vines per hectare.

Each vine on the Petaluma vineyard has a large soil volume from which to obtain scarce nutrients and moisture. The resulting vines are very hardy, but meagre, bearing small bunches of berried fruit, often showing lesions on the skins of the berries due to sun exposure and the limited available moisture. These vines hang onto their leaves and keep functioning under most difficult climatic conditions and compensate for their lack of physical beauty by imparting to the fruit abundant aromatic Riesling flavour.

The development of this aromatic flavour is the criterion by which picking time is decided. Samples taken weekly, in the early morning cold of the vineyard, are scientifically representative of each of the small subsections of the vineyard. These samples are processed under refrigerated conditions to give clear green juice samples which are assessed for flavour development by the winemaker. Optimal flavour develops quite late in the season, and picking of the Petaluma vineyard usually begins when most others in the area have finished.

The grapes are harvested by hand and carefully placed in the specially designed containers for the 140 kilometres journey to Piccadilly, a township at the highest point of the Mount Lofty Ranges, close to Adelaide, where they are delivered to the Petaluma winery. The grapes are gently crushed by a roller and the must is immediately chilled to 5°C after crushing. The must is pumped into a Willmes Tank Press filled with carbon dioxide, and is recombined with the stalks. Cold juice is run away from the press by gravity to insulated, refrigerated settling tanks, where clarification occurs by natural settling at −2°C for a period of five weeks.

At the end of the vintage, each tank of clear wine is carefully separated from the solid material at the bottom of the tank and is stored at −2°C until its turn for fermentation. The juices are carefully evaluated for flavour and only the best tanks are selected for Petaluma. The juices are warmed to 10°C and inoculated with a pure culture of Petaluma's own yeast strain, selected from the Clare vineyard. Fermentation proceeds for 30 days at this low temperature in a sealed tank.

When all of the sugar has been consumed, the wine, turgid with yeast, is chilled again in its fermentation tank to −2°C. Natural settling allows a clear wine to be separated from the yeast solids and this is then ready for immediate bottling. The care exercised in this process, and the use of natural settling over a period of time under very cold conditions in a carbon dioxide atmosphere means that the wine has a floral, aromatic and grapey flavour, very reminiscent of the fresh fruit off the vine. Time is the added element required to bring the wine to its peak of quality, complexity being conferred by chemical changes which are catalyzed by time in the bottle.

The late picking of the Rhine Riesling, together with the warmth of the vineyard on summer afternoons, result in a level of alcohol of between 12 and 12.5 per cent by volume, a pH level of 3.1, a total acidity of 7.0 grams per litre, and only 100 parts per million total sulphur. The cold nights of the vineyard protect the grape's natural acid and these natural chemical preservatives, together with the abundance of fruit flavour, give the wine longevity.

This kind of attention to detail was the reason that one of the greatest wine successes of the 1970s was the rise of the Petaluma label, although they only made a few releases. The name—that of a town in the Sonoma Valley of California—was adopted by Brian Croser, who became attached to it while studying at Davis University. Croser, one of the most gifted and highly trained of the younger generation of Australian winemakers, returned from the United States to head the Thomas Hardy winemaking team. The 1975 Siegersdorf Riesling, a winner of many prizes, was one of his early creations.

Croser then left to found the Oenology School of the Wagga Wagga College of Advanced Education. During that time he obtained batches of fruit from various locations and his various creations included a Rhine Riesling in 1976, a Traminer in 1977, and a Chardonnay in 1977 and 1978 which received national acclaim.

In 1978, he contracted to build a winery at Piccadilly in the hills outside Adelaide which Croser considered to be ideal for the

PETALUMA

1979 RHINE RIESLING

750ml

PRODUCE OF AUSTRALIA BOTTLED AT PICCADILLY SA

The 1979 Petaluma Rhine Riesling was outstanding and has won many prizes. Petaluma in the Clare Valley is a new estate which has attained a very high reputation, particularly for its German-style white wines.

The Petaluma vineyards in the Clare Valley (BELOW) give a low yield. This is partly because the shallow and skeletal nature of the shale-based soil (BOTTOM) means that the vines have to be well spaced. Petaluma plants only 800 vines per hectare.

Brian Croser (LEFT) is the winemaker at Petaluma. He is in his laboratory in the winery at Piccadilly in the hills outside Adelaide. The winery and its equipment (BOTTOM LEFT) are very modern, having only been built in the late 1970s.

These young Riesling grapes (LEFT) are in superb condition, just approaching maturity. The 1979 Rhine Riesling vintage was outstanding.

growing of high quality Chardonnay. The 1979 vintage resulted in an outstanding Rhine Riesling from Clare, voted by many panels and magazines as the best of its variety of the year.

A partnership was then formed to consolidate the Petaluma label. The wine will never be available in other than limited quantities. Eventually the company will produce about 6,000 cases of Riesling, 5,000 cases of Chardonnay and 5,000 cases of Cabernet Sauvignon.

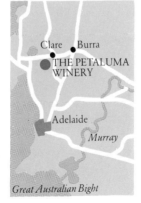

The Petaluma vineyard is in the Clare Valley in the northern part of South Australia.

BROWN BROS

Brown Bros is a family concern, with extensive vineyards at Milawa. The jewel of their collection is their remarkable Noble Riesling, which is Rhine Riesling made aristocratic by the effects of noble rot. Deep coloured and truly luscious, it is a wine of complexity, body and persistence on the palate.

Some of the most interesting sweet wines to have evolved in recent years in Australia have been late-picked Rhine Rieslings. The name Rhine is used in Australia to differentiate this type of true Riesling from Hunter Riesling, which is Sémillon, and Clare Riesling, which is Crouchen. The confusion is unfortunate, but since people in the Hunter Valley and Clare have used the misnomers for over a century, there seems to be little rush to change.

Rhine Riesling makes the finest aromatic, flowery, and fruity dry whites of Australia. Most of these come from South Australia, in particular from the Clare and Watervale districts and the Eden Valley which is among the hills on the Eastern side of the Barossa Valley. However, there are pockets of quality Rhine Riesling production elsewhere.

As demand for dry whites grew, a greater interest in sweeter whites developed. This resulted in a wide range of 'late-picked' styles. Some of the material was genuine enough, two or three extra weeks on the vine concentrating the sugar content of the grape. In other cases, concentrated grape sugar is added, and the result can be a wine of 'lolly-water' character. 'Lolly', short for 'lollipop', is the colourful Australian slang term for something which is plain and sweet in character.

Though some pleasant, authentic wines have been made, some of which—including Lindeman's, Buring's, Stanley, Yalumba and Kaiser Stuhl—have considerable character, especially when matured in bottle for a few years, two companies have consistently produced outstanding wines. One—Brown Bros in Milawa—is an isolated family company which occupies a small corner of a relatively small wine district; the other—Orlando—is one of Australia's largest, a subsidiary of one of the world's multi-national companies. Yet both companies enjoy the skills of superb winemakers, men completely dedicated to producing the finest they are capable of.

Brown Bros forms a closely knit family unit as the father, John Brown, casts a paternal eye over everything, while his son John Jnr makes the wine, another son, Peter, tends the vineyards, a third, Ross, sells the wines, and the fourth, Roger, researches and propagates new vinestocks. They make a vast range of top varietals, and have long produced the most consistent dry Rhine Rieslings in the district. But the late-picked wine of noble rot character is their triumph.

In only a few areas of Australia are weather conditions satisfactory sufficiently late in the season to allow noble rot to develop on the ripened grapes. For this reason the luscious, aromatic, sweet wines with the character produced by *botrytis cinerea* are rarely made.

One vineyard fortunate to be affected is the Milawa vineyard of Brown Bros. Noble rot was first recognized in 1934, although at that time the grapes were not handled separately. The next known occurrence was in 1962, when the first Milawa Late Picked Rhine Riesling (now known as Noble Riesling) with botrytis character was made. This has been followed to some extent in most years since 1970.

Each year an area of about 10 acres is left for harvesting late in the season in the hope that noble rot will visit. Without noble rot this area provides grapes which are used for Late Picked Rhine Riesling, a wine which is still luscious but without the special character imparted by the *botrytis cinerea* or noble rot.

Picking of normal Rhine Riesling takes place about the second week of March depending on the season, and the noble rot affected grapes about three weeks later.

During the 1978 vintage, for example, the progress of the mould was watched very carefully; it first appeared about six days before picking commenced. During the week before the mould became evident, the weather was wet and cool with noticeable humidity, giving conditions under which mould was able to develop and spread rapidly.

So extensive was the mould growth that, instead of selecting the affected bunches for picking as is the practice in Europe, bunches were removed in the few unaffected areas, after which the arduous task of gathering the sticky, highly developed bunches proceeded. So fragile were the bunches that picking went very slowly to avoid knocking grapes to the ground and losing them.

Problems did not finish in the vineyard. In the winery, the skins were found to be so broken down that they could not be separated from the juice and it was not possible to take an hydrometer reading. Calculation from the alcohol strength after fermentation indicated that the sugar was about 18° or 20° Baumé at the time of harvest. Separation was effected by adding pectic enzymes to the syrupy mush to deposit as much solid matter as possible and allow a cap to form during fermentation. It was then possible to press the remaining solids without difficulty.

Normally, fermentation lasts about three weeks, after which the wine is cold stabilized. At an early stage it shows considerable complexity and intensity of flavour. Noble Riesling is generally matured in the bottle for several years and released about five years after vintage. The late-harvested Riesling of both the 1979 and 1980 vintage were affected by noble rot. When mature, the wine has a deep golden colour, with no hint of green which, to some, has an appearance of oxidation. There is no evidence of this on the nose, which is rich and full. The flavour is intense and complex, redolent of the variety, and the finish long. The wine appears more luscious than all but the finest German *Auslesen* and it has even been jokingly referred to as 'a Sauternes made with Riesling,' and after about five years in the bottle, it is a lovely wine.

MILAWA

Special Limited Production

SPAETLESE
RHINE RIESLING

Vintage 1973

The magnificent aged Rhine Riesling bouquet sets the pattern for a full yet delicate flavour. The luscious sweetness followed by typical Rhine Riesling character is enhanced by the late harvest, giving a wine of great complexity and style.

GROWN & BOTTLED BY
Brown Bros
Milawa Vineyard
Victoria

738ml PRODUCT OF AUSTRALIA E.2644

The late harvest Rhine Rieslings produced by Brown Bros have gained the estate a high reputation. This is one of the few areas in Australia where weather conditions allow the development of noble rot.

John Brown (BELOW) *is the*
winemaker at Brown Bros.
He carries on the family
tradition of making high
quality wines. He is tasting
some of the 1976 Riesling.

The cellars (BELOW) *were*
extended during the 1950s
and again during the 1970s.
The larger casks are used for
red wines.

The vineyards at Milawa
(CENTRE) *are about*
200 metres above sea level
The climate in this area of
north-east Victoria is liable
to severe spring frosts. Al-
though noble rot is relatively
rare in Australia, these
Riesling grapes (LEFT) *have*
been affected. Both the
Spaetlese Rhine Riesling 1973
and Noble Riesling 1974
were affected by noble rot.

The first wine was made here
in 1889 and the first winery
built in 1900. The modern
equipment and technology used
at Brown Bros today are
aimed at producing as high
quality wines as possible. The
estate (LEFT) *buys in grapes*
from selected growers as well
as using their own produce.

The grapes are placed in the
hopper (ABOVE CENTRE)
when they are brought in from
the vineyards. The hopper
leads to the destemmer. In
the background is the Coq
press.

CALIFORNIA: INTRODUCTION

The wine industry in California had, in effect, two birthdates. The first was at the end of the eighteenth century, and the second after the repeal of Prohibition in 1933. The rebuilding process after Prohibition was not accomplished rapidly. Until after the Second World War, production was concentrated mainly in simple table and appetizer wines for which the bland Thompson grape, which can also be used as fresh fruit, was ideal. However, Americans returning from Europe after the War had developed a taste for more sophisticated table wines and so consumers soon began to demand higher quality home-produced wines.

Before the Second World War there were few acres devoted to fine varietals such as Chardonnay or Cabernet Sauvignon. However, in the 1960s taste and consumer demand was sufficient to create a 'wine boom', which continued into the 1970s. For example between 1970 and 1974, 200,000 acres of new vineyards were added to the existing strains of vines and, by the mid 1970s, the acreage of wine grapes exceeded that of raisin grapes for the first time since Prohibition. Demand for wine began to accelerate rapidly in the late 1960s culminating in a 25 per cent increase in 1971 alone. The number of wineries also increased rapidly. In 1966 just before the wine boom, there were 227 licensed wineries in California; by 1977 there were 352, an increase of 125. These new wineries brought with them new winemakers; many were trained in oenology, while others came from different careers.

The consumer demand for California wines of high quality could be seen in several ways. Firstly, the rate of consumption of table wine increased rapidly. Secondly, consumers rapidly accepted wines coming from new and previously untried areas. There are also an almost cult-like seeking out of wines made by new winemakers. Finally, there was an unexpected and as yet unexplained shift to the use of white wine as an aperitif.

During the 1970s California wines also gained international recognition of their quality. In 1976 a tasting took place in Paris in which five French white Burgundies were set against five California Chardonnays and five *grand cru classé* Bordeaux wines against five California Cabernet Sauvignon wines. Californian wines finished first in both categories, thus irrefutably establishing California as capable of producing quality equal to the best Europe has to offer. California has always had the potential to produce great wines. The state is 700 miles long and thus stretches across a wide range of latitudes which provides a broad diversity of microclimates in which are found soils, solar exposures, climatic conditions, and combinations of these factors, which ought to match any found in the main wine regions of Europe.

The Napa Valley is the best known wine producing area in California, dating from the mid nineteenth century. The Valley is about 35 miles long and between one and five miles wide. It is bounded on both sides by mountains of the Coast range. The fertile soil of the valley floor is of volcanic origin. The soil is, however, uneven in composition and this, together with the valley's variations in

Vintage chart
Vintage charting for California is difficult for several reasons. First, in contrast to the practice in Europe where vineyards are specialized, growers in California frequently plant several varieties in the same vineyard or at least in the same viticultural area. Second, several viticultural areas now recognized as the source of fine, even superb, wines have been producing in this century for far too short a time to have compiled a lengthy record. Third, California's entry into the world arena of truly fine wines on a broad scale is a relatively recent one, dating, perhaps, from the mid 1960s.

Napa County

1970 Despite low harvest due to bad spring weather, a superb vintage for reds, especially for Cabernet Sauvignon and Zinfandel. Whites merely average.

1971 Heavy crop but only average quality wines, both reds and whites.

1972 Except Chardonnay, overall quality below average, and almost a complete failure for Cabernet Sauvignon.

1973 Good maturity at harvest with excellent sugar/acid ratios, largest harvest ever in Napa Valley. Good to excellent red wines especially Cabernet Sauvignon and Zinfandel. Whites average, except for excellent Chardonnay.

1974 Another great year for reds, especially Cabernet Sauvignon and Zinfandel. Whites generally well above average. Excellent Chardonnay.

1975 Harvest two to three weeks late with reduced crop Above average quality reds, with some excellent wines from top vintners. Cabernets were harder, with more backbone and solid structure than the 1974s. Outstanding whites and Chardonnays.

1976 Drought produced a very small crop and juice yields well below normal. But high grape quality with excellent sugar/acid ratios. Good to excellent reds frequently with intense perfume, robustly full-body and almost concentrated flavours. Excellent Zinfandels for long ageing; some Cabernets of great intensity and good flavour. Whites also were good to excellent. Chardonnays were full and well balanced, holding well.

1977 The second year of drought required irrigation in spring to restore soil moisture; mild summer and autumn. Reds and whites of good to excellent quality. Cabernets are showing potential to the 1974s; Pinot Noirs best of the decade. Chardonnays rich and well-balanced, as were Johannisberg Rieslings.

1978 Heavy winter rain, largest ever Napa harvest, generally excellent. Superb whites, especially Chardonnay and Sauvignon Blanc. Cabernet quality again high. Reds generally appear to have the fullness of the 1974s.

1979 Lush growth and heavy yields, early ripening of late varieties. Whites generally excellent. Chardonnays of excellent balance, with potential to develop into superb wines.

1980 Coolest growing season of the decade and large crops. Chardonnays and Pinot Noirs have the potential to be among the best ever. Cabernet Sauvignon had good sugar/acid ratios, but may be a bit hard.

Sonoma County

1970 Very low crop producing superb red wines. Cabernet Sauvignons fine with balance, finesse and excellent ageing ability. Zinfandels also excellent. Good white wines.

1971 Late harvest. Below average quality red; whites only slightly better.

1972 Cool growing season meant that red wines suffered. Cabernet Sauvignon and Zinfandel were virtually no better than *vin ordinaire*; Pinot Noir only slightly better. Good whites with excellent Chardonnays.

1973 A crop of excellent quality. Cabernets and Zinfandels are particularly fine and are holding well with good fruit and acidity; Chardonnays also excellent.

1974 Ideal weather resulted in excellent quality vintage with reds better than the whites. Cabernet Sauvignons and Zinfandels the best since 1970. Main white varieties were better than average,

temperature, marine influence on climate and in terrain, provide a wide range of combinations, many of which still have to be exploited by the area's winemakers.

South of San Francisco is Monterey County, a relatively recent grape-growing and winemaking region. In 1960 there were only 35 acres of grapes growing in the county; the total today exceeds 40,000 acres, most of which were planted between 1970 and 1974. The area receives less than 10 inches of rainfall in an average year; so irrigation from the Salinas river was necessary to ensure good conditions for planting vines.

Sonoma County is viticulturally two separate regions. One is the Sonoma Valley in the southern part of the county, the second is the Russian River Valley. The 23 mile long Sonoma Valley is shaped like a long, curving funnel with a broad mouth opening into a northern extension of San Francisco Bay and the narrow end pointing towards Santa Rosa. The soils are extremely varied. Growing season temperatures are coolest near San Francisco Bay and become steadily warmer further south reaching Region III of the University of Davis classification near Santa Rosa. The microclimates of Sonoma are dominated by the Pacific Ocean, Coast range of mountains and the great interior valley of California. The mountains, up to 6,000 feet in height, act as a barrier to ocean fogs. Geologically, Sonoma is similar to Napa. Rainfall is heavy in winter. Soils range from clay loam to sandy loam, most are well drained and thus provide excellent growing conditions for grapes.

UNITED STATES – CALIFORNIA

Climatic regions of California
Region I Up to 2,500 degree-days of sunshine, conditions like those of the Rhine, Mosel and Champagne
Region II 2,501 to 3,000 degree-days, similar to Bordeaux
Region III 3,001 to 3,500 degree-days, similar to northern Italy and the Rhône Valley
Region IV 3,501 to 4,000 degree-days, similar to central Spain
Region V More than 4,000 degree-days, similar to North Africa

The University of California at Davis developed a system of classifying the state's wine-growing regions based on 'degree-days'. A 'degree-day' is a unit for calculating how long the temperature remains over 50° F between 1 April and 31 October.

---

although not as fine as the reds. Some lightly botyrized Rieslings had good balance and intriguing complexities.

1975 Cabernet Sauvignons and Zinfandels were good, but more austere than those of the previous vintage. Their firm structure should mean long life in the cellar. Equally good whites matching 1974 in quality. Johannisberg Riesling even better, especially with the late-harvest styles.

1976 Low juice yields and reduced crop. Cabernets are intense with flavour, loaded with extract and will benefit from lengthy ageing. Zinfandels have intense aroma and flavour, needing age. Excellent Chardonnays with balance and finesse. Fine Sauvignon Blancs and Johannisberg Rieslings.

1977 Despite drought, outstanding Zinfandels and Pinot

Noirs with good balance and structure which means that they will age well. Cabernets with potential to develop. Chardonnays rich and full. Johannisberg Rieslings may be the best of the decade.

1978 Chardonnays are probably the best of the decade with superb balance and richness, needing up to five years of age for full development of their potential. Sauvignon Blancs are also fine with fullness of aroma and flavour.

1979 Whites are better than reds, which are of only average quality or slightly better. Chardonnays have excellent balance and show potential to develop nicely.

1980 Early varieties ripened well and their wines may achieve excellence of quality. Late varieties, particularly Cabernet Sauvignon, were relatively high in acidity and can be expected to be a trifle hard.

Monterey County

1975 Harvest began two to three weeks late and was not completed until December. White wines were good to excellent, particularly Chardonnays, Chenin Blancs and Johannisberg Rieslings. Pinot Noirs were full-bodied, exhibiting good varietal character on the nose, but lacking a little in fruit. Cabernet Sauvignons from the warmer regions developed nicely, with medium body, good acid and sound structure.

1976 A difficult wine year in Monterey County. Production was low and quality only average for both reds and whites.

1977 Quality and yield were both high. Red wines were particularly fine, especially Pinot Noir, Zinfandel and

Cabernet Sauvignon. Above average quality whites but lacked the finesse and attractiveness of the reds.

1978 White wines were generally excellent, particularly Gewürztraminers and Chardonnays, as were Pinot Noirs, Zinfandels and Cabernet Sauvignons.

1979 A mild growing season produced heavy yield; despite September heat wave followed by a cold rainy spell, the harvest was completed two weeks early. Excellent Cabernet Sauvignons which promise to develop suppleness with adequate ageing. Excellent quality Rieslings and Chardonnays.

1980 Too early to evaluate.

Temecula

This is a very recent viticultural area in which Callaway is the main vineyard.

1977 Interesting Petite Sirah with upper Rhône Valley characteristics.

1978 Callaway produced interesting Santana and Xenos wines from dehydrated white Riesling grapes.

1979 Long harvest. Chenin Blanc has intense aroma and flavour.

1980 Early season and short harvest, no botrytis. Sauvignon Blanc developing into fruity, well-rounded wines.

FREEMARK ABBEY

The best known wine from Freemark Abbey is Edelwein. Produced from botrytized Riesling grapes when weather conditions permit, it rivals its German <u>Beerenauslese</u> and <u>Trockenbeerenauslese</u> counterparts in lusciousness and elegance of style. Cabernet Sauvignons are usually big and fruity when young. Good recent vintages include 1973, 1974, 1975, 1977 and 1978.

Freemark Abbey occupies the cellar level of an old stone building below a candle factory, gift shop and restaurant. The building dates from about 1895. The quarters are cramped even for a winery of Freemark's relatively small annual production of 23,000 cases, in spite of the fact that a separate structure to house the bottling plant, offices, laboratory, case storage and a visitors' centre was built alongside the old building in 1973. However, the winemaking equipment which is crowded in that relatively small area in the cellar is adequate for Freemark Abbey's purpose, which is to make a limited list of varietal wines very well.

The list is currently limited to only four varieties — Cabernet Sauvignon (including the label-designated single vineyard wine called Bosché), Chardonnay, Riesling and Petite Sirah. Cabernet Sauvignon and Chardonnay make up 80 per cent of their total production. Both the regular Cabernet Sauvignon and the Cabernet Bosché are blended with small amounts of Merlot. Botrytized Riesling is produced in years in which weather conditions allow for a significant vineyard infection by noble rot. A light sweet style is produced in other years. Production of Petite Sirah is limited primarily to that which can be made from about 15 tons of selected grapes purchased from the Fritz Maytag vineyard on Spring Mountain.

The site of the present Freemark Abbey was first occupied as a winery in the mid 1880s. The present stone building was constructed in the 1890s by the second owner, Antonio Forni. After several changes of ownership, it was purchased around 1939 by Albert M Ahern who changed the name to Freemark Abbey. Ahern coined the name from portions of the names of his two partners and his own nickname—Charles Freeman, Mark Foster and Abbey Ahern. The trio continued to make the wine there until the mid 1950s.

The present seven owners formed a limited partnership in the 1960s. They acquired the basement area of the old building in 1965 and reactivated the winery and the name in 1967. Three of the partners are grape growers who together own 6,000 acres of prime varietals in the eastern part of the Napa Valley near Rutherford. Freemark Abbey has the choice of the best grapes which are grown on the partners' property.

It is thus not surprising that winemaker Larry Langbehn's view is that winemaking begins in the vineyards. Therefore he spends much of his time in the vineyards owned by the partners. His concerns range from the relatively simple matter of training the vines to the more complex matching of grape varieties to soils and terrains. He is also concerned with such questions as pruning and cluster thinning.

Freemark Abbey selects grapes from six vineyards, three of which are owned by the partners Lorrie Wood, Chuck Carpy and William Jaeger. A fourth, Red Barn Ranch, is the most southerly and is owned by the winery. The Cabernet Sauvignon grapes for Freemark Abbey's famous Cabernet Bosché wine come from an independent vineyard called John Bosché's Vineyard. Freemark Abbey selects grapes from a total of approximately 850 acres.

The soils on the vineyards owned by the three vineyardist partners differ widely from one location to another. East of the Napa River, soil structures change rapidly, from deep black to red gravelly, and the different soils flow into each other with few clear demarcations between soil types. Individual plots show distinct differences, however, particularly those in the Red Barn vineyards, which consist of several different plots in physically different locations. For example, the Merlot from Chuck Carpy's ranch, about 9 acres, is harvested in two sections up to three weeks apart. This is because of a soil change from a light red soil, which easily matures fruit, to a deep rich soil on which the vines require a longer period of time to ensure that the fruit ripens.

Chardonnay grapes from the various vineyards show individual characteristics, which is indicative in a general way that soils and microclimates make a difference in grapes. These vineyards are all on the valley floor, so it is unlikely that solar exposure differs very much from one location to another.

These differences in vineyard plots result in differences in the wines produced from the grapes grown on each. Freemark Abbey's winemaker, Larry Langbehn, and Bradford Webb, the consulting oenologist and one of the partners in the winery, have noted differences and similarities between barrels of both Cabernet Sauvignon and Chardonnay in the winery. They find that blending results in wines which are more complex and better tasting than any of the individual lots. Therefore, except for the Cabernet Bosché, they prefer to blend lots to take advantage of the improvements which result from doing so.

At harvest, care is taken to ensure that only fully ripe grapes with a good balance of sugar and acid go to the winery. To this end, one of the partners, Carpy, is in charge of picking and scheduling the arrival of grapes at the winery during the harvest season.

When the grapes arrive at the winery, they are put through a Healdsburg stemmer-crusher which destems the grapes before crushing them. Straight liquid sulphur dioxide at levels of 50 to 150 parts per million, depending on the temperature and condition of the fruit, is added at the crusher to stun wild yeast and to prevent oxidative browning of the white musts. Chardonnay is crushed directly into a fermenter and allowed to remain in contact with the skins for some period of time in

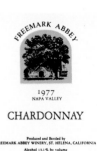

FREEMARK ABBEY

1977
NAPA VALLEY

CHARDONNAY

Produced and Bottled by
FREEMARK ABBEY WINERY, ST. HELENA, CALIFORNIA
Alcohol 13.1% by volume

The 1977 harvest produced some rich and well-balanced Chardonnays. Freemark blend different lots of grapes from various vineyards for both their Chardonnays and Cabernet Sauvignons.

order to extract phenolics and flavour from the pulp and the skins. The length of time varies with each lot and depends on the winemaker's judgement. Subsequently, the free-run juice is racked to another tank, the pomace is pressed and the press juice is blended with the free-run before fermentation is begun.

The Riesling is crushed directly to the press where free run and press juices are separated. The free-run juice is chilled to 45°F, and allowed to settle for two days before being racked off the lees and fermentation initiated. All the wines are fermented in stainless steel. Temperatures are in the range of 45°F to 60°F for the white wines and 67°F to 77°F for the reds. Inoculation to commence fermentation is made with Montrachet yeast for both white and red wines.

As fermentation proceeds, a vigorous pumping over is employed to break up the cap on the red must and to mix it thoroughly into the fermenting juice. Late in the fermentation, at low Brix or after they have gone completely dry, the reds are inoculated for malolactic fermentation. When the winemaker determines that the proper amounts of colour and tannin have been extracted, the red pomace is pressed. Free-run and press wines are kept separate.

After fermentation, the wines are clarified as soon as possible. The whites take priority. The Chardonnay must be clarified, placed in 60 gallon cooperage for wood development and bottled prior to the next vintage. The red wines are on a different schedule. They get approximately two years of barrel ageing. However, the length of time in the wood may vary; the purpose of wood ageing is to balance fruitiness with wood character. Both reds and whites are filtered and fined as necessary. Gelatin fining is used for the red wines. Bentonite and a commercial fining agent called Sparkloid are used to fine the whites. In 1979, a centrifuge was used for the first time in clarifying the white wines.

Although Freemark Abbey has experimented with the use of Limousin and American oak for ageing, at present all the 60 gallon barrels—over 1,000—are provided by a single cooper, Demptos, from the oak of a single forest, Nevers. The only additional wooden cooperage in the winery consists of three 2,300 gallon American oak tanks, which are used primarily for Riesling.

The wines are sterile bottled by a bottling machine rebuilt especially for Freemark Abbey's needs. As the bottles enter the machine, air is removed and replaced by nitrogen at atmospheric pressure. The wine is released into the bottles near the bottom, displacing the nitrogen and providing an oxygen-free fill. The head-space is filled with additional nitrogen just before the bottles are corked.

Although Freemark Abbey concentrates most of its attention on making fine Chardonnays and Cabernet Sauvignons, the winery was catapulted to national and international attention in 1975 when the Freemark 1973 Edelwein received a Grand Prize at the Los Angeles

County Fair judgings. This 1973 Edelwein was one of the first of the new wave of botrytized Rieslings to be produced in California. The grapes were heavily infected with botrytis and the must weight was in the neighbourhood of 35° Brix. The finished wine was low in alcohol, with high residual sugar. The aroma was intensely fruity, intensely Riesling, with overtones of peaches or apricots and honey. The flavour repeated the intensity of the aroma with its mouth-filling complexity. A truly elegant wine, it was easily the match of the best of the *Beerenauslesen* of Germany.

This wine prompted Freemark Abbey to change the style of their Riesling. Prior to 1975, Freemark's Rieslings had been bone-dry in the Alsatian style favoured by California winemakers before the 1970s. With the winery's success in producing Edelwein, the decision was made to discontinue production of the bone-dry style, and to make botrytized Riesling only when weather conditions permit, or a light fruity style in the years in which the noble rot does not develop.

A second wine for which Freemark Abbey is justly noted is the Cabernet Bosché. The Cabernet Sauvignon grapes for this wine are grown entirely in John Bosché's vineyard. They make a big, fruity, deeply coloured distinctive wine like Beaulieu's Private Reserve or Heitz's Martha's Vineyard, though less aggressive with distinctive notes of mint or eucalyptus. This, too, is an elegant wine.

Current production at Freemark Abbey is at its maximum capacity of 23,000 cases annually. Output could not be expanded without changing the entire operation; therefore, production will, for the foreseeable future, remain at this level. Limitations of space may place restraints upon volume, but certainly not upon quality.

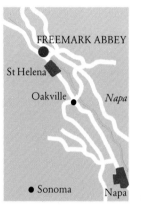

Grapes for Freemark Abbey's wines come from six vineyards. Freemark Abbey is based at St Helena. Their renowned Bosché vineyard is in the central part of the Napa Valley.

Freemark's winemaker is Larry Langbehn (LEFT). He is checking the apparatus which automatically monitors the sulphur dioxide level in the fermenting must.
He attaches great importance to matching grape varieties and soil conditions.

CHATEAU MONTELENA

Production is concentrated primarily on two varieties with smaller amounts of two others. Cabernet Sauvignons have been rich and full with excellent ageing potential. Chardonnays have shown restrained varietal fruitiness and have generally developed full body with bottle age. Zinfandels have been lighter in style and, showing considerable fruit, have been ready to drink when young, like Beaujolais.

Chateau Montelena is the most romantic in appearance of all the new wineries which have begun producing in the Napa Valley since 1970. The building, which resembles a medieval castle, was built as long ago as 1882.

Imported cuttings were planted in the vineyards around the building and wine-making began there in 1886. After Prohibition began in 1919, the winery lay in deteriorating idleness until it was reactivated in 1972 by Lee Paschich, James L Barrett and Ernest W Hahn. In the interval, the building and grounds had been owned by a Chinese couple, the Yort Franks, who created the present spectacular setting—a five acre lake, surrounding four small islands, on one of which is a Chinese style gazebo, and three of which are connected with the others by zig-zag bridges which skim just above the surface of the water. The scene is completed by a genuine Chinese junk.

The three partners invited Milenko Grgich to join them as winemaker. Born into a large Croatian winemaking family, Grgich was trained in oenology at the University of Zagreb in Yugoslavia. He emigrated to California in 1958, and worked at various vineyards. In 1976, Milenko Grgich and Chateau Montelena shot to international attention when his 1973 Chardonnay finished top in the first Paris comparative tasting of Californian and French wines. In 1977, Grgich left Montelena to open his own winery. His successor was Jerry Luper, whose 1973 Edelwein, a late-harvest Riesling, produced at Freemark Abbey, had won the Sweepstakes award in the Los Angeles County Fair judgings in 1975. Luper is an oenology graduate from Fresno State University.

Chateau Montelena's production is concentrated primarily on only two varieties—Cabernet Sauvignon and Chardonnay. The annual crush is approximately 325 tons of Cabernet (45 per cent) and Chardonnay (55 per cent). In addition, about 50 tons of Zinfandel are crushed each year, and smaller amounts (25 tons) of Riesling have been vinified in the past.

Montelena aims to produce mainly high quality Cabernet Sauvignon from grapes grown in their own vineyards. There are 92 acres of vineyards at the Chateau. In 1978, approximately 135 tons of Cabernet Sauvignon were produced. In 1979, the 60 acres which have been planted to Cabernet Sauvignon produced only 90 tons of grapes, an unusually low yield. When vines are mature, 3 tons per acre or 180 tons per year are expected. Of the 92 acres, 10 have been planted to Zinfandel and the balance (22 acres) is being planted to the Chardonnay variety.

Although the location of Chateau Montelena, in a Region III at the foot of Mount St Helena north of Calistoga, is not generally considered to be the best area in the Napa Valley to plant Chardonnay, steps are being taken to reduce the effective temperatures in the vineyard during the growing season.

The Chardonnay being planted in Montelena's vineyards is the newest Chardonnay clone developed by the University of California at Davis. It is certified virus-free, and is expected, on the basis of what is known of the clone, to produce grapes of high varietal character in spite of a projected yield of 4 tons per acre and in spite of the relatively warm climate.

Chateau Montelena will continue to buy Chardonnay grapes from vineyards of proven quality in the southern part of the Napa Valley. Cabernet Sauvignon, Zinfandel and some Johannisberg Riesling grapes are also purchased; however, the volume of purchases has been declining as Montelena's own vineyards have started to produce.

Ideally, the winery would prefer to use only grapes grown on Chateau Montelena's own vineyards. Such a practice would increase Luper's control over the vineyards and permit him to direct cultural practices there. However, Luper has a good working relationship with the vineyardists from whom Montelena purchases grapes. He works closely with each vineyard during the growing season and at harvest to ensure that the grapes are delivered to the winery at full maturity in prime condition.

The grapes are picked by hand as much as possible. Although Luper prefers hand-harvested grapes, he is not opposed to machine-harvesting so long as the quality of the machine-picked grapes meets his standards.

One important quality factor stressed by Luper is that some potassium metabisulphite is put in the bottom of every gondola in which the grapes are transported to the winery. This releases sulphur dioxide to protect the grapes from wild yeast fermentation and from oxidation. There is no field crushing; the grapes are crushed at the winery, usually within an hour after the gondolas are filled. The grapes are received into a stainless steel hopper, and a continuous helical screw conveyor feeds them to a stemmer-crusher. In the crushing area, metabisulphite is used to maintain minimum sulphur dioxide levels to protect the grapes. Inside the winery, liquid sulphur dioxide replaces the metabisulphite in order to keep potassium levels low.

White grapes are pressed immediately after crushing without skin contact for a longer period than that required to complete the pressing of a lot of grapes. A pneumatic bladder press is used because bladder presses are faster than other batch presses, the press cake is thinner and they extract fewer phenols and less bitterness.

Luper personally supervises pressing at Montelena. In his opinion, pressing is never a routine operation. Every vineyard is different from other vineyards, and the grapes from each vineyard differ from year

The wines of Chateau Montelena grabbed the world's attention in 1976 when the 1973 Chardonnay came top in the first comparative Paris tasting of French and Californian white Burgundy-style wines. In 1977 the Napa Valley produced rich and well-balanced Chardonnays.

to year. He seeks to extract the maximum juice from the white grapes with the least pressure and with the least maceration of the skins. This personal involvement in pressing limits the effective maximum size of Montelena to the handling of approximately 400 tons per year.

The pressing cycle as conducted by Luper results in high quality juice being extracted at relatively low pressures. Only the last press is made at relatively high pressure, and the last press yields only 1 to 2 per cent of the total volume of juice extracted from a batch of grapes. Therefore, the press' juice is blended with the free-run juice prior to fermentation.

From the press, the must goes into stainless steel fermenters, where it is chilled to 45°F and settled by gravity for 36 to 48 hours. The settled juice is then racked to

another tank where it is fermented on Montrachet yeast at 55°F. Montrachet yeast is preferred by Luper for fermenting white wines, because it will carry on a vigorous fermentation at low temperatures. At 55°F, fermentation takes approximately four weeks.

All fermentation at Chateau Montelena is carried out in temperature controlled stainless steel; there is no barrel fermentation. Luper favours fermentation in stainless steel, followed by ageing in small wood barrels to develop subtleties in the wine.

The white wines are settled by gravity, racked twice, fined with bentonite and filtered to Limousin barrels for six months of wood development. Some of these barrels are new, but most have been previously used.

Throughout the handling of the Chardonnay, great care

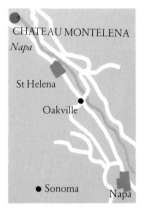

CHATEAU MONTELENA
Napa

St Helena

Oakville

● Sonoma   Napa

	Harvest date	Sugar at harvest	Reducing sugar	Residual sugar	Total acid	Recommended ageing	Optimum period for consumption	Winemaker's comments
SONOMA **Cabernet Sauvignon** 1973				0.12%	0.62%	Up to 8 years		A classic wine, big rich mouth-filling, given extended ageing in French and American oak.
NAPA & ALEXANDER VALLEYS **Chardonnay** 1974				0.11%	0.7%	2–4 years		A big and full-bodied wine with a delicate balance of wood complexities gained from ageing in Limousin oak.
NAPA & ALEXANDER VALLEYS **Zinfandel** 1974				0.12%	0.6%		1978–1980	The wine has fruit and spice in the nose and on the palate. Rich and long-lasting finish.
NORTH COAST **Cabernet Sauvignon** 1975	15–27 Oct	23° Brix	Dry		0.61%		1981–1985	The wine receives 2 years wood ageing and one year in bottle before release.
NAPA & ALEXANDER VALLEYS *Late Harvest* **Johannisberg Riesling 1975**				13%	0.74%	1–2 years		Late harvest style was affected by a good deal of botrytis.
NAPA & ALEXANDER VALLEYS **Chardonnay** 1976	1–7 Sept	23.5° Brix	Dry		0.84%		1979–1981	Wine with good body, intense varietal character and a rare degree of richness and complexity.
NORTH COAST **Zinfandel** 1976	14 Sept–14 Oct	23.5° Brix	Dry		0.59%		1980–1983	Drought produced low yield with concentration, intensifying flavours of nose, taste and finish.
NAPA VALLEY **Chardonnay** 1977	3 Sept–2 Oct	23.5° Brix	Dry		0.73%		Early 1980s	Dry season produced a wine with concentrated varietal character.
NORTH COAST **Johannisberg Riesling** 1978	mid Sept	24.2° Brix	1.7%		0.79%		1979–1981	A very full-bodied Riesling with intense fruit flavours. Good as aperitif or with a meal.

The Chinese gardens, lake and
summerhouse (RIGHT) were
built by Chinese owners.
The estate building is,
however, medieval in style
(INSET LEFT), and the
medieval atmosphere is
carried over into the figure of
a knight (INSET RIGHT).

is taken to prevent oxidation. Carbon dioxide and nitrogen are used for purging tanks and barrels. In addition, free sulphur dioxide levels are monitored continuously.

Following the period of barrel ageing, the wines are reassembled from the barrels. The wines from two or three different vineyards may be blended. Then the wines, filtered with cellulose polishing filter pads, are bottled by an automated bottling line under what are essentially anaerobic conditions, permitting the addition of relatively low amounts of sulphur dioxide at bottling.

Riesling receives no wood development, but is bottled directly from stainless steel after fining and filtration. The latter is essential; the Riesling contains approximately one per cent residual sugar at bottling. Chardonnay receives up to one year of bottle age at the winery before being released into retail channels. Thus, Chardonnay is approximately two years of age at the time of its release.

The red grapes are crushed to a stainless steel fermenter. Fermentation is initiated by the addition of Burgundy yeast, which Luper prefers because it ferments sluggishly at the relatively high controlled temperatures—of between 75°F and 80°F after pumping over—at which fermentation proceeds. This yeast also permits a longer period of skin contact, up to 8 days, during fermentation. As there are two completely separate chilling systems at Montelena, temperatures of both reds and whites can be accurately controlled.

During the alcoholic fermentation, vigorous pumping over takes place twice per day. The cap is totally broken up and an effort is made to make sure that every skin in the cap is washed. The length of time taken for pumping-over varies, depending on the stage of fermentation.

At the completion of the alcoholic fermentation of the Cabernet Sauvignon, draining and pressing follow, with the press wine being blended with the free-run. Malolactic fermentation follows.

On the completion of malolactic fermentation, the Cabernets are fined and filtered before being racked to small French and American oak barrels where the wine spends two years maturing. Again, as in the case of the Chardonnay, most of the barrels have been previously used, although a small proportion are new. After barrel ageing, the Cabernets are assembled for bottling and given a final polish filtration as the wines are put into bottles. An additional 18 months of bottle ageing precedes release. Luper aims to extract as much as possible from the Cabernet grape during fermentation, but then to age the wine at the winery so that it leaves the winery ready to drink.

By contrast, Chateau Montelena's Zinfandel is made in a fresh, fruity style. Therefore, malolactic fermentation is inhibited. The wine receives one year of wood ageing in previously used French oak. It is given millipore filtration at bottling. Aged for one year in the bottle at the winery, the Zinfandel is released just before it is three years old.

While the wines are in wood, a statistically representative sampling is made of each lot of barrels once each month; these barrels are topped up, and the wine tasted and analyzed in the laboratory. As the barrels are in permanent stacks, all this work is done with the barrels in place. However, as a consequence of the fact that the wines are filtered before being placed in the barrels, no racking is necessary once the wines are in wood.

The annual production at Chateau Montelena is approximately 20,000 cases. The small size of the winery, plus the emphasis on quality and personal attention to each stage of the winemaking process by Luper, make it impossible to increase production beyond that figure. However, these are the very factors which assure high quality.

Jerry Luper is the winemaker at Chateau Montelena (RIGHT). He graduated in oenology from Fresno State University. This institution, together with the University of California at Davis, has produced many of the top winemakers in the revitalized California wine industry.

Chateau Montelena has 60 acres of Cabernet Sauvignon, 22 acres of Chardonnay and 10 acres of Zinfandel. These Zinfandel grapes (ABOVE) just before the vintage are in good condition with evenly ripened berries.

HEITZ WINE CELLARS

The best wines are the Cabernet Sauvignons, especially those produced from Martha's Vineyard grapes. Big, dark and intense, their minty (but not necessarily varietal) style, with its undercurrents of eucalyptus, has attracted a legion of afficionados. The Chardonnays, which also have a following, are crisp and clean. Good recent vintages include 1968, 1971, 1973, 1974, 1975 and 1977.

The Heitz Wine Cellars are set in a small valley on Taplin Road near Joseph Phelps. Both the family's home and the production facilities are here. There are three parts to this winery. There is a tasting room on Highway 29 just south of St Helena and north of Louis Martini's winery, while on Taplin Road are the old stone winery built in 1898, and a new concrete-block and wood structure built in 1972 to house offices and additional ageing facilities. White wines are aged in the old stone cellar and the red wines are matured in the newly constructed building.

Joe Heitz first came to California in 1944 while serving in the US Air Force. Stationed near the Italian Swiss Colony winery, near Fresno, he was introduced to winemaking while working as a cellarman there. He then studied and taught winemaking, as well as working extensively in various vineyards. In 1961, the Heitz family bought a small winery and eight acres of Grignolino grapes, on the site of the present tasting room south of St Helena, and moved back to the Napa Valley to make wine. In 1964 he acquired the old winery at the end of Taplin Road and moved the winemaking activities there.

In addition to the 20 acres of a selected strain of Grignolino planted around the winery, Heitz cellars now have 30 acres of Chardonnay. They continue to buy other grapes from growers with whom they have built up a relationship over the years. The vineyards of these growers are located mainly on the floor of the Napa Valley between St Helena and Yountville. Heitz prefers, as he puts it, not to spread himself too thin, but to concentrate his efforts on winemaking. He does, however, work closely with the growers, particularly at harvest time, to ensure that grapes come to the winery at balanced maturity. At present all grapes are harvested by hand, although Heitz is not against accepting machine-harvested grapes.

As the grapes enter the winery they are crushed in a stainless steel crusher-stemmer. Potassium metabisulphite is added to the must in sufficient quantity to generate 100 parts per million of sulphur dioxide. Red grapes then go directly to the fermenters. White musts are allowed to settle overnight; by the time they are inoculated with yeast on the following day to begin fermentation, the sulphur dioxide levels will have fallen to 60 or 70 parts per million. Montrachet yeast is used for both red and white wines, and both reds and whites are fermented in temperature-controlled stainless steel tanks, the whites at 50°F, and the reds at 70°F.

A one ton Willmes bladder press is employed on both red and white pomace. Each batch is given two or three gentle pressings, but the final pressing is never more than 60 pounds per square inch.

Red wines are usually fermented on the skins for approximately one week. When they reach between 5° and 6° Brix, the free-run wine is racked off the pomace and the pomace is pressed.

Heitz Cellars do not own a centrifuge, thus the wines are clarified by racking, fining and filtration. Heitz believes that fining and filtration are interchangeable with centrifugation. When the fermentation is finished, all wines go into 1,000 or 2,000 gallon American white oak tanks for preliminary settling and clarification. Riesling and Gewürztraminer spend approximately six months in 1,000 gallon tanks. There they receive several rackings, are fined with bentonite to remove excess protein and filtered just before bottling from these large tanks by the summer following fermentation. Chardonnay also spends about three to four months in 1,000 gallon American white oak tanks where it is given bentonite fining and a rough filtering. As soon as the previous vintage of Chardonnay is bottled, the current vintage is transferred into the 60 gallon Limousin barrels which have just been emptied. The small barrels are mainly Nevers and Limousin, except for some limited amounts of Yugoslavian. Chardonnay receives one full year of ageing in small oak barrels.

The white wines are not cold stabilized as a separate step in preparing them for bottling because fermentation at cold temperatures and the fact that Chardonnay spends 1½ winters and the other whites spend one winter in a naturally cold cellar usually eliminates the need for it. In addition, Heitz feels that his customers are sophisticated enough to recognize and accept tartrate crystals in a bottle of his wine as a natural phenomenon.

Zinfandel, Grignolino, Barbera and Burgundy spend 1½ to two years in 1,000 or 2,000 gallon American oak. While there, they also are given two to three rackings, gelatin fining to remove excess tannin, and, if the wine calls for it, rough filtration before it goes into 60 gallon barrels for an additional year of wood ageing. Cabernet Sauvignon spends 1½ years in the large tanks undergoing 'rough' clarification, including bentonite fining if necessary; then it is transferred to the small, 60 gallon vessels for an additional two years. At the end of small wood ageing, the wine is reassembled in the larger containers to blend out any differences between barrels. The wine is then fined, filtered and bottled. By this time, Heitz's Cabernet Sauvignon will have spent a total of 3½ years in wood.

Malolactic fermentation is not induced by inoculation. The bacillus is present in the winery and, if a red shows a tendency to undergo malolactic fermentation, it is encouraged to do so spontaneously. The rosés are usually the first to be bottled, and are frequently in the bottle by the March following fermentation.

The grapes from each vineyard are vinified separately, but individual vineyards are given credit on the Heitz label only if the wines produced from the grapes

Martha's Vineyard is acknowledged to be one of the finest in California. Heitz always specify the name of this vineyard on the label.

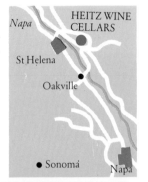

grown in that vineyard are somehow special or if the vintage from that vineyard is especially fine. Martha's Vineyard is the one vineyard consistently to receive identification on Heitz labels. Martha's Vineyard is located against the foothills on the west side of the Napa Valley just south of Oakville. The vineyard is about 15 acres in size and planted entirely with Cabernet Sauvignon. In the hands of Joe Heitz, grapes from Martha's Vineyard produce intense, minty, darkly coloured, very distinctive wines, which are known and are eagerly sought out by wine cognoscenti the world over.

In addition to the wines which are produced by Heitz Wine Cellars, there are several wines which may have been purchased from another winery for bottling under the Heitz label. Others may be subjected to various cellar treatments such as ageing, blending and clarifying to perfect them before they are bottled. A portion of the generics are either labelled as 'Selected by' or 'Perfected and Bottled by' Heitz Wine Cellars.

Heitz's international reputation and fame as a wine-maker rests upon two factors. He is noted for the acuteness of his palate, on which his unquestioned skill as a blender rests. Second, his Chardonnays and Cabernet Sauvignons are acknowledged to be consistently among the finest in the whole of California.

Heitz is one of the top Napa Valley wineries. The old stone winery (ABOVE) is on Taplin Road with a more modern tasting room on Highway 29 just south of St Helena.

This specially commissioned picture of the Heitz estate (BELOW LEFT) appeared on the Heitz 1974 Napa Valley Cabernet Sauvignon from the famous Martha's Vineyard.

Joe Heitz (LEFT) owns the Heitz Cellars. He bought a small winery and eight acres of Grignolino vines in 1961. The concern has expanded since then. Heitz first gained a reputation for his Cabernet Sauvignon, and more recently his Chardonnay has come to the world's attention.

It is important to remove any unripe or rotten grapes (FAR LEFT), so that only the best and most healthy grapes are used in the wine.

Initial fermentation of the must is fairly rapid and a large amount of carbon dioxide is produced (LEFT). Heitz ferments in stainless steel tanks.

The rugged terrain of the Napa Valley (LEFT) is testimony of the region's volcanic geological origins. Most of the vines in this area are planted on the valley floor, the hillsides remaining more wooded.

White oak tanks with capacities of 1,000 and 2,000 gallons (ABOVE) are used for clarification of all wines, both reds and whites. The Cabernet Sauvignon spends about 18 months in large tanks for rough clarification and is then put into small barrels.

The care and attention which are lavished on all aspects of winemaking at Heitz are accompanied by research and experiment in the laboratory (ABOVE). Joe Heitz's son David is carrying out fining tests on an experimental wine.

JOSEPH PHELPS VINEYARDS

The white wines are typified by the extreme delicacy of their bouquet and flavour. They show the hand of a master winemaker who has adapted German winemaking techniques to California. The late-harvested Johannisberg Rieslings reveal extreme lusciousness. The newer red wines are developing both class and style. Good recent vintages are 1975, 1977 and 1978.

Joseph Phelps Vineyards was founded in 1973 by a Colorado builder who discovered the challenges and satisfactions of making wine during the construction of wineries for other new entrants to the world of winemaking. The first wine, a Johannisberg Riesling, was produced in 1973, using equipment leased at another winery. It was released a year later during the construction of Phelp's own winery on Taplin Road just down the hill from Heitz Cellars. The new winery was completed in time for the 1974 vintage.

The winery is two separate buildings which are linked together at the upper levels. Crushing, pressing and fermenting take place in a northerly building. The southern building houses small wooden barrels of both French and American oak for ageing, and the bottling room. The small 60 gallon barrels are set in individual racks to facilitate movement and cleaning. There are also several fibreglass tanks from Germany in this room which are unique in Californian winemaking.

The building is impressive, but even more impressive is the vineyard programme under the supervision of winemaker Walter Schug, who is assisted by vineyard manager Bulmaro Montez. Schug is a graduate of the Institute of Oenology and Viticulture at Geisenheim in the Rheingau, and had worked for several producers in the Rhine before going to the United States in 1961. After arriving in the Napa Valley, he worked for two major California wineries before joining the new Joseph Phelps Vineyards as winemaker in 1973.

Like most of the new breed of winemakers, Schug believes that great wines begin in the vineyard with great grapes. To this end he is directing a complex programme of planting. In his work Schug is fortunate to have the full support and participation of winery owner Phelps and a well-qualified team of assistants. Phelps is no absentee owner, but takes an active interest and role in all the winemaking decisions.

The rolling hills which surround the winery afford Phelps Vineyards the opportunity to experiment with matching varieties to varying solar radiation exposures. Schug and Phelps believe that solar radiation is a major factor in determining microclimate and is more crucial to the development of balanced maturity of grapes than is heat summation during the growing period. For example, Chardonnay is planted on eastward sloping backlit hillsides to reduce the amount of radiation it receives. Except for making sure that each site is well drained, there is less emphasis upon matching varieties to soil types. Schug notes that the soils on the Phelps ranch are extremely varied because of the hilliness of the terrain, and that there are frequent changes in soil type within a relatively few feet. Vineyard practices attempt to encourage each vine to produce at its natural optimum level and the vines are not artificially stressed. A permanent overhead sprinkler system provides frost protection.

In choosing the varieties to plant on the home ranch, the prime consideration was whether that variety would do well in that location. Varieties include Gewürztraminer, Sauvignon Blanc, Johannisberg Riesling, Scheurebe, Cabernet Sauvignon, Zinfandel and Syrah Noir. The Syrah Noir is the latest approved French clone of the grape which is used in producing the great wines of Hermitage in the Rhône Valley. In total 20 acres are planted on the Phelps property near the winery. Scheurebe is a Riesling and Silvaner cross developed at the State Research Institute at Alzey in Rheinhessen in 1916. The Phelps property has the only acreage of Scheurebe growing in California.

In addition to the 170 acres planted on the hills around the winery, Phelps owns 50 acres near Yountville, and also grows grapes on leased land on the Silverado Trail near Oakville. Grapes are also purchased from several growers in specific locations selected for the variety and quality of the grapes.

In Schug's view, the winemaker must not only identify with each grape variety, he must treat each load of grapes received at the winery individually to determine how to help them to vinify themselves correctly. This means that practically every batch is vinified separately which requires extra equipment and extra effort. However, Schug is quick to point out that his techniques allow him to conduct constant research into the effect of microclimates upon grape and wine quality.

Although each batch of grapes is treated individually there are some general practices which provide an insight into Schug's winemaking style. All harvesting is done by hand. All white grapes are destemmed as they arrive at the winery. The Chardonnay must usually receives some skin contact. Gewürztraminer may, if it appears that it will improve the finished wine. Riesling musts are given skin contact only if they are late harvest. Skin contact varies from three hours to overnight.

Sulphur dioxide is introduced during crushing, wherever possible as gas, or as potassium metabisulphite in a solution sprinkled over the grapes in the crusher. Levels depend upon the quality and condition of the fruit. Usually, reds

will receive only 60 parts per million in order not to interfere with malolactic fermentation later. Whites receive 75 parts per million at the time of crushing, mostly as an anti-oxidant. Subsequent to fermentation, the whites are adjusted to 30 parts per million of sulphur dioxide for the bottling process.

Pressing takes place in one of the two Willmes tank presses recently acquired by the winery, to give the pomace the lightest degree of pressure possible. The Willmes

Late harvest Johannisberg Rieslings are a speciality of Joseph Phelps Vineyards. The 1978 harvest was the largest ever in the Napa Valley.

tank press is incapable of exceeding a pressure of 30 pounds per square inch. The result is a low level of solids in the press juice.

All white juice, including press juice, is put through a centrifuge before fermentation is initiated, to reduce the volume of solids in the must. Schug and Phelps believe that solids not only interfere with the rate of fermentation, they result in 'off' tastes in the wine. Hydrogen sulphide may develop from sulphur dust on the grapes. Fruit fibre may add a tea-like taste; shredded leaves, the taste of alfalfa. Solids also make it more difficult to control the temperature of fermentation. Centrifuging also removes wild yeasts from the must.

Fermentation temperatures vary with the grape variety. Riesling, Gewürztraminer and Scheurebe are fermented at 45°F–50°F; Chardonnay, at 55°F to 60°F, to encourage enzyme action for extra complexity.

When fermentations of the Rieslings require 25 days or more, a mid fermentation racking at 10° Brix removes yeast lees and prevents 'off' characteristics developing from the expired yeast cells.

After each white wine completes fermentation, it is filtered as soon as possible and racked to the German ovals. These casks have been steam leached to ensure that they will impart no wood flavour to the wine. Sauvignon Blanc goes entirely into 650 to 12,000 gallon ovals. Rieslings go into larger 1,500 or 2,000 gallon ovals to allow some slight oxidative development without picking up any woodiness.

The early harvest Riesling, for example, requires some time in these ovals to develop the crispness for which it is noted. Gewürztraminer also goes into the 1,500 gallon ovals for oxidative development.

The ovals are also used for holding Chardonnay in preparation for placing them in the small cooperage or for storing any Chardonnay which has matured early in the barrel. The ovals provide a neutral container in which to hold those wines while the balance is catching up. The different sizes of these ovals allow the winemaker to assess variations in the body and flavour from different vineyards and to select the size appropriate to a particular wine.

Riesling and Gewürztraminer master blends for bottling are made at the end of January or the beginning of February. The large tanks are then filled with Sauvignon Blanc which remains in wood until June or July of the year following fermentation before bottling. Chardonnay is placed in 60 gallon Limousin barrels to develop a slight wood character. Chardonnay is also bottled in June or July. White wines are fined with isinglass and heat-stabilized with bentonite before bottling. All the white wines with residual sugar are sterile bottled.

One of the specialities at Joseph Phelps Vineyards is the production of late-harvest style Johannisberg Rieslings. Every vintage produces grapes which have been affected by botrytis. From these have been made Riesling wines in the German style ranging from *Spätlese* to *Trockenbeerenauslese* quality.

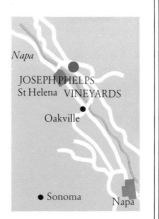

The Joseph Phelps winery is in the middle of vineyard land on the Taplin Road in the Napa Valley near Heitz cellars. The rolling hills which surround the winery give the vine growers the opportunity of matching the grape varieties to soil and solar conditions, which vary with altitude.

Joseph Phelps started off as a builder of wineries and became interested in establishing his own. The winery itself (LEFT) is constructed from wood and consists of two separate buildings which are linked at the upper levels.

The stainless steel screw in the hopper (BELOW) propels the grapes—in this instance Cabernet Sauvignon—gradually to the destemmer-crusher. The screw has to be made of stainless steel to avoid possible metal contamination.

Young vines (RIGHT) are protected with plastic from frost or animals. This practice is ideal in a dry climate, such as California, but is not always desirable in wetter climates as humidity can build up under the plastic and damage the vines.

Walter Schug, an oenologist trained at the Geisenheim academy in Germany, is the winemaker at Joseph Phelps, He is filling a stainless steel vat with must (RIGHT) and working with a centrifuge (CENTRE). At Joseph Phelps the white must is centrifuged to remove solids and wild yeasts.

Sauvignon Blanc juice (RIGHT) is fermenting in one of the large stainless steel fermenting tanks. On completion of fermentation, the white wines are filtered, and racked to the large German ovals (CENTRE).

After destemming and crushing, red musts are inoculated with Bordeaux yeast and are fermented to dryness on the skins at 80°F. The really big reds, such as the Eisele Vineyard Cabernet Sauvignon, are allowed to remain on the skins for several days following completion of fermentation. At 10° Brix, the fermenting must is inoculated with malolactic starter. Vigorous pumping over occurs three times per day for fifteen minutes each time. After fermentation, the red wines are pressed in the Willmes tank presses and the press wine is blended into the free-run wine. The wines are then placed in the large Yugoslavian

oak tanks and monitored until malolactic fermentation has been completed. They are then filtered into small oak barrels for wood ageing which varies from eight months to two years, depending upon the body of the wine.

The small barrels are re-used; when new barrels are purchased, they are simply added to those already in use. New Nevers barrels are used first for Cabernet Sauvignon, then for Merlot, and are simply handed down to varieties which do not require much wood.

At the end of the ageing period in small barrels, the red wines are blended in the Yugoslavian oak tanks and

allowed to sit for two additional months before being bottled. Just before bottling, the red wines are fined with egg whites and given a coarse pad filtration.

Relatively early bottling is favoured by Joseph Phelps Vineyards for their red wines. Schug and Phelps prefer the fruity character which results from early bottling over the lower fruitiness resulting from longer wood ageing.

Experimentation with various blends of Cabernet, Merlot and Cabernet Franc to be released under the Insignia label is continuing from harvest to harvest. The Insignia label is to be used for the best claret-style wines.

The 1975 harvest produced an outstanding Merlot. So the Insignia wine for 1975 was mainly Merlot with Cabernet Sauvignon added to provide the wine with a structural framework, while in 1977, the blend was 50 per cent Cabernet Sauvignon, 30 per cent Merlot and 20 per cent Cabernet Franc. The blends may change from year to year depending on the grapes.

Schug describes himself as a jack-of-all-trades. He actively oversees and participates in every operation at Phelps, from the vineyard planting programme to the bottling of the finished wines.

The terrain around Joseph Phelps is hilly (ABOVE). *All picking is done by hand* (INSET). *In addition to the vines close to the winery, the estate has vines elsewhere in the Napa Valley, Yountville and on the Silverado Trail near Oakville.*

ROBERT MONDAVI WINERY

The Cabernet Sauvignons are developing elegance and intensity, especially the 'Reserve' wines. The Chardonnays show excellent varietal character, with good fruit, balance and complexities, but may be overshadowed by the Fumé Blancs, which gain depth and their own unique complexities with each new vintage. Johannisberg Rieslings appear to be taking on a Germanic subtlety of aroma and flavour

When Robert Mondavi left the family-owned Charles Krug Winery in 1966 to open his own winery at Oakville, he built the first entirely new winery to be constructed in the Napa Valley since the end of Prohibition. It has remained one of the most modern and progressive wineries in California because the Mondavis are constantly experimenting with new techniques and trying out new equipment.

From the beginning, the winery has used stainless steel, temperature-controlled tanks for fermentation before they came into the widespread use they now enjoy in California. Mondavi has aged red wines, and some whites, in small, 60 gallon, mainly French barrels on a larger scale than had been previously the case in California. Other techniques include clarifying both juice and wines with a centrifuge and introducing large rotating tanks for draining white musts and for fermenting some red wines.

Although the initial size of the winery was on a relatively small scale, expansion was rapid. In 1966 the storage capacity of the winery was 100,000 gallons; it is now 1,800,000 gallons. Fermenting capacity has increased from 75,000 gallons to 1,200,000. Two centrifuges now sit side by side at the end of a bay of stainless steel fermenters. The vineyard has been increased from 12 to 600 acres. Despite its apparently large size, the winery still has the capacity to handle small lots of grapes and to produce some wines on a small scale. Flexibility has been maintained and quality has not been sacrificed.

When entering the fermentation wing of the winery for the first time, a visitor sees that virtually every piece of equipment is either stainless steel or oak. From the time the grapes arrive at the crusher until the wine made from those grapes is put into barrels, casks or tanks for ageing, the grapes, must and wine are put in stainless steel.

Harvesting is done by hand into 10 to 30 pound containers and these containers dumped into two or four ton gondolas. As each gondola is filled, the grapes are protected from oxidation by a layer of carbon dioxide gas under a tarpaulin cover. As the gondolas arrive in the crushing area, they are emptied into a stainless steel hopper from which the grapes are fed to a stainless steel stemmer-crusher in a concrete-lined pit below ground level. As the grapes are destemmed and broken open, the stems are expelled from one end of the stemmer-crusher cylinder into a conveyor which transports them into a truck for return to the vineyard where they are ploughed under as mulch. The must is pumped from the hopper below the crusher cylinder by a specially designed, French manufactured must pump.

There are two complete and entirely separate, parallel sets of crushing equipment. Thus both red and white grapes can be crushed simultaneously. Each variety—

and indeed each batch—of grapes is treated individually.

Cabernet Sauvignon is typical of the way red wines are produced at Robert Mondavi Winery. The pursuit of excellence by any winemaker begins in the vineyard. Oakville was chosen as the best location in the Napa Valley for Cabernet Sauvignon. After the Robert Mondavi Winery was built at Oakville in 1966, 600 acres to the west and south-west of the winery were planted to Cabernet Sauvignon. In the western Oakville area, the soils are deep and well-drained, and the climate is cool enough so that the grapes retain high natural acidity as they reach maturity, but warm enough to bring the grapes to full maturity in which the skins are soft but chewy and in which the tannins have softened and lost any aggressive harshness.

At proper maturity, the grapes are turgid, are soft but not flabby and, in the case of Cabernet Sauvignon, have reached sugar levels of 23.8° to 24.7° Brix when they are brought to the winery. They are destemmed, then crushed. The free sulphur dioxide level is adjusted to 30 parts per million at the crusher.

The must goes directly to a stainless steel fermenter where it is inoculated with French Red yeast to begin fermentation. Any necessary acid corrections will be made in the fermenter with tartaric acid. During alcoholic fermentation, the temperature rises gradually to 88°F at the end of the fermentation. At the temperatures at which the primary fermentation proceeds, it is completed within four to six days. Pumping over is undertaken four times each day during fermentation, with the length of the pumping over determined by the size of the fermenter. Malolactic fermentation also begins during the primary alcoholic fermentation; if it does not commence spontaneously, an inoculation to induce it will be made. At the completion of malolactic fermentation, any additional acid corrections are made on the basis of taste, as well as after consideration of sugar and pH levels. Free sulphur dioxide is adjusted to 25 parts per million at this stage.

At the completion of alcoholic fermentation, continued skin contact will depend on the vineyard from which the grapes came and on the judgement of the winemaker, and varies with each season. This period varies from seven days to two weeks, with 12 days being the average. Each fermenter is sampled twice each day and the fermenting must is tasted for such elements as the development of varietal character, the identification of any potential defects and to determine skin contact time. Added skin contact gives the wine added flavour and phenolics from the skins which will allow it to develop fullness and richness with age.

Next, the young wine is drained and pressed. The free-run is kept separate from the heavy press wine. The young wine will be settled and racked three or four times before it goes into barrels for wood

The 1974 vintage produced exceptionally fine red wines, particularly Cabernet Sauvignon, while the 1978 vintage favoured the whites especially the Sauvignon Blanc.

development.

Individual lots and individual vineyards are kept separate. The body and the flavour of the wine will determine the type of barrel in which the wine is placed. After the young wines are racked into barrels, they are checked frequently and topped up over a one or two week period, then tightly bunged and stored in the 'bung-over' position which permits as much as six months' time to elapse between toppings. Free sulphur dioxide levels of 25 parts per million protect the wines during their development in barrels.

During barrel ageing, frequent tasting continues, to determine how much time the wine should spend in the barrel, to choose which of the barrels are suitable for the reserve blends, which exhibit normal varietal characteristics and which are not developing varietal character.

When, in the winemaker's judgement, the wines have received the appropriate amount of wood development, they are prepared for bottling. If necessary they are fined with egg white one or two months before bottling. The wines may also be centrifuged or filtered prior to bottling. Tasting determines which, if any, of these processes is necessary.

The reserve Cabernet Sauvignons chosen by tasting during the barrelling period receive additional barrel and bottle age before being released.

Small amounts of Cabernet Franc and Merlot are produced for blending with the reserve Cabernet Sauvignons. The percentage of each in the final blend varies with the temperature conditions during the growing season. In warm years, Cabernet Franc adds firmness to the wine and enhances the cleanliness of bouquet; in cool years, the Merlot contributes firmness.

For the winemaker, Tim Mondavi, the Cabernet Sauvignon should exhibit the character of the land on which the grapes are grown and carry the imprint of the winemaker's style. The Cabernet Sauvignon's style is of vitality and interest, of good bouquet and mouth-feel, with an oak character which harmonizes with and enhances the wine without dominating it, a supple but firm wine with enough tannic structure to age well. The ultimate test of the success is the finished wine on the palate of the consumer.

The white wine for which the Robert Mondavi Winery is perhaps best known is the Fumé Blanc first produced in 1967. The wine is made primarily from the Sauvignon Blanc grape, although approximately five per cent Sémillon is usually added. The grapes are usually harvested at 23.5° Brix. On arrival at the winery, they are destemmed, crushed and then put in the fermenter. Free sulphur dioxide is adjusted to 30 parts per million. The must receives six to eight hours of skin contact in the fermenter, depending on the degree of ripeness and condition of the grapes. Subsequently, the free-run juice is drained and settled. The light press juice is centrifuged before being

Robert Mondavi is 'nosing' one of his wines (LEFT). Mondavi has undoubtedly done a great deal to spread the reputation of the Napa Valley wines both within the United States and throughout the world.

blended back into the free-run. The heavy press juice is fined and racked to a separate tank for fermentation.

Any necessary acid correction—depending on sugar, acid and pH levels in the must—will be made prior to fermentation by the addition of a combination of malic and tartaric acids. Inoculation for fermentation is made with French White yeasts. Approximately 10 per cent of the Fumé Blanc will undergo complete alcoholic fermentation in small oak barrels. The remainder will be fermented in stainless steel at 60°F until the reducing sugar level reaches 8° Brix, at which point the temperature will be allowed to rise to 68°F as the alcoholic fermentation is completed. Between 10 and 20 per cent of the tank-fermented Fumé Blanc will be induced to go through malolactic fermentation.

At the end of fermentation, the wine is racked, sulphur dioxide levels are adjusted to 25 to 27 parts per million, the wine is chilled and allowed to settle. Periodic rackings serve to clarify the wine naturally, and tastings determine when the wine is ready to go into barrels for oak development. The length of barrel ageing varies from four to 10 months, depending on the year and the barrels into which the wine is placed. The average is six or eight months. Chardonnay receives treatment during fermentation similar to that given the Fumé Blanc. Three styles of Johannisberg Riesling were made in the vintages of 1977 and 1978; rains prevented the production of all three styles in 1979. The three styles are botrytized, late-harvest and normal.

Robert Mondavi Winery is very much a family enterprise. In the exercise of their winemaking skills, the Mondavis not only have available the latest technology in the winery, but also have access to the grapes produced in the vineyards owned by the winery plus output of an additional 1,000 acres which are under long-term contract to the winery. In the Mondavis' view, fine wines begin in the vineyards. The major factors in achieving quality in the wine are understanding the vineyards which provide grapes, the ability to choose the vineyards which are

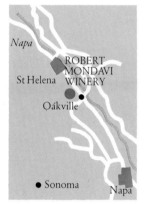

The Robert Mondavi winery is in Oakville in the Napa Valley. The estate has over 600 acres of vineyards to the west and south-west of the vineyard, planted with Cabernet Sauvignon vines.

A typical comparative tasting session is in progress at the Mondavi Winery (BELOW). Regular tastings are held with the winemaking staff in a continuous search for quality. These tastings are also used for training staff.

Bob Mondavi's son Tim is tasting a white must from one of the stainless steel fermenting vats (BELOW CENTRE).

producing the best grapes, and to provide in the winery the conditions which will permit those grapes to develop their fullest potential. Other important factors are climate, soil, proper drainage, and correct choice of grape variety for the conditions. Then the appropriate yield for each vine in the vineyard must be adjusted to the growing conditions. In the Mondavi-owned vineyards, yields range from $1\frac{1}{2}$ to four tons per acre. The grapes must be fully mature at harvest. Full maturation is judged by the grapes' skin condition, softness of tannins and the sugar-acid ratio. With the correct balance of these factors, wine can be made in a 'natural' way with a minimum of movement, fining and filtration, to ensure that the maximum in quality of aroma and flavour can be extracted from the grapes.

Flexibility is another major factor in the art of winemaking at the Robert Mondavi Winery. Flexibility is needed in the vineyard where the best grapes in a particular harvest must be chosen. Flexibility in the cellar involves choosing the techniques most appropriate to the development of each lot of grapes and wine. For example, different yeasts may be utilized in fermenting different varieties. Another important area of flexibility is in the selection of the best lots and specific barrels to go into the reserve programme and how they are used.

A further important aspect of the Mondavi approach to winemaking is summed up in one word—dedication. To the Mondavis, dedication means living with the wine they make from vineyard through fermentation, maturation and bottling—even to the selling of the wine to the consumer. To this end they seek to imbue the winemaking staff with the same dedication they feel. By placing each member of the winemaking staff in charge of a portion of total production and limiting his or her direct responsibility to a manageable portion, the Mondavis can duplicate within their own relatively large estate the conditions which exist within smaller wineries and secure that attention to detail and that quality of wine for which the small wineries are known.

A glass 'serpent' is inserted into the top of a vat to help guard against oxidation (RIGHT).

This view of the magnificent ageing cellars at the Mondavi winery (ABOVE) shows the large quantities of French oak barrels which Robert Mondavi uses. These are used mainly for the red wines, but sometimes for white wines as well.

The Mondavi Winery at dusk (RIGHT) reveals the startling white buildings which led the way to much of the distinctive winery architecture now often found in the Napa Valley.

SPRING MOUNTAIN VINEYARDS

Meticulous attention to detail has produced superb Chardonnays and Cabernet Sauvignons. The Chardonnays, based on solid varietal character, display considerable complexity, resulting from partial barrel fermentation. Cabernets, with depth and richness, are full-fruited. Sauvignon Blancs combine sound structure with crisp fruitiness. Good recent vintages include 1973, 1974, 1975 and 1978.

Much of the energy devoted to the pursuit of excellence in California wine in the 1970s came from men such as Michael Robbins, the managing partner of Spring Mountain Vineyards. Robbins epitomizes the new breed of amateurs—in the true sense of the word—whose enthusiasm and dedication to quality were the hallmark of the decade.

Robbins entered the wine industry after a successful career in commercial and industrial real estate. Spring Mountain Vineyards began in an old Victorian house north of St Helena which Robbins bought in 1962 and spent several years restoring. As the remodelling proceeded, the basement below the living quarters was bonded as a wine cellar and equipped as a small winery. Spring Mountain was founded there in 1968. The first wines were in leased tanks at established wineries in the Napa Valley and first released in 1970. They were immediately acclaimed as excellent.

In 1974, Robbins bought the old Tiburcio Parrott Estate, called Miravalle, situated several hundred feet above the floor of the Napa Valley on a south eastern slope of Spring Mountain, and constructed a new winery there. The old house on the property was completely restored. The old winery building was torn down and replaced by a structure, the architectural style of which matches that of the residence. The new winery building, like the old, sits in front of a 90 foot long tunnel dug into the mountainside by pick and shovel in the nineteenth century. Because the winery was completely rebuilt, it was possible to lay out the equipment first, and design the building around it. In addition, the multi-level building and the winery curve around the side of the mountain. They are dug into the slope, in such a way that gravity assists in moving must from the crusher-stemmer to the fermenting area on the lower level and wine from the bottling tanks to the bottling line, also at ground level.

The crusher-stemmer is of Italian style. Instead of the usual blades, there are spokes whose length is precisely calculated to remove grapes from their stems without shredding the stems. Inside the winery is an impressive array of jacketed, temperature-controlled stainless steel fermenters, seven oval casks, and 30 different 1,000 gallon upright oak tanks and 900 60 gallon barrels which are all made of French oak.

Robbin's view on wine quality is that great wines begin with great grapes and require an almost fanatical attention to detail throughout the process from harvesting to bottling. For him and his Spring Mountain wines, that attention to detail begins in the vineyard. The grapes, at balanced maturity, are harvested by hand into one-ton bins to minimize juicing and oxidation and to decrease the time between picking and crushing. Spring Mountain

switched to this method from field crushing because Robbins felt that field crushing led to an excessive amount of skin contact for the white wines. Four varietals are produced—Chardonnay, Sauvignon Blanc, Pinot Noir and Cabernet Sauvignon.

The Chardonnay begins fermentation in the stainless steel fermenters and is finished in small French oak cooperages. After crushing and a light pressing, the must is transferred to thermostatically-controlled tanks and is inoculated with Champagne yeast when the temperature has been brought down to 60°F. As fermentation progresses, the temperature is lowered to between 45°F and 47°F so that sugar is reduced at the rate of 1° Balling per day. When the fermenting must has reached 10° or 12° Balling, it is moved into 60 gallon barrels to complete the fermentation process.

Sauvignon Blanc is also cool fermented, then transferred to the German ovals for ageing until April. Cabernet Sauvignon is fermented to dryness in the stainless steel fermenters at temperatures between 85°F and 90°F, with the temperature allowed to rise toward 90°F at the end of fermentation to increase colour extraction. The fermenters are filled only to 50 to 60 per cent of capacity to increase surface area of the must relative to volume. Vigorous pumping over, using a three inch line to ensure a thorough mixing of the cap occurs for a total of one hour per day. These practices enhance colour and flavour extraction from the skins. About midway through the primary alcoholic fermentation, the must is inoculated with malolactic starter and malolactic fermentation is usually completed at about the same time as alcoholic fermentation. Following fermentation, the Cabernet remains on the skins for up to 10 more days for added flavour and tannin extraction, before being transferred to the 1,000 gallon Yugoslavian uprights for approximately one year. These tanks are relatively tall with domed heads. Their height is an aid in natural clarification of the wine. The domed heads reduce the chances of the wine oxidizing as steady absorption into the wood ensures that the dome is kept tight. In August of the year following fermentation, the Cabernets are returned to stainless steel for blending and fining with egg white before being racked into 60 gallon French oak barrels for a further year of ageing.

Pinot Noir is produced on a smaller scale than the other wines, and might be regarded as experimental. However, several vintages have been produced, and Robbins has planted Pinot Noir grapes on 10 of the 20 acres in another of his properties, the Soda Creek vineyard. The Pinot Noir is fermented in stainless steel at relatively high temperatures, which are allowed to reach 90°F as the fermentation nears completion. Stems are added to the fermenter, to give tannin, in a ratio which varies with

SPRING MOUNTAIN
1975
Napa Valley
CABERNET SAUVIGNON
MADE AND BOTTLED BY
Spring Mountain Vineyards • St. Helena, California
ALCOHOL 13.0% BY VOLUME

The 1975 vintage gave good quality red wines, although the Cabernet Sauvignon was a little harder than the 1974.

the ripeness of the fruit. Following the completion of fermentation, Pinot Noir receives cellar treatment very similar to that given to the Cabernet Sauvignon.

The Pinot Noir is made from the clone known as Pinot Noirien Petit, believed to be a descendant of the one given by Louis Latour to Paul Masson in 1898, and to come either from Romanée Conti or from the Corton Grancey (Latour) vineyards on the Côte de Beaune.

All wines are fermented on a pure yeast starter, wild yeasts on the grapes having been neutralized with sulphur dioxide at the time of crushing. None of the wines is cold stabilized, but they are treated with Bentonite or gelatin for heat stability. If acid corrections are necessary they are made with tartaric at the time of crush or with citric or malic acid following the crush. The white wines are filtered before bottling, but Robbins prefers not to filter his red wines.

Each lot of each variety is fermented, stored and evaluated separately until the master blend of that variety is made. Subsequently, each *cuvée* is re-evaluated just before bottling, which permits adjustments to be made in the blend. So far, no single vineyard wines have been made.

Approximately 100 of the 265 acres surrounding the house and winery on Spring Mountain can be planted with grapes. It appears that Spring Mountain will eventually be able to supply a large portion of their grape needs from their own vineyards even when both wineries reach their anticipated annual production of 23,000 cases each. A number of grapes, however, will continue to be purchased from growers.

To ensure that the winery can continue to operate throughout the critical period of the harvest there are two Willmes presses, two multi-purpose must pumps and two identical refrigeration units, each of which is capable of supplying the cooling and refrigeration requirements of the winery. The two presses not only allow for flexibility, but also provide back-up in case of breakdown of one or the other. Similarly, the two must pumps provide back-up capability as well as flexibility. Each can be used for pumping over or for moving the pomace.

The winery building is designed so that the fermenting room on the ground level can be sealed off from the ageing cellars. Likewise, the first year red wine ageing cellars on the second level can be sealed off from the ground floor cellars. Thus the temperature can be allowed to rise on the second level or in the fermentation room during malolactic fermentation, without raising the temperature in the barrel rooms on the ground level. The barrels are château models, identical to the barrels used by the top classified growths of Bordeaux and Burgundy, from which

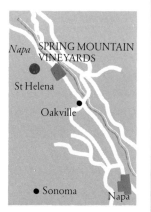

The winery is situated on Spring Mountain Road which leads out of St Helena in the Napa Valley. Some of the grapes used in Spring Mountain wines are grown in the estate's own vineyards, but most are bought in, so the vineyards used are rather scattered.

This stained-glass window (ABOVE) is in the Spring Mountain cellar, part of which is in a 90 foot long tunnel hewn into the mountain rock.

The area around the Spring Mountain winery (LEFT) is heavily forested. The land near the winery building has been cleared and is being prepared for planting vines.

This view of the winery (BELOW) shows the proximity of the stainless steel fermenting vats and the horizontal press.

Michael Robbins supervises every aspect of the winemaking process at Spring Mountain. He is following the progress of the Chardonnay fermentation in stainless steel vats (BELOW). Fermentation is completed in French oak barrels.

The Chardonnay grapes are put into the hopper (LEFT) by a rotating fork-lift truck. The grapes are in excellent condition. Grapes are mainly harvested by hand. Spring Mountain abandoned their earlier practice of putting the grapes straight into a field crusher because this led to excessive contact between the skin and juice. Field crushing is a Californian practice, not used in Europe.

Kept in the beautifully ventilated cellars (LEFT) is a fine collection of French oak barrels which Michael Robbins uses for both red and white wines.

regions they are obtained.

Robbins' own vineyards are severely pruned to reduce yields. The vines on the Wildwood property produce approximately 2.5 tons of Chardonnay and 2.25 tons of Cabernet Sauvignon per acre. Immature bunches are thinned before harvesting begins, to ensure that only fully ripened fruit will be picked. Robbins lists four natural factors which he believes affect yield and grape quality—microclimate, soil, drainage and solar exposure. He notes that microclimates differ throughout the Napa Valley, ranging from some which closely resemble those of the coolest regions of France to some which resemble the northern regions of Italy. He says that

microclimate is much more than a question of heat summations utilizing the system developed at Davis; it is also a question of the number of hours each day of solar radiation, not merely heat, and the exposure of the vines to that solar radiation. He adds that complex, well drained, sparsely bearing soils produce more complex and more flavourful wines than do heavy soils giving high yields.

Robbins attributes the great improvement in the quality of California wines during the 1970s to the fact that the new amateurs began winemaking with open minds about how wine should be made. They are flexible, willing to experiment, keen to apply the latest technology and able to blend all this with tradition. Better wines are the result.

The estate house (LEFT) is built in a rather ornate, late nineteenth century style and has been carefully restored in recent years. The other buildings on the estate are built in the same, somewhat flamboyant style.

STAG'S LEAP WINE CELLARS

Noted for powerful Cabernet Sauvignons with deep colour, intense flavour and full body, Stag's Leap also produces equally rich, full-flavoured and full-bodied Chardonnays. Recent noteworthy vintages include 1973, 1974, 1975, 1977 and 1978.

For Warren Winiarski, proprietor of and wine-maker at Stag's Leap Wine Cellars, the major prerequisite for making fine wines is a 'model of excellence.' He begins each wine with an image of what he would like to make, a form he would like to achieve. With this ideal in mind, he allows the grapes to point out the direction in which it lies. To Winiarski this is the art of the winemaker—allowing the grapes to show the winemaker what can be done with them, what wine can be made, and, following that route, ending as near as possible to the initial ideal.

This approach has served him well. He and his wife Barbara left teaching jobs at the University of Chicago in 1964 to settle in the Napa Valley with the intention of becoming winemakers. Warren worked with Lee Stewart at the old Souverain and with Robert Mondavi, before, in 1972, building Stag's Leap Wine Cellars. The site he chose lies alongside the Silverado Trail, below the rocky outcrops on the eastern ridge line of the Napa Valley known as Stag's Leap. It was the Cabernet Sauvignon from his third harvest, the vintage of 1973, which won the Paris tastings of 1976 in competition against four classified growths of Bordeaux and five other California Cabernet Sauvignons. Subsequent vintages have maintained or improved on that standard. Winiarski's method requires that he spends a great deal of time in the vineyards during the growing season and at the harvest and that his wine-making technique be flexible and adaptable to the qualities of his grapes.

Stag's Leap Wine Cellars own 45 acres of vineyards around the winery which are planted in Cabernet Sauvignon and Merlot. These vineyards are planted on well-drained loam which contains about 15 per cent gravel, and slope slightly downward from the eastern foothills to the Silverado Trail. The temperature during the growing season makes the region a low II or a high I according to the system of heat summation and classification developed at the University of California.

Chardonnay, Johannisberg Riesling and Pinot Noir grapes are purchased. Winiarski makes frequent observations of the development of the grapes in the vineyards from which he purchases grapes, as well as in his own vineyards. He pays particular attention to the size of the berries, to the distribution of the fruit on the vine and, as the harvest season approaches, the pattern of ripening of the fruit. He knows that there is no absolute uniformity even in a small vineyard, and that the pattern and degree of maturity vary from vine to vine, from bunch to bunch on the same vine, and even within the same bunch.

Ideally, each berry should be identical and perfectly mature at the same moment. Although this is not possible, it is necessary to get as close to uniformity as is possible,

Winiarski feels, and this requires making many adjustments in the vineyard during maturation of the fruit. At harvest, the maturity of the vineyard must be gauged accurately by the winemaker.

In a normal year, each vineyard is picked in five or six stages. Allowances are made for small differences from vine to vine in order to equalize out these differences and to achieve as much uniformity as possible in the fruit when it arrives at the crusher.

Some vineyards are picked only in the morning and some vineyards are picked only in the afternoon. Winiarski has found that must crushed at one temperature in the morning turns out to be different from must crushed in the afternoon at another temperature. Skins break down more rapidly, and the behaviour during fermentation of the tiny bead of fuzz-like material on the inner surface of the grape skins varies with the temperature at which the grapes were harvested. Some adjustments can be made in the temperature by the use of temperature-controlled fermenters. For example, if a must is too warm, the jacketed tanks can be used to cool it; however, Winiarski stresses that what has taken place in the field has already had its effect on the composition of the grapes when they arrive at the crusher.

Wild yeasts are sterilized at the crusher by the addition of 50 parts per million of sulphur dioxide to the red musts and slightly more to the white. Sulphur dioxide levels are kept low in the red musts to facilitate malolactic fermentation following the end of the primary alcoholic fermentation.

Flexibility in technique is again required when it comes to the addition of a pure yeast inoculum to start the alcoholic fermentation. Both the amount of the inoculum and the temperature at which the must is held while these yeasts are growing into populations which can successfully conduct the alcoholic fermentation are variables, and are adjusted to levels which Winiarski judges to be appropriate for each lot of grapes.

At Stag's Leap Wine Cellars, even the vigour and amount of crushing or maceration to which grapes are subjected is varied according to the size and condition of the berries. This variation in crushing technique determines the degree to which the skins of the grapes are turned inside out or are merely bruised. It is accomplished

by the use of a variable speed crusher. Although the crusher also destems the grapes, stems may be added to the red musts during fermentation.

The choice of fermentive yeasts has been and is subject to experimentation at Stag's Leap Wine Cellars. In general, Champagne strains have been utilized. However, some red Bordeaux strains have also been used. Assmannshausen has been employed for Pinot Noir, and a strain from Geisenheim has been found to produce a more desirable

The 1973 Cabernet Sauvignon of Stag's Leap won the international tasting of 1976 in Paris. The 1977 vintage is showing potential to equal the excellent quality of the 1974 vintage.

floral character in the Riesling. However, Stag's Leap is still experimenting to find the best yeast or yeasts.

Fermentation temperatures are also subject to variation in accordance with growing conditions in vineyards, condition of grapes at crush, and temperatures at crush. In general, temperatures for the white musts vary between 45°F and 55°F and those for the red musts between 75°F and 85°F. Other factors in the choice of fermentation temperatures for the red musts include the anticipated length of the fermentation and the anticipated length of contact with the skins after the wine is dry. Winiarski stresses that close monitoring of the progress of fermentation is necessary.

The length of the fermentation of the red wines and the degree of maceration of the fruit during fermentation will depend on the condition of the fruit in a particular year and, specifically, upon the image which Winiarski has developed of the natural tendencies of the grapes in that year. For example, if the wine is tending naturally towards austerity, he will allow the wine to develop austerity rather than try to force it to softness. A contrast which illustrates exactly this point is that between the austere, somewhat steely, Napa Valley Cabernet Sauvignons of 1975, and those of 1974 which are soft and aromatic. Neither vintage could have been forced into becoming what the other became.

Winiarski regards the treatment of the cap during red wine fermentation as critical in achieving the best wine possible. Pumping over during the course of fermentation is ordinarily done twice each day. However, the duration of each pumping over, its vigour, and degree of completeness are crucial in the development of complexity, depth and richness in the finished wine. Management of the cap must be done with restraint in California because of the richness of California grapes, and modified in accordance with the material in the fermenter.

The length of time that the red musts are in contact with the skins also varies with the nature of the material. Some wines are left for relatively long periods of maturation on the skins after the wine has 'gone dry.' In other instances, skins and wine are separated while there is still unfermented sugar. Separation of the wines from the skins is followed by pressing in a Willmes ram press, which Winiarski feels gives him better control over pressing than other types. The proportion of the press wine added to the free-run varies each season.

After separation and pressing, the wines are inoculated with the strain of lactobacillus designated as ML-34 by the University of California at Davis to induce malolactic fermentation. Malolactic is usually completed by mid November. Acid corrections, if necessary, are made in the red musts with tartaric acid following the completion of the malolactic fermentation.

Natural clarification by settling and racking follows

Stag's Leap is situated on the Silverado Trail on the eastern side of the Napa Valley, one of California's main wine-producing areas.

Most of the vines (LEFT) are planted on the floor of the Napa Valley. The natural terrain on the slopes of the hills is woodland and scrub.

*Skilful cellar work is needed
so that the wine in stacked
barrels (BELOW) remains in
good condition particularly
during racking and topping up.*

until mid January. Detartration is also accomplished by
allowing the outside cold in late November and December
to chill the wines and precipitate excess tartrates. After
light gelatin fining, the red wines and the Chardonnay go
into small Nevers barrels for wood development. The red
wines receive 12 to 15 months in small oak, and Chardonnay
spends six to 12 months in the small barrels. The length of
wood ageing each wine receives depends on the nature of
the wine and the age of the barrels. The purpose of the
time in wood is to enhance the tastes of the wines, not to
overwhelm a particular character with wood flavour.
Because Winiarski feels that Johannisberg Riesling is not
enhanced by wood development, his Rieslings are not

Careful records have to be kept especially during fermentation (BELOW). Here Warren Winiarski is updating the fermentation charts.

aged in small barrels. The development of the wines in wood is followed closely by frequent examination, tasting and topping up. Once the wines are in the barrels, they are not moved until the final assemblage shortly before bottling.

In making the final blending, Winiarski takes advantage of the fact that different lots have developed differently in the cellar, and of the fact that barrel age and even position of the barrel in the stack make for differences in the wine. Wines at higher elevations in the cellar are at slightly higher temperatures than those at lower elevations, for example, and this helps in the creation of wines of complexity. Winiarski looks for both unity and 'tenseness', hoping to satisfy the palate with the unity of his wines but to stimulate some interest by the contrast of opposing elements, such as hardness and softness, or suppleness and tannin. In fact, unity is created by the contrast and 'tenseness' of the different elements.

If the attention to detail, and the accommodation of techniques to the demands of the grapes are successful, the image with which Winiarski began will be achieved: wines of richness, depth and complexity, with a certain degree of opulence, but also a modest display of restraint. Such a wine finished first in the Paris tasting in 1976. Similar wines continue to be produced by Stag's Leap Wine Cellars each year.

The red must has to be pumped over (LEFT) during fermentation to ensure that skin-juice contact is maintained. This tank of Pinot Noir is being pumped over after fermenting for 14 days. Pumping over takes place at intervals during fermentation.

Warren Winiarski (LEFT) is the winemaker at Stag's Leap. He is transferring carbon dioxide from a tank containing Chardonnay to another containing Pinot Noir. This has been fermenting too long to be able still to produce its own carbon dioxide.

STERLING VINEYARDS

Sterling wines are typified by their character and depth of flavour, which need time to develop. The Cabernet Sauvignon 1974 is just beginning to show a glorious rich nose and a powerful, well differentiated taste. The Merlot is always dark and sappy with fruity opulence, while the Chardonnay is powerful and withdrawn, understated at first opening, but becoming outstanding with bottle age.

Sterling Vineyards sits at the top of a wooded hill at the north end of the Napa Valley, just a few miles south of Calistoga. The winery buildings' architecture is Mediterranean in style. The buildings were obviously designed to accommodate the winery. Both buildings and equipment follow the gentle slope of a ridge running north to south. The winery was laid out to take advantage of the slope of the ridge so that gravity would help to move the grapes, must and wine through the various stages in winemaking from the crusher-stemmer through the fermentation room and the large, 3,000 gallon, French oak tanks used for preliminary clarification and blending, to the ageing cellars filled from floor to ceiling with 60 gallon French barrels neatly stacked in metal racks. The buildings were obviously also designed to cater for visitors.

Sterling Vineyards date from 1964 when the owners of Sterling International, a San Francisco paper company, began buying land in the Napa Valley and planting grapes. By 1968, a winery was being planned. The first wines were produced in 1969 to 1971 in a small group of structures at the bottom of the hill, while the hill top winery was being designed and built. Sterling had hired Richard Forman as their first winemaker. After touring wine regions in Europe in 1968 and 1969, he worked on the construction of the winery. The equipment was installed and the building erected around it. The first wine produced on the hill top was the vintage of 1972. In 1977, Sterling was purchased by The Coca-Cola company of Atlanta. Michael Stone, who had been General Manager of Sterling Vineyards, became its President. After the 1978 crush, Forman left Sterling to start his own winery, and Theo Rosenbrand succeeded him as winemaker.

Rosenbrand joined Sterling in time for the 1979 harvest after 22 years in Beaulieu Vineyards. Although he had no training or experience in winemaking, he was hired by Beaulieu in 1956 to work in the cellars. He advanced to positions of increasing responsibility, working on the way with some of the best winemakers in California, until he was named winemaker.

Under the guidance of Rosenbrand, Sterling are reducing their number of major grape varieties to four—Cabernet Sauvignon, Merlot, Chardonnay and Sauvignon Blanc. However, some Sémillon will be produced for blending with the Sauvignon Blanc. The last vintage of Pinot Noir was made in 1978 and of Zinfandel in 1979. Chenin Blanc and Gewürztraminer will not be made after the 1980 vintage. Acreage has been acquired on Diamond Mountain, at the northern end of the Napa Valley in the western hills, for planting in the four main varieties. These vineyards are located on steep terraced hillsides. The purchase and planting of the hillside vineyards

brings the total planted acreage owned or controlled by the winery to 500. When the vines planted on Diamond Mountain are mature (the earliest plantings were made in 1976), Sterling will be capable of producing for themselves all the grapes needed for the maximum capacity of 100,000 cases.

Approximately 300 acres of the Sterling vineyards were originally planted by the founders in the 1960s on the flat, alluvial, relatively rich soils of the valley floor. Yields and quality are good, but the feeling at the winery is that the vineyards on Diamond Mountain will produce higher quality grapes when the vines reach full maturity.

The soils are deep and well drained, so that in some areas the four year old vines still require some irrigation. The soils were created by alluvial action. The rough nature of the terrain results in a wide variation in exposure to solar radiation. As a consequence, it has been possible to match varieties to sites on the basis of the amount of solar radiation received as well as to drainage. For example, Sauvignon Blanc, Cabernet Sauvignon and Merlot have been planted on sunny, well-lit slopes, and Chardonnay, on the less sunny, back-lit slopes. Although these vineyards are steep and difficult to cultivate and harvest, there is apparently little danger of frost damage at the elevations at which they are located, approximately 1,000 to 1,500 feet above the floor of the valley.

The vines around the winery are cluster-thinned in the spring to reduce yield; in some instances as much as 50 per cent of the clusters have been dropped. Harvesting is entirely by hand. Vineyards are sometimes selectively harvested in blocks. The winery equipment includes a Healdsburg stainless steel crusher-stemmer, a Willmes tank press, stainless steel fermenters, 3,000 gallon tanks of French oak, and small barrels for ageing, also made of French oak. The fermenters are unique in that there are two separate cooling jackets on the exterior. One extends in a band from the bottom of the tank about three-quarters of the way to the top. The second is a relatively narrow strip, approximately 12 inches wide, located near the top of the tank. The purpose of the second band is to provide for uniform temperature control throughout the fermenter and to eliminate the possibility of hot spots near the top.

White musts go directly to the press after crushing, without skin contact. It has been the practice to ferment Chenin Blanc, Gewürztraminer and Sauvignon Blanc in the stainless steel fermenters at temperatures of 50°F to 60°F. At these temperatures, the length of fermentation varies from 10 days to three weeks.

After overnight settling of the must in stainless steel, the Chardonnay is placed in Limousin barrels for fermentation; the Chardonnay is, at present, entirely barrel-

Sterling's Reserve Cabernet Sauvignon is made from selected grapes and built for long life.

fermented. Some Sauvignon Blanc was barrel-fermented in 1979 as an experiment. Rosenbrand sees the chief problem with barrel fermentation as temperature control, which must be weighed against the advantage of added complexity which is traditionally attributed to barrel-fermentation. All white wines receive some exposure to wood. Chardonnay remains in the barrels for an additional four to five months. About 50 per cent of the Sauvignon Blanc spends some time in 60 gallon barrels. The balance of the white wines receive wood exposure in the 3,000 gallon tanks.

The premier Sterling wine is the Reserve Cabernet Sauvignon. Vinification and maturation techniques employed in producing this wine differ from those utilized in making the other red wines. In fact, a separate *chais* was completed in 1979 to house the Reserve Cabernets during the two years they spend in 60 gallon Nevers barrels.

All Cabernet Sauvignon and Merlot musts are separately vinified by vineyard lots. Fermentation takes place in the stainless steel fermenters at temperatures of 75°F to 85°F. Skin contact varies from five to 10 days. Following primary fermentation, the young wines, still separated by vineyard lots, are moved to 3,000 gallon tanks to undergo malolactic fermentation. Reliance is placed primarily upon spontaneous initiation of the malolactic fermentation; however, selected strains of malolactic bacillus taken from the winery during previous malolactic fermentation are cultured in the laboratory and may be added to provide a boost to the spontaneous development of the secondary fermentation. Following the completion of malolactic

fermentation, selections are made of the best of the Cabernet Sauvignon lots for the Reserve programme. The best of the Merlot is also selected for blending with the Cabernet. Some Merlot is added to the Cabernet in making up the Reserve blends, the exact percentage varying according to the judgement of the winemaker. Blending of the wines from separate vineyard plots is undertaken to add complexity to the wines. For this reason, no single vineyards are identified on the Sterling Vineyards label. All are Napa Valley appellation.

The wines are then racked to 60 gallon Nevers barrels which are stored in the bung-up position, with glass bungs inserted to allow any carbon dioxide left from malolactic fermentation to escape from the barrel. At the end of three months, the wines are racked for the first time, bunged tightly and stored in the bung-over position and, thereafter, they are racked quarterly. Just after the second racking, the Reserve wines receive fining with egg white in the barrel. As the programme develops, one third of the barrels in the Reserve ageing programme will be new, one third will be two years of age, and one third will be four years of age. After two years in wood, the Reserve Cabernets are bottled and receive one to two years' additional bottle age before being released.

The balance of the red wines, the Cabernet Sauvignon and Merlot not selected for the Reserve programme, undergo malolactic fermentation and clarification in the 3,000 gallon tanks. Then they receive one year of barrel ageing before they are filtered and bottled. These barrels are also stored in the bung-over position. This practice

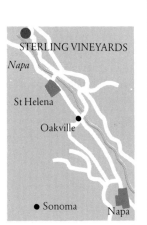

Sterling Vineyards are based in Calistoga in the Napa Valley. The Napa Valley narrows at its northern tip near Calistoga and the climate is very hot.

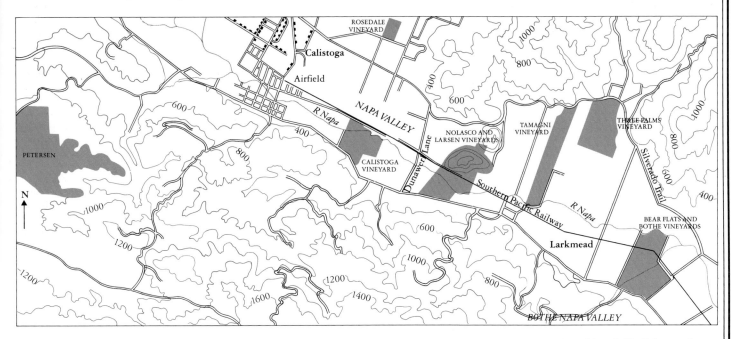

Sterling's vineyards are planted on a variety of terrains. Those planted on the alluvial soil near the Napa River need to be well drained and pruned back to control the harvest. The vines planted on the gravelly outcrops of the valley bear fewer grapes but attain excellent fruitiness. The hillside plantings are at an altitude of several hundred feet where soil and climatic conditions differ from those at lower altitudes. The 150 acre Petersen vineyard has well drained sandy soil and is planted with Pinot Noir, Cabernet Sauvignon and Merlot. The 13 acre Rosedale Vineyard is planted solely with Cabernet Sauvignon. The rich clay-loam soil of the 62 acre Calistoga Vineyard suits the Chardonnay, Sauvignon Blanc, Sémillon and Gewürztraminer grapes grown there. The Nolasco and Larsen Vineyards straddle the Napa River. On the east bank the white varieties Chardonnay and Sauvignon Blanc are grown and on the west bank Pinot Noir and Zinfandel grow. The Tamagni Vineyard slopes from the Silverado Trail down to the river. Both red and white varieties grow in the Three Palms Vineyards. Similarly in the Bear Flats and Bothé Vineyards, Pinot Chardonnay grows on the upper sandy soil and Cabernet and Merlot on the lower loam slopes.

After harvesting, the grapes are put into the hopper (BELOW) which feeds them through into the crusher. The largest of the stainless steel fermentation tanks (BELOW RIGHT) has a capacity of 10,000 gallons. Each grape variety is fermented separately, so several different sizes of tank are needed. Stainless steel can be cleaned easily and the fermentation temperatures accurately controlled.

Large oak tanks (BELOW) are used for the red wines. The tanks have a capacity of 3,000 gallons and are made of French oak.

This aerial view shows the vineyards (BELOW) near the winery. The design of the gleaming white winery (INSET) is based on that of an early Californian mission. The building was completed in 1973.

At Sterling the Cabernet Sauvignon and Sauvignon Blanc wines are aged in barrels made from Nevers oak and the Pinot Noir and Chardonnay (BELOW) are aged in Limousin oak.

The cellar foreman at Sterling is checking the progress of the Chardonnay (BELOW). He takes the sample using a glass pipette. The wines are still young as the bungs have not been turned to the side.

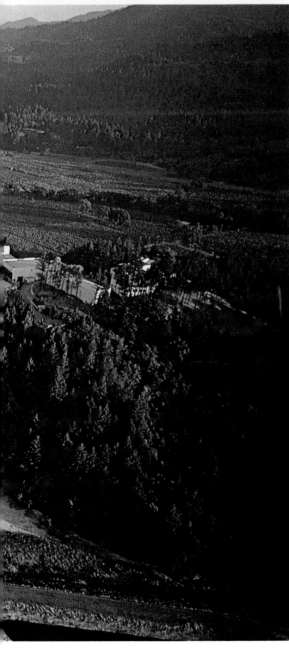

reduces the frequency with which the barrels must be topped. Experiments conducted at the Monterey Vineyard have shown that tightly bunged barrels produce a vacuum in the barrel, and that contrary to the teachings of conventional wisdom, no air enters the wine through the staves or the head of the barrel. The practice of ageing red wines in small cooperages in the bung-over position was initiated in California at Beaulieu Vineyards in 1958. Forman adopted this practice with the very first vintage of red wines to go into barrels at Sterling. Rosenbrand has not only continued the practice, but has also extended it to white wines being aged in small cooperage.

Both red and white musts have 50 to 60 parts per million of sulphur dioxide in the form of liquid sulphur or meta-bisulphite metered into the must lines as they leave the crusher-stemmer. Both reds and whites are fermented with Montrachet and other selected yeast strains. The press used, a new model Willmes tank press, not only permits the gentlest pressing possible of the pomace, but also introduces less air into the press wine than do other methods. A pre-press cycle, at pressures which never exceed five pounds per square inch, increases the yield of free-run quality juice. The pre-press cycle is followed by a pressing cycle at higher pressures. The white press juice is centrifuged before fermentation and is fermented separately. Free-run juice is centrifuged selectively as necessary. For example, in 1979 only the Gewürztraminer and the Sauvignon Blanc free-run juice were centrifuged before fermentation.

The Coca-Cola company's purchase of Sterling Vineyards has added impetus to the development of new vineyards, made for increased specialization in grape varieties and resulted in increased emphasis on the Reserve Cabernet programme. More importantly, Coca-Cola has done nothing to interfere with or dictate winemaking technique at Sterling. Rather, it has given talented winemaker, Theo Rosenbrand, freedom and encouragement to produce the finest wines from Sterling's vineyards.

The tasting rooms (ABOVE) are used by guests and visitors. Shelter from the extreme summer sun outside is important, especially when tasting red wines which are not chilled and are adversely affected by heat.

CHATEAU ST JEAN

Chateau St Jean is noted for white wines precisely labelled by the name of the vineyard which produced the grapes. The Chardonnays easily match Burgundy's best. However, more attention has focused on botrytized Johannisberg Rieslings. Cabernet Sauvignons from Wildwood Vineyard, are rich and robust, developing finesse and complexities with age.

Chateau St Jean is the realization of a dream to enter the premium California wine field shared by three grape growers with extensive vineyards in the San Joaquin Valley. The three are brothers Edward and Robert Merzoian and Edward's brother-in-law Kenneth Sheffield. Having made a decision in the early 1970s to found a winery dedicated to the production of the finest possible wines, they hired Richard Arrowood as winemaker and charged him with the responsibility of designing and developing the winery. They also gave him the freedom to purchase the finest equipment and to make all the winemaking decisions. Sheffield served as President of Chateau St Jean for its first three years.

The 250 acre Goff Estate at the foot of Sugarloaf Ridge, near Kenwood in the Sonoma Valley was purchased in 1973 as headquarters for the new enterprise. In 1974 the first wines produced were made in leased equipment at another Sonoma County Winery. When they were released, it was to critical acclaim which immediately established Arrowood's reputation as a winemaker of the first rank. Subsequent vintages have enhanced that reputation.

Working entirely with purchased grapes, Chateau St Jean was among the first to give credit on the label to the grower who produced the grapes. This practice not only requires that each lot of grapes is crushed, fermented, aged and bottled separately, but also that each vineyard is closely supervised from pruning in the spring through cluster thinning during the growing season to harvesting in the autumn. Some 30 different vineyards in three different counties are under contract to and supervised by Chateau St Jean. Contracts with their growers are unusual, in that they call for a set fee for a sugar range, and specified minimum sugar and acid levels. Growers are also compensated for cultural practices which reduce yield but enhance quality. This increases the cost of grapes to St Jean above typical market levels, but because the vineyard feels that the finest quality wines begin with the best grapes available, the owners and the winemaker think that the results are worth the added cost.

Winery construction began in 1975. The first stage was finished in time for that year's crush, and the final stage is almost completed. The winery was designed to accommodate a large diversity of small, separate lots of grapes resulting from the policy of crediting the vineyards which produced the grapes. The stainless steel, temperature-controlled fermenters range in size from 500 to 3,500 gallons. As a consequence of the fact that the white musts are fermented at temperatures of between 45°F and 47°F, fermenters, once filled, contain fermenting juice for considerably longer than the three or four weeks usual in California. This increases the number of fermenters required to accommodate a given volume. The owners,

especially Sheffield, must be given credit for the support they have given this programme and their courage in making the financial commitment to acquire the necessary equipment to make wine in small, separate lots.

All grapes are harvested by hand. As they arrive at the winery, the grapes are passed through a stemmer-crusher which removes the stems before the grapes are crushed by a pair of nylon rollers. Carbon dioxide is used throughout the system, from the crusher-stemmer to the bottling of the finished wine, to prevent exposure of grapes or juice to oxygen. As the grapes are crushed they are blanketed with carbon dioxide. Between 60 and 70 parts per million of sulphur dioxide are added to the crusher to stun wild yeast naturally present on the skins of the grapes and to inhibit polyphenyloxidase browning. The white must, still under its blanket of carbon dioxide, goes to a dejuicing tank to increase the yield of free-run juice. Because most of the varietal flavours and aromas are contained in the skins of the grapes, the juice is allowed to remain in contact with the skins for some time. After several hours or several days, depending on the variety, the free-run juice is drawn off the pomace to a fermentation tank. There it is inoculated with a yeast culture, a Geisenheim strain for the Johannisberg Rieslings and a Champagne strain for the other whites. Pressing of the pomace is accomplished in a bladder-type press at no more than 45 pounds per square inch. The press juice is fermented separately but may be returned to the wines made from the free-run juice if Arrowood feels that this will enhance the varietal character of the wine. All white juice is centrifuged before fermentation is initiated. Removal of some of the insoluble solids from the juice results in a cleaner, more uniform fermentation.

Chardonnay, Pinot Blanc and Fumé Blanc (dry Sauvignon Blanc) are fermented in stainless steel at approximately 47°F until between 10 and 12° Brix is reached. They are then transferred to oak barrels to complete fermentation to dryness at approximately 60°F. The Fumé Blanc goes into American oak, and the Chardonnay and Pinot Blanc into 60 gallon French barrels. Arrowood feels that this partial fermentation in oak results in better extraction of the wood's desirable flavouring elements than does mere ageing in oak following fermentation.

The drier Gewürztraminer, Johannisberg Riesling and Muscat Canelli are fermented in stainless steel for periods of eight to 16 weeks. The so-called late harvest Riesling and Gewürztraminers which have been affected with botrytis and which are high in sugar content may ferment for as long as eight months in the stainless steel fermenters.

Although the production of late harvest style wines accounts for a relatively small portion of total output at Chateau St Jean, it was these wines which helped to bring the wine-making talents of Arrowood to

The Wildwood Vineyard from Chateau St Jean produces high quality wines. This vineyard has some of the oldest Cabernet Sauvignon vines in California growing in it.

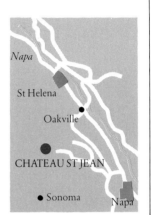

Richard Arrowood (BELOW) *is the Chateau St Jean wine-maker. He was one of the dynamic, well-trained oeno-logists to come to the fore in California in the 1970s.*

Arrowood was one of the first winemakers to explore the possibilities of producing German style late harvest Johannisberg Riesling wines in California.

international attention. He was among the first in California to make use of grapes which had been affected by botrytis mould to produce Riesling wines in the German *Qualitätswein mit Prädikat* style. His first Johannisberg Riesling at Chateau St Jean, produced in 1974, was *Spätlese* quality and the botrytis character was clearly and cleanly present in the nose and on the palate. Subsequent vintages have yielded both Riesling and Gewürztraminer wines of various *Prädikat* levels, including at least one at the *Trockenbeerenauslese* level. Prohibited by the American federal agency, which grants approval for wine labels, from using German terms to describe his botrytized wines, Arrowood identifies them by such terms as 'Individual Bunch Selected' or 'Individual Dried Bunch Selected' on the label.

These botrytized wines begin with vineyard practices designed to encourage the development of the mould. Where it is known that botrytis will develop, growers are asked not to spray certain rows with mould-retarding chemicals. At harvest time the pickers are paid by the hour and instructed to pick only the 'rotten' fruit, which is contrary to what they are usually instructed to do. Fields may be gone over three times. The most heavily affected fruit is usually picked first; this yields the *Trockenbeeren-auslese* quality fruit. The final picking may yield only *Auslese* levels of sugar and mould. Chateau St Jean has even laid down its own standards for each of its five *Prädikat* quality wines, in terms of minimum must sugar, residual sugar in the finished wine, and minimum botrytis infection, all in correspondence with the relevant German equivalents.

Juice yields are predictably low sometimes reaching 40 gallons per ton compared with the more usual 150. The grower is therefore compensated for the risks involved in producing these special wines. He is paid on the basis of a normal crop from that section of his field which has been set aside for the production of late harvest wines, even though only half that many grapes may be harvested. The quality of these wines has been recognized throughout the

world. In California they have received Gold Medals in every competition in which they have been entered.

Chardonnay, Pinot Blanc and Fumé Blanc are aged in French barrels following fermentation, for an additional three to six months, before they are bottled. None of the Gewürztraminer, Johannisberg Riesling or Muscat Canelli wines spends any time in wood. They are bottled directly from stainless steel tanks whenever Arrowood feels that they are ready. Cold fermentation means that the white wines are relatively cold stable when fermentation is complete. Holding at temperatures below 35°F for 10 to 20 days prior to bottling completes tartrate stabilization.

Chateau St Jean produces only three red wines which account for less than 10 per cent of total production. They are Zinfandel, Merlot and Cabernet Sauvignon. All are fermented with Pasteur Institute yeast, to dryness on the skins in stainless steel tanks at 75°F to 85°F. Pumping over occurs two to three times daily until fermentation ends. Subsequently, depending upon the quality and the condition of the grapes when they arrive at the winery, the young wines may be left in contact with the cap for an additional three to six days, then pressed. The press wine would then be separated from the free-run and transferred to small oak barrels for at least six months of wood ageing. Zinfandel goes into American oak, Merlot and Cabernet Sauvignon into French oak. After bottling, the red wines receive additional ageing in glass at the winery before being released.

In the future, the only red wines which will be produced by Chateau St Jean are those which can be produced from the Cabernet Sauvignon grapes of Wildwood Vineyard, which is reputed to have some of the oldest Cabernet Sauvignon in California growing on the property. Wildwood has produced consistently successful wines in the hands of Arrowood. The French barrels at St Jean are mainly Nevers and Limousin, but there are also barrels from Allier and Tronçais. They are replaced after 36 to 48 months of use. If acid corrections are necessary on any of the wines, Arrowood prefers to make them with juice, but will use tartaric acid if the corrections cannot be done only with juice.

The bottling line at Chateau St Jean was purchased from Seitz in Bad Kreuznach, Germany. This system permits bottling under completely sterile conditions. As the bottles enter the bottling room they are given a cool rinse to rid them of any cardboard dust. They are then filled with sulphur dioxide gas to sterilize them. This gas is removed and the bottles transferred by conveyor to the bottling machine where they are filled under an inert carbon dioxide atmosphere. The corker jaw is heated to 260°F to prevent any bacterial or yeast contamination. The wines are sterile filtered as they are transferred from storage tanks or barrels to the bottling tanks in a room adjacent to the bottling room. The bottles are filled from these tanks. Wines which are expected to be aged in the

The vineyards of Chateau St Jean are widely scattered. The grapes used to be entirely bought in from 30 vineyards in three counties. St Jean always shows the name of the vineyard from which the grapes came on the label. St Jean now has 105 acres around the winery planted with several varieties of white grapes.

In the dejuicing tanks (BELOW) *the pomace gathers in the bottom of the tanks, while the free-run juice is drawn off. The pomace is then put into the press.*

bottle for more than three years are closed with a 50 millimetre cork; those which are to be consumed within three years receive a 45 millimetre cork.

About 105 acres of the property around the winery have been planted with the white varieties of Chardonnay, White Riesling, Gewürztraminer, Sauvignon Blanc, Muscat Canelli and Pinot Blanc. The latter was acquired from French cuttings propⁿ gated at the University of California at Davis and was chosen for its Chardonnay-like character.

There are five different soil types on the property, and an effort has been made to match grape varieties to soils, to humidity conditions and to exposure to solar radiation. Approximately half the vineyards are planted on Sugarloaf Mountain above the chateau, and the other half on level ground adjacent to the chateau. The upper vineyards are generally free of fog. They are planted in Chardonnay, Pinot Blanc and Gewürztraminer. On one particularly rocky section Sauvignon Blanc is grown. The lower vineyards are planted with Chardonnay, Pinot Blanc, Gewürztraminer, and Muscat Canelli. One area, where morning fog combines with afternoon sun, appears to be conducive to the development of botrytis, Riesling is planted. Even when these vines are mature, Chateau St Jean will continue to rely upon grapes purchased from the 30 growers presently under contract for approximately 70 per cent of its grape needs and will continue the policy of identifying vineyards on the labels.

The frost-protection system (RIGHT) *consists of a series of pipes which spray water on the vines. When the grapes arrive at the winery, sulphur dioxide* (RIGHT CENTRE) *is added to kill the wild yeasts present on the grape skins. Later, a selected yeast culture will be added to start fermentation.*

Chardonnay grapes (RIGHT) *are tipped into the hopper and propelled into the dejuicing tank. This increases yield of the free-run juice and allows some contact between the skins of the white grapes and the juice, bringing out maximum flavour and aroma in the wine.*

The elegant white main building (RIGHT) *of the Chateau St Jean estate stands in large grounds at the foot of Sugarloaf Ridge in Sonoma, California.*

Chardonnay pomace has been emptied into the horizontal pneumatic press (BELOW) which presses the grapes extremely gently, at pressures of no more than 45 lbs per square inch. This is vital for white grapes.

The free-run juice is taken off the pomace (BELOW) and the solid matter which remains behind is put into the press.

The white wines are fermented largely in stainless steel, but are transferred to oak barrels to complete the process. French oak is used for the Chardonnay. The wines gain flavour from com-completing fermentation in oak (BELOW).

PAUL MASSON VINEYARDS

Masson's relatively young vineyards at Pinnacles have produced several fine white wine vintages. Chardonnays and Gewürztraminers are best. Rieslings have been rather flowery, and Sauvignon Blancs, a little too varietal. At present the balance of Masson's wines are sound but do not compare with these four vintage-dated, estate-bottled varieties.

Paul Masson is one of the 10 largest wineries in the United States, with shipments totalling five million cases annually. It also claims to be the oldest winery in continuous operation in California. The history of Paul Masson begins in 1852, when Etienne Thée, a *vigneron* from Bordeaux, purchased 350 acres of land five miles east of Los Gatos and planted vineyards. A compatriot, Charles LeFranc, became Thée's partner in 1857 and his heir at Thée's death. Subsequently, LeFranc was succeeded by his son-in-law, Paul Masson.

Paul Masson changed the name of the firm to Paul Masson Champagne Company, built a mountain-top winery called La Cresta above Los Gatos, and survived Prohibition with continuity unbroken by producing medicinal sparkling wine under government permit. At the end of Prohibition he sold La Cresta to Martin Ray.

Martin Ray operated Paul Masson until 1942 when he sold the winery to the Canadian-based House of Seagram, one of many large distillers who rushed to buy California wineries at the beginning of the Second World War as a means of producing some alcoholic beverage to keep open distribution channels. At the end of the war, when their home stills were not required for the production of industrial alcohol, many sold out, but Seagram retained a majority interest in Paul Masson. Until 1971, the minority shareholders (Alfred Fromm, Franz Sichel and Otto Meyer) ran the Paul Masson vineyard and guided the growth of the winery.

They built the showcase winery, completed in 1959, at Saratoga. In the early 1960s they began, with Mirassou, the test plantings which were to lead to the explosive increase in plantings in the early 1970s. They planted vineyards at Pinnacles and, in 1966, built the winery at Soledad to produce wine from these vineyards. Paul Masson was ready for the 'wine boom' of the late 1960s. As a consequence of the information gathered in the trial plantings, each variety was planted in small plots, matching the variety to soil and microclimate conditions in the new vineyards.

In 1971, Seagram bought out the interests of the minority shareholders, combined the marketing of Paul Masson wines with Seagram's Browne Vintners imported lines and began to expand production of Paul Masson wines. A third winery facility—Paul Masson Sherry Cellars—was built in 1974 at Madera to house their dessert and aperitif wine production. The headquarters winery at Saratoga is now devoted to finishing, blending and bottling. The winemakers are Joseph Stillman and Buddy Masuda.

Combined storage capacity of the Paul Masson wineries is now 28 million gallons. Of that, two million gallons are in the form of 6,000, 12,000 and 25,000 gallon wooden cooperage. Two buildings house 10,000

American and 6,000 French oak barrels. The American barrels are replaced after two or three years of use. About 250 new French barrels are purchased each year, and introduced into the ageing programme. Paul Masson now owns over 5,000 acres of vineyards.

The largest Paul Masson vineyard is at Pinnacles, where they made the first Monterey County plantings in 1962. There are now 1,800 acres in two vineyards designated Pinnacles I and Pinnacles II. In addition, there are 1,700 acres in Greenfield and at San Lucas. The latter are planted in red grapes which were mature enough for their first harvest in 1978.

Microclimates vary within the vineyards, as do the soil structures. Most soils are decomposed granite, but there are patches of gravelly, sandy loam and coarse sandy loams. Vineyard slope is approximately four per cent, except close to the foothills of the Gavilan Mountains where the slope increases to 10 per cent. The combination of soils and terrain slope produces good drainage in the vineyards.

As Paul Masson's Vineyards in Monterey County come into full bearing, it will be possible to adopt a Monterey County appellation for most of the Paul Masson table wines. At the end of the 1970s, all Paul Masson white wines, with the exception of Chenin Blanc and the generic Chablis were produced entirely from Monterey County grapes. The red wines are generally blends of Sonoma County and Monterey County reds.

Rainfall in Monterey County rarely exceeds 10 inches per year. Therefore, the vineyard supplements the natural rainfall with irrigation of 10 to 11 inches per year by means of permanently installed sprinklers. Soil moisture is very carefully controlled by means of devices placed in the soil. Monterey County is virtually frost-free, so sprinklers are not needed for frost protection. Nitrogen is the only fertilizer applied.

All vines are trained on two- and three-wire trellises for easier cultivation and to increase leaf exposure to solar radiation. All grapes are harvested by hand and transported to the winery in two-ton gondolas.

As the grapes arrive at the crushers, each load is inspected; sugar, acid and temperature readings are taken. Sulphur dioxide is added as the grapes are fed to the stemmer-crushers up to levels of 80 to 100 parts per

million. White grapes are crushed into Potter tanks for separation of juice from the skins and for juice clarification. The Potter tanks are conical stainless steel tanks, inside of which is a cylindrical stainless steel screen running down the centre. The free-run juice thus separated by passage through the skins and the screen contains one per cent or less of solids. This permits inoculation of the free-run for fermentation without further mechanical separation or settling. The must is chilled to 45°F in a

The Pinnacles Vineyard Gewürztraminer is perhaps the finest wine produced by Paul Masson Vineyards. The label bears the signature of the winemaker, Joseph Stillman.

stainless steel fermenter and inoculated with a strain of yeast developed and cultured in Paul Masson's own laboratories. Fermentation proceeds slowly at temperatures of 55°F to 60°F.

The drained white must is sent through a Coq dejuicer to a Coq press; the press juice is centrifuged. The press wines, both red and white, are never added to the varietal wines but go into the generics produced by Masson.

After fermentation is complete, all white wines are centrifuged and rough-filtered to remove yeast and solids. By the February following harvest, all the lees from the white wines have been sent to the distillery for distillation into 190° neutral spirits. Of the whites, only the estate-bottled Chardonnay and Sauvignon Blanc receive wood ageing. These two wines go into Limousin barrels for up to 26 weeks. The balance of the white wines is held in stainless steel tanks under temperature control at 50°F

until they are cold-stabilized, bentonite-fined and filtered before bottling.

Red wines are also fermented in stainless steel. Fermentation temperatures are controlled between 75°F and 80°F. Occasionally, Pinot Noir is fermented at 85°F for greater colour and flavour extraction. As in the case of the white wines, the reds are inoculated for alcoholic fermentation with a yeast culture developed in the Paul Masson laboratories. During alcoholic fermentation, pumping over is undertaken two or three times daily. The fermenting must is sprayed lightly over the cap and allowed to percolate through. At the end of alcoholic fermentation, the reds are transferred to large redwood tanks to undergo malolactic fermentation. The malolactic microflora are present in the redwood tanks, having been introduced when they were new, so no additional inoculation is necessary. However, the wines are cross-inoculated between tanks to ensure that the malolactic is completed smoothly. At the end of malolactic, the red wines are filtered. Acid corrections with tartaric acid are made, if necessary, when the red wines are blended. All red wines receive some wood ageing in small oak cooperage. Gelatin fining precedes bottling.

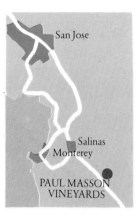

Paul Masson vineyards

Other vineyards

The Paul Masson vineyards are in Monterey County, mainly around Soledad, Greenfield and south of King City. At Soledad are the famous Pinnacles vineyards where the first vines were planted in 1962.

The imposing facade of the old Masson winery (ABOVE) *came from the St Patrick's Church in St José which was severely damaged in the 1906 earthquake.*

Both red and white wines receive cold stabilization. In the winery at Soledad is a cold room where the temperature is maintained at 25°F. Wines are pre-chilled and then pumped into stainless steel holding tanks in the cold room, where they are held for two weeks. At the end of that time the whites are transferred to stainless steel and the reds moved back into redwood tanks.

Bottling takes place under conditions designed to eliminate the possibility of dissolved oxygen in the bottled wines. A vacuum is created in the head space above the surface of the bottled wine just before the cork is inserted. All red wines are bottle-aged for six to 12 months at the winery before being released.

With the maturing of the Pinnacles vineyards, Paul Masson has made two substantial departures from past policy. The first was to begin a programme of vintage dating. The second was to initiate an estate-bottled programme. No legal definition exists at the moment for the term estate-bottled either in California or federal legislation but in 1983, new federal regulations will take effect, giving the term rough equivalency to that of chateau-bottled wine.

Prior to 1977, Paul Masson did not vintage-date any of their wines, preferring to blend old and new vintages to maintain consistency. But as the vines in the Pinnacles matured, Masson took the opportunity to select certain varieties from the best harvests for bottling as vintage-dated estate wines. The first of these wines to be released, in April 1977, were labelled as Pinnacles Selections and included a 1975 Johannisberg Riesling, a 1975 Chardonnay, a 1976 Gewürztraminer, and a 1974 Johannisberg Riesling sparkling wine. The releases in this series have continued with subsequent vintages.

The estate-bottled programme will remain relatively small. It is expected to involve approximately 35,000 cases per year out of a total production of 6,000,000 cases. Red wines will be included in the programme in the future.

In addition to Paul Masson's size and self-sufficiency in supplying their own grape needs, two other characteristics of this winery are striking. The first is the experience of the staff; the general manager has had 30 years experience, the winemaker 22, and the assistant winemaker eight years. The entire production staff are also technically orientated.

The production staff from cellar and laboratory supervisors to winemaker includes chemical engineers, microbiologists, chemists and oenologists. In addition, there is a separate research department, consisting of a director and four staff members with backgrounds in chemistry and engineering, and a technical services group whose responsibility is improvement of winemaking skills.

Flexibility at Paul Masson, another factor in the vineyard's success, is manifested in several ways. There is the flexibility of the winemaking personnel and their willingness to try new techniques and research.

White grapes (BELOW) *are transported from the scattered vineyards in 2 ton gondolas. To prevent oxidation in transit, the grapes may be given a light dusting with sulphur dioxide before leaving the vineyard.*

The free-run red juice (LEFT CENTRE) *is being pumped out and the solid matter caught by a screen. This is made of stainless steel to avoid possible metal contamination.*

White grapes are unloaded into the stainless steel hopper (LEFT) *which gently propels the grapes into Potter tanks for the separation of juice from skins and for clarification.*

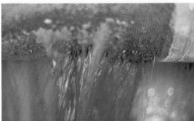

Paul Masson has a storage capacity of over 20 million gallons in stainless steel (FAR LEFT). *Capacity of the storage tanks ranges from 6,000 to 25,000 gallons. In the modern sterile bottling plant* (LEFT), *a vacuum is created in the headspace above the bottled wine before the cork is inserted.*

This large wooden barrel (LEFT) *shows the date of 1852 when the winery which is now Paul Masson was first established. Paul Masson was the son-in-law of the partner of the first owner, Etienne Thée, who came from Bordeaux.*

The two cellar buildings contain 10,000 American and 6,000 French oak barrels (ABOVE). *The estate-bottled Chardonnay and Sauvignon Blanc are put into French Limousin barrels for about six months.*

Joseph Stillman (ABOVE) *is the highly experienced winemaker at Paul Masson. His signature appears on the label of all the top Pinnacles selection wines.*

CALLAWAY VINEYARD AND WINERY

Emphasis is primarily upon crisp, dry white wines meant to accompany foods. Callaway also produce reds, which in recent vintages have become less aggressive, tannic and darkly coloured and hence more pleasantly varietal. The Petite Syrah is reminiscent of an upper Rhône, and the Cabernet of a well made Bordeaux. The Chenin Blanc Sweet Nancy is luscious, showing complexities of fruit and botrytis.

The first phase of the winery at Callaway Vineyard and Winery was completed in time for the 1974 crush, the first to be built near Temecula, after the planting of vineyards began there. Ely Callaway began planting his vineyards in 1969 on the rolling hill to the east of Temecula, a region of 2,000 acres which is now a recognized appellation because of Callaway's wines. Experimental wines, made at the University of California from the first small crop in 1972, were so encouraging that larger quantities were crushed in 1973 at a Napa Valley commercial winery and the decision was then made to go ahead with construction of Callaway's own winery.

The varieties selected for Callaway's vineyards were Cabernet Sauvignon, Zinfandel, Petite Sirah, White Riesling, Chenin Blanc and Sauvignon Blanc. Callaway planted 140 acres to these varieties. Callaway Winery uses all the production of Callaway's vineyards plus the output of selected independently owned vineyards nearby. Callaway's vineyards are managed by John Moramarco, part of an old Southern California grape-growing and winemaking family. Moramarco oversees the crews which farm the Callaway vineyards and does all the final pruning. He not only prunes personally, but also cluster-thins after the grapes are set, in order to limit yield and, by giving each vine individualized pruning, to permit each vine to produce only the amount of fruit it can mature properly. Yields, as a consequence, are low, averaging around four to five tons per acre. In addition, Moramarco oversees and his crews do the work on the independent vineyards from which, as is the practice in much of California, Callaway buys grapes.

Vineyard soils are mainly decomposed granite, well-drained but lacking in fertility. With no history of phylloxera in the Temecula area, the vines are all planted on their own roots.

In terms of the system of heat summation employed for classifying by the University of California, the vineyards are in Region III. However, they lie directly in the path of the wind which sweeps over the area through Rainbow Gap in the coastal mountains about 23 miles to the west, on its way to the inland desert sea, the Salton. The wind is a natural air conditioner; the wind-chill factor and the moisture in the wind probably reduce the effective temperatures at the vine to those of Region II or cooler. Average rainfall is between 15 and 18 inches annually, confined primarily to the winter months. Irrigation is used to supplement the low natural rainfall and to control the timing of water reaching the vines. The combination of these factors results in fruit of intense varietal characteristics, which are captured in the finished wine by the same attention to detail in the winery as in the vineyard.

Grapes are harvested by hand, using scissors rather than knives for the removal of the grape clusters from the vine in order not to break open the grapes before they reach the crusher. Picking is done into two ton gondolas, which are only half filled. Each variety in each section of the vineyards is picked according to the maturity of the grapes.

As the grapes reach the winery the gondolas are dumped into a stainless steel hopper which feeds them to a German made Amos destemmer before they are passed to soft corrugated rubber rollers for crushing. As the must passes out of the crusher and into a hopper below, it is covered with a blanket of carbon dioxide to protect it from oxidation. Red must then goes directly into stainless steel, temperature controlled fermentation tanks inside the winery building. White must is transferred from the hopper to a dejuicer; free-run juice is fed directly to a fermentation tank; the pomace goes to a Coq continuous press. Press juice is fermented separately from the free-run juice. The entire process of destemming, crushing, dejuicing and pressing is completed within approximately three minutes. The maximum elapsed time from the harvesting of the grapes until the must is in a fermentation tank is 90 minutes.

White must is centrifuged before fermentation and again at 7° Balling for removal of yeast sediment, to reduce 'yeasty' flavours. Fermentation temperatures are between 45°F and 50°F for white and 65°F for red wines.

White musts are inoculated with yeasts obtained from Geisenheim in Germany. Red musts are fermented on a single strain of yeast, Assmannshausen, also obtained from Geisenheim. Malolactic fermentation in the red wines is encouraged during the primary alcoholic fermentation, but usually occurs following the completion of alcoholic fermentation. During the alcoholic fermentation, vigorous pumping over to break up the cap thoroughly is undertaken twice daily for 20 minutes each time. Following the completion of both fermentations, the red wines are left in contact with the skins so that total skin contact time averages three weeks.

Acid corrections are usually unnecessary for either red or white wines at Callaway. On those occasions when the winemaker Steven O'Donnell feels an increase in acidity would enhance the quality of one of his wines, he makes an addition of a quantity of tartatic acid.

The ageing cellars are separated so that a temperature differential of approximately 8°F can be maintained between the white wine cellar and the red wine cellar. The former is kept at 52°F and the latter at 60°F.

Just before bottling, all Callaway table wines are lightly fined and filtered. The fining agent for the white wines is protein material from the flotation bladder of the Black Sea sturgeon. Fresh egg whites are used for the red wines. Fining and filtration are undertaken only as a final polishing.

CALLAWAY
Vineyard & Winery·

Estate Bottled
Vintage 1977

TEMECULA, CALIFORNIA
"Sweet Nancy"
CHENIN BLANC HARVESTED LATE
Grown, Vinified and Bottled by Callaway Vineyard & Winery
Temecula, California   ALCOHOL 10.5% BY VOLUME ·

This late harvested Chenin Blanc wine has good aroma, complexity, richness and balance. Noble rot affected the berries.

When the Callaway Winery was founded, Karl Werner was winemaker and designed the winery. All the ageing barrels in the winery were initially of German origin. Not only were casks of 164 gallons and of 337 gallons purchased, but Werner had 60 gallon barrels constructed for ageing the red wines. The winemaker at Callaway now is Steve O'Donnell, a qualified oenologist, who joined Callaway in 1977. O'Donnell has made few equipment changes, but he has added a major collection of French and American cooperage for ageing the red wines and Chardonnay.

Seven varieties of grapes are made into wine by O'Donnell. Up to a dozen wines are made from these varieties. The grape varieties are White Riesling, Sauvignon Blanc, Chardonnay, Chenin Blanc, Zinfandel, Cabernet Sauvignon and Petite Sirah. Two wines are made from the Sauvignon Blanc grape—Fumé Blanc and Sauvignon Blanc. The essential difference between these two wines is that the Sauvignon Blanc is selected by daily tastings during fermentation to identify those tanks with the most varietal character. Fermentation in these tanks is then halted at 0.7 to 0.8 per cent residual sugar. This wine receives two to three months of ageing in German casks. The Fumé Blanc consists of the balance of the Sauvignon Blanc which is allowed to ferment to dryness and receives longer ageing in the German ovals.

The 1979 vintage shows clearly the direction O'Donnell is taking with the wines at Callaway. O'Donnell is exploring the hypothesis that the length of time a grape hangs on the vine affects flavour and aroma and that maturity may be as much a question of this as it is of sugar-acid ratios. The 1979 White Riesling was harvested at 18° Brix and 1.3 per cent total acidity. Here O'Donnell's goal was to produce a dry wine with all the desirable varietal characteristics of White Riesling but without bitterness in the finish, which is typical of many fully dry Riesling wines in California. Research showed that there is Muscat in the parentage of the Riesling grape, and that wines made from fully mature Riesling grapes exhibit the same bitterness of finish—intensified in Riesling grapes grown on their own *vinifera* roots as they are at Temecula— as do wines made from fully mature Muscat grapes. O'Donnell also observed that German Riesling wines usually do not possess the same bitterness of finish as do California Rieslings, even when made without residual sugar. This made him decide to harvest the 1979 Riesling according to the German growing season, which is much shorter than the Californian. The Riesling was therefore harvested at 18° Brix and 1.3 per cent total acidity. After fermentation, the wine had 0.85 per cent residual sugar, 0.7 per cent total acidity and 10.3 per cent alcoholic content. After maturing in German cooperage for 10 weeks, the wine exhibited the desired characteristics. The wine had the light colour of a Mosel, but had begun to develop the desirable fruitiness of aroma and flavour of the

Riesling grape, but without the bitterness of a Riesling wine made from fruit which has been allowed to hang on the vine through a longer growing season.

The 1979 Chenin Blanc went to the opposite end of the spectrum. As the harvest began, the grapes were at 21° Brix. One third of the way through the harvest, the weather turned cool and harvesting was halted temporarily with the consequence that approximately $3\frac{1}{2}$ weeks elapsed between the beginning and the end of the harvest. During that period of time, the sugar level remained at 21° Brix, but flavours and aromas became much more intense. When harvesting was resumed, the first lots to be harvested had developed a slight pineapple character in aroma and flavour. The final lots were harvested at 20° to 20.5° Brix and had developed a distinct butterscotch aroma and flavour. Each lot was fermented separately, and each wine exhibited the characteristics observed in the grapes. During the additional time the grapes were hanging on the vine, they developed a richness and intensity, plus some added flavours—in short, a complexity not usually found in Chenin Blanc wines. The master blend of the separate lots combined these pineapple and

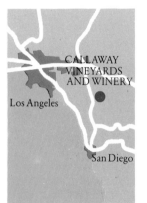

Temecula south of Los Angeles would be too hot for vine growing without the cooling influence of the winds which blow from the Pacific through the Rainbow Gap in the coastal mountain range. Temecula is a very recent grape growing area, and Callaway is the main winery in the area.
The Rainbow Gap can be seen behind the extensive Callaway vineyards (LEFT).

Snow-capped mountains form the backcloth to the long rows of the Callaway vineyards in winter (BELOW). The area is very dry and the meagre rainfall has to be supplemented by artificial irrigation.

butterscotch flavours into a harmonious whole. At 11 per cent alcohol, 0.9 per cent residual sugar and 0.7 per cent total acidity, the wine is well balanced and richly aromatic.

Dry white wines amount to approximately 85 per cent of Callaway's total wine production. However, three dry red wines are produced from the three varietals grown. O'Donnell's style of winemaking is also apparent in his treatment of the red varieties produced at Callaway. As he phased the German barrels out of the wood ageing programme, he began to phase French and American barrels in and to moderate the aggressive character exhibited by the first vintages of Cabernet Sauvignon and Petite Sirah. O'Donnell's first vintage as winemaker was 1977. His 1977 Cabernet Sauvignon exhibits fully the varietal character of Cabernet Sauvignon—at 12.4 per cent alcoholic content, the wine is fruity and softly astringent. It spent seven months in American oak barrels. In the 1978 and 1979

vintages O'Donnell has reduced the alcoholic content of the Cabernet Sauvignon with the 1979 vintage containing less than 12 per cent.

The 1977 Petite Sirah exhibits the peppery character associated with the wines of the northern Rhône. It is a complex wine which is developing additional complexity with age and it has a finesse which is characteristic of O'Donnell's way with red wines.

In addition to the dry table wines produced at Callaway, there are several speciality products. The first of these to be made was a botrytized Chenin Blanc, called Sweet Nancy, named after Callaway's wife. The pinnacle thus far reached in the production of this wine was achieved with the 1978 vintage. The grapes were harvested at 48.7° Brix. The juice fermented at between 36°F and 37°F for 15 months before it stopped of its own volition, at 10 per cent alcoholic content, 16.5 per cent residual sugar and over

This vine on the estate's 20 acre experimental plot has both white (Chenin Blanc) and red (Zinfandel) grapes growing on it (BELOW). The bunches are in perfect condition with grapes of equal size.

Inside the winery (LEFT) the Callaway winemaker Steve O'Donnell and the vineyard manager John Moramarco are discussing carbonic maceration of the Zinfandel grapes which will be used in making the speciality, early-release wine called Noël which is sold at the Christmas following the vintage.

Callaway is fairly German orientated in its approach to winemaking. In the winery (LEFT) are stainless steel fermentation tanks with German white oak barrels in the foreground.

1.0 per cent total acidity. The complex richness of Chenin Blanc varietal aromas and flavour blend well with the pronounced honey flavours of the botrytis. The wine is liqueur-like, and makes an excellent dessert served without accompaniment, or perhaps with a fully ripe pear. Unfortunately, the quantity is extremely limited. Botrytis mould occurs in Callaway's vineyards each year, thus it is possible to produce a Sweet Nancy each year, as well as occasionally to make botrytized Rieslings such as the 1978 Santana and Xenos.

The limitation of the number of varieties and the number of wines produced at Callaway Winery is indicative of and consistent with the objectives of Ely Callaway for the winery which bears his name, goals which are understood and shared by winemaker Steve O'Donnell. Those objectives are simply to produce a pleasing mealtime wine, a dry wine to accompany food.

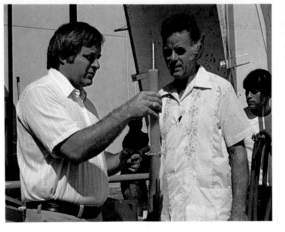

Steve O'Donnell and John Moramarco (LEFT) are checking the density (sugar level) of a must after crushing using a hydrometer.

GLOSSARY

Small capitals indicate cross-references.

Foreign words are indicated by the letters *Fr* (French), *Ger* (German) and *It* (Italian) after the headword.

Acetic acid Traces of acetic acid are present in all wines, but usually in amounts too small to be detected by taste or smell. Sometimes, usually as a result of bad handling, the acetobacter microbe attacks the wine causing it to acetify and giving it the vinegary taste and smell characteristic of sour or spoiled wine. Acetic acid is a volatile acid.

Acidity There are several kinds of acid present in wines. These acids are important. Not only do they help the wine keep well, they also give it an edge without which it would taste flat. There are three kinds of acidity involved in the winemaking process: volatile, fixed and total acidity. Volatile acidity includes acids such as acetic acid which are not present in the fresh grape juice but are produced during vinification; fixed acidity includes the natural fruit acids—tartaric, malic and citric acid; together volatile acidity and fixed acidity make up total acidity.

Alcohol The sugar in grapes is converted by yeasts into ethyl alcohol (C_2H_6O) and carbon dioxide gas—the process is called ALCOHOLIC FERMENTATION. Alcohol is a colourless, volatile liquid which is toxic. Taken in small quantities it has a pleasing and soporific effect.

Alcoholic content Methods of measuring the alcoholic strength of wines and spirits vary from country to country. A percentage per volume scale is now generally used in Europe. See conversion tables (p.253).

Anthocyanines These are chemicals which result from the breakdown of the tannin and colouring substances in wine. They sometimes cause haze and deposits which must be stabilized by coagulation and precipitation with FINING AGENTS such as gelatin, egg whites etc.

Appellation Contrôlée (AC) (*Fr*) The highest rank in the APPELLATION D'ORIGINE system. These two words on a label guarantee the origin and production standards of the wine.

Appellation d'Origine Contrôlée (AOC) (*Fr*) This is a classification of French wine which is regulated by a body known as the Institut National des Appellations d'Origine des Vins et Eaux-de-Vie (INAO) which was set up in 1935. This organization dictates not only the name which a wine carries, but also regulates very strictly the viticulture and vinification procedures: type of grape; degree of ripeness when picked and so on. The Appellation Contrôlée wines are the most strictly controlled. The second category are the Vins Délimités de Qualité Supérieure (VDQS) wines. Both AOC and VDQS wines are VQPRD or Vins de Qualité Produits dans une Région Déterminée, following these are the Vins de Pays and lastly the Vins de Table.

Aroma The aroma of a wine is that part of its smell which is derived from the grape and should be distinguished from the bouquet which develops in the bottle. It is most pronounced when the wine is young. Certain varieties of grape such as the Muscat are easily identified by their smell, although much of this disappears during the fermentation process. Aroma tends to fade as the wine is aged and in good wines the bouquet should take over.

Assemblage (*Fr*) A term which describes the blending of wines of different origin, characteristics or age. Fine wines are not usually blended, but wines from different casks may be 'assembled' to avoid cask-to-cask differences. The vin de presse and the vin de goutte of a particular vineyard are often blended.

Auslese (*Ger*) The third of the German QUALITÄTSWEIN MIT PRÄDIKAT wines—the categories above it are Beerenauslese and Trockenbeerenauslese. Auslese grapes are picked late in the harvest when they will be ripest. Only the very ripest bunches are selected so Auslese wines will be sweeter and more expensive than the Spätlese from the same cellar and the same vintage. It must by law have an Oechsle reading of 97°.

Balling Hydrometer scale used in America to measure the approximate sugar content of unfermented grape juice—from this the probable alcohol content of the finished wine can be calculated.

Barrel The containers in which wine is matured, stored and, very rarely, shipped. They come in a variety of shapes and sizes, and with local names characteristic of particular areas. They are usually of wood and often of oak.

District	Name	Contents in litres	Contents in cases
Beaujolais	Pièce	212	23
Bordeaux	Barrique	225	25
Burgundy	Pièce	228	25
Rhine	Halbstück	600	68
Alsace	Foudre	1,000	108

Barrel ageing Wines are aged in barrels. During this time the wine clarifies and matures. Sometimes malolactic fermentation takes places at an early stage. Wood is used because it allows small amounts of air to reach the wine. The wood itself imparts character to the wine so the type and age of timber used is important. Wines are often rotated between new and old barrels.

Barrique (*Fr*) One of the many French words for a BARREL.

Baumé (*Fr*) A hydrometer scale for measuring the density and hence the sugar content of grape juice—from this the probable alcohol content of the finished wine can be calculated.

Beerenauslese (*Ger*) The second highest category of German wine which is even rarer and more expensive than Auslese. The grapes are harvested late and only the ripest from each bunch are used so the sugar content of the unfermented juice is high. Under the German wine law of 1971 Beerenauslese must have an Oechsle reading of at least 120°.

Bentonite (Aluminium silicate) A clay used as a fining agent to remove excess protein.

Binning The laying down of bottled wine for ageing. The bottles are stored on their sides so that the wine is in contact with the cork ensuring an airtight seal. Temperature, humidity and other factors are important if the wine is to age well.

Blending See ASSEMBLAGE

Body The wine's weight in the mouth due to its alcoholic content and other physical components.

Bonde de côté (also bonde à côté) The 'bung-over' position in which many wines in barrel are stored. When the bung is at an angle, less wine evaporates, and therefore the practice makes economic sense to many winemakers.

Botrytis cinerea The 'noble rot', a parasitic fungus or mould which attacks grapes. In German it is called Edelfäule and in French pourriture noble. The fungus penetrates the skin of the fruit causing it to wither. As the grape dries the sugar and flavour become concentrated.

The wine produced from these grapes is sweet and high in alcohol. Botrytis cinerea is responsible for some of the finest of the world's sweet white wines such as Sauternes, the Beerenauslesen and Trockenbeerenauslesen of the Rhine and Mosel, and the late harvested Rhine Rieslings of California.

Bottling, cold sterile process Bottling under cold sterile conditions is desirable for nearly all fine wines which are expected to age in the bottle. Other processes such as pasteurization or hot bottling ensure totally sterile wine and are ideal for wines which will be sold and consumed quickly. In the cold sterile process no heat is used, the bottles are sterilized with sulphur dioxide and filled with wine under an inert carbon dioxide atmosphere to prevent recontamination.

Breed A term used to describe a wine which is distinctive and distinguished—these qualities stemming from such factors as the wine's origin, its soil and the winemaker's skill.

Bouquet This term describes the smells given off by a good mature wine once it has been opened. It results from the slow oxidation in the bottle of alcohol and fruit acids into esters and aldehydes. Wines which are high in acids tend to have a more powerful bouquet than those which are lower in acids. Bouquet should not be confused with AROMA.

Bung This is the stopper for the barrel in which wine is aged. It may be of wood, earthenware or glass.

Cap In red winemaking the solid matter of the grapes which floats to the top of the fermenting vessel. It consists of skins,pips and sometimes stalks. This cap or, in French, chapeau must be broken up so that the skins are kept in contact with the fermenting juice in order that the maximum amount of colour can be extracted. This is achieved by the process called REMONTAGE.

Cask See BARREL

Cave (*Fr*) A cellar—this can mean a place for storing wine, sometimes though not always underground. It can also refer to the collection of wine which it contains.

Cellar See CAVE

Cellar master In France called the chef de cave, the cellar master is responsible for the cellar, its contents and their condition.

Centrifuge A spinning device which pushes the particles in a liquid outward from the centre by centrifugal force thus separating out the particles in suspension. It can be used on must or wine.

Cépage (*Fr*) A grape variety. Pinot Noir, Chardonnay or Cabernet Sauvignon are examples of cépages.

Chais A building for storing wine, usually above ground. It should be distinguished from a *cave* or true cellar, although in practice both terms are interchangeable.

Chapeau See CAP

Chaptalization The process of adding sugar to the must before or during fermentation in order to increase the potential alcoholic content. This practice is forbidden in Italy but is permitted, under very strict controls, for some wines in certain circumstances in Germany and France.

Château (*Fr*) A wine estate—a wine with a château label should be a wine made entirely from the grapes grown on that property. Although château-labelling on most wines is a reliable designation of origin and authenticity, some trade marks use the word château without justification.

Château-bottled A wine made and bottled on the estate on which the grapes were grown. The terms mis du château or mis en bouteilles au château are a guarantee that the wine is unblended with wines from other properties, but it is not necessarily a guarantee of quality.

Chef de cave Cellar master.

Chef de culture Manager of the vineyards at a property.

Chromatographic analysis The analysis of substances in solution by passing it over an absorbing material so that the constituent parts separate into bands of different colours.

Classed growth In French cru classé. A property which has been awarded an official classification. The Bordeaux Classification of 1855 was developed for the Paris Exposition Universelle of 1855. A committee of Bordeaux winebrokers worked out a classification based on soil, prestige and price for the wines of Médoc, Sauternes, and one Graves—Haut Brion. Relatively few changes have been made to this general classification.

Climat Burgundian term for a vineyard—equivalent of cru in Bordeaux.

Clodosporium cellare A fungus present in some cellars—said to give a 'sealing wax' tone to wine.

Clone A group of plants which have been produced vegetatively from a single plant.

Clos (*Fr*) A walled vineyard, or one that was once walled, especially in Burgundy. The word clos may not appear on a wine label unless the vineyard actually exists and produced the wine. The vineyard must also be surrounded by an enclosure, unless the existence of the clos has been recorded for more than 100 years.

Commune (*Fr*) A township or parish, an administrative unit consisting of a village and the surrounding land. Many commune names are adopted as official APPELLATION CONTRÔLÉE place names in which case the wines produced within the area must be of a similar standard and quality.

Cooperage General term for wooden casks, barrels or vats used for storage in a particular cellar or winery. Also means storage capacity and the repair of these containers. Used especially in the United States.

Côte (*Fr*) A slope with vineyards. When used to describe a wine it usually, but not always, refers to wine of lesser quality than a clos or a château. It can also describe a whole area i.e. Côte d'Or, Côte de Beaune, Côte de Nuits.

Cru (*Fr*) The French term for a growth or a crop of grapes. When applied to a vineyard it means the vineyard and the wine it produces. Most classifications of French wine divide the wines and vineyards into crus. The crus classés or classed growths of the Médoc are divided into five categories: premier cru; deuxième cru; troisième cru; quatrième cru and cinquième cru. Below the crus classés are the crus bourgeois, which include 18 crus exceptionnels, and bourgeois growths. This system has been adopted with modifications in other areas. Other Bordeaux regions, for example, have only two categories of classification: premier grand cru classé and grand cru classé. It should be remembered that a troisième cru is not a third-class wine—it is one of the finest wines.

Cuvage (*Fr*) Vatting—putting fresh grape juice into fermenting vats after crushing.

Cuve (*Fr*) Wine vat, tank or cask, especially a large one in which wines are fermented or blended.

Cuvée (*Fr*) From cuve—a term meaning the contents of a vat, or all the wine made at one time under the same conditions. It may also refer to a particular batch or blend of a blended wine.

Cuvier (*Fr*) A winery.

Débourbage (*Fr*) In the making of white wine the practice of delaying the fermentation of the newly pressed juice for a day or so allowing the juice to 'stand' and clarify. The juice can then be drawn off from the coarse sediment. Centrifuges are sometimes used at this stage in modern wineries. The onset of fermentation is postponed by keeping the juice at low temperatures or by treating it with metabisulphite. Fermentation may then be initiated by the introduction of a culture of selected yeasts.

Décuvage Devatting—in the making of red wine the fermentation vats must be emptied in order to separate the fermented grape juice from the skins or marc. This occurs at the end of fermentation after the cap of skins has sunk to the bottom of the vat.

Denominazione di Origine Controllata (DOC) (*It*) The second level of wine established under the 1963 Italian wine laws and broadly equivalent to the French APPELLATION CONTRÔLÉE. It is a guarantee that the wine comes from a specific vineyard and that that vineyard's production is controlled by law. The top level of wine under the 1963 laws is the denominazione di origine controllata e garantita (DOCG), so far only one has been awarded to Brunello di Montalcino.

Domaine (*Fr*) A single property or estate which can be made up of several separate vineyards.

Edelfäule (*Ger*) See BOTRYTIS CINEREA

Egrappage (*Fr*) The removal of the stalks from grapes before they are pressed or placed in the fermentation vat. Whether the grapes are destemmed or not depends on the type of wine, the region and the traditional practices of the winemaker. The stalks contain tannin which can render the wine harsh but they also give the wine lasting qualities. Egrappage also allows the wine to attain higher levels of alcohol for the stalks tend to absorb a certain amount of alcohol during fermentation. Originally the stems were removed manually but this operation is now done mechanically in a machine called an 'égrappoir' or destemmer. Egrappage is usually combined with crushing in an 'égrappoir-fouloir' or stemmer-crusher.

Einzellage (*Ger*) Under the 1971 German wine laws, Einzellage is used to describe a single vineyard with strictly defined boundaries and a legally approved name. Most Einzellagen have evolved from a consolidation of several small vineyards.

Eiswein (*Ger*) A rare German wine made from the first pressings of frozen grapes. Grapes harvested after a severe frost will be frozen, the less ripe grapes will be frozen solid, the ripe ones will be only partially frozen. The first run of juice will be from the ripest and therefore only partially frozen grapes. It is a very rare and expensive wine. See QUALITÄTSWEIN MIT PRÄDIKAT.

Estate-bottling Virtually the same as château-bottling and a guarantee that the wine is unblended with wines from other properties. It is also to some extent a guarantee of quality. Terms used on the label to indicate estate bottling are: 'mise du domaine', 'mise du propriétaire', 'mis en bouteilles par le propriétaire' and 'mise à la propriété'.

Estery A sweet, fruity smell which results from the slow reaction between the acids in a wine and the alcohol.

Extract Non-volatile, soluble solids which give wine substance, body and depth. In must and sweet wine, sugar forms the major part of the soluble solids but in a dry wine the extract consists of non-sugar solids.

Fat Term used to describe a wine with a full body, high in glycerol and extract.

Fermentation, alcoholic The process by which sugar is converted into alcohol and carbon dioxide by a series of complex chemical reactions between the sugars in the grape juice and zymase, an enzyme carried in the yeast cells. Today the fermentation process is carefully controlled.

Fermentation, malolactic A secondary fermentation during which the rather green, appley malic acid, a fruit acid, is converted into the milder lactic acid and carbon dioxide. This secondary fermentation can take place straight after the alcoholic fermentation, some time later or, in certain conditions, it may proceed at the same time. It is usually considered desirable in red wines and in white wines where the acidity is too high.

Filtration The clarifying of wine prior to bottling by passing it through a filter—there are many kinds of filter. Many German wines are sterile filtered to remove all bacteria so that the wine can be bottled earlier.

Fining The traditional method of clarifying wine. An organic agent such as gelatin, egg white (for fine red wines) or isinglass (for white wines) is added to the wine causing the particles in suspension to coagulate and settle at the bottom of the cask as sediment or lees. Clay products such as bentonite are also used as fining agents. Fining stabilizes the wine and can improve the flavour if, for example, excess tannin is removed. Nowadays it is being replaced by filtration and centrifuging, though fining is still used for the finest wines.

Finish A wine tasting term which describes a wine's end taste. A wine with a good finish has a firm distinctive end and some length on the palate.

Foudre (*Fr*) A barrel.

Foulage (*Fr*) The process of lightly crushing the grapes in order to start the juice running. After crushing, grapes for red wine are pumped into the fermentation vats. Crushing and destemming are often combined in one operation in a stemmer-crusher or 'égrappoir-fouloir'.

Free-run wine Also known as vin de goutte, this is the juice which runs off from the crushed grapes before pressing. It is usually of superior quality. The juice extracted by pressing is known as PRESS WINE.

Frost Many of the fine wines of the world are produced at the northern limit of the vine where frost is a hazard. The period of greatest danger is in spring when the buds have just begun to form. A severe frost can destroy most of the vine and affect not only that year's crop but also the following year's vintage. Growers try to minimize frost by putting heating apparatus among the vines or by spraying water over them—the water freezes and protects the vines.

Generic A name which describes a type of wine, for example, sparkling wine, rosé etc. It can also refer to a region, generic Bordeaux or Burgundy, for example. In the USA and Australia, European wines are 'reproduced', and so the names Sauternes, Chablis and Rhine wine appear on the labels. However, the origin of such wines must be clearly stated and their sale is not allowed in EEC countries.

Glycerol A thick, colourless, sweet-tasting chemical, $CH_2(OH). CHOH. CH_2(OH)$. It is an important component of wine.

Goût de terroir A distinctive earthy taste which some wines have and which is related to the soil of the particular vineyard.

Governo (*It*) A winemaking process practised in the Chianti region. Four different kinds of grapes are fermented together. A small proportion of the vintage, about 5 per cent, is put aside before fermentation. These grapes are left to dry, usually on straw trays or hanging in a well-ventilated attic. These raisinated grapes are added to the wine which has already fermented. The dry sweet grapes cause fermentation to start again. This second fermentation proceeds very slowly and gives the wine more body and a higher glycerine content which is ideal for a Chianti Classico intended for ageing.

Grafting In the late 1800s almost all the existing European vineyards were destroyed by the aphid PHYLLOXERA which was introduced accidentally from America. Almost all European wines now come from species of VITIS VINIFERA grafted onto phylloxera-resistant American rootstocks. Grafting involves taking a twig (scion) from the desired variety and inserting it into a notch cut in the stump of the 'stock'.

Grand vin (*Fr*) A 'great wine', it is not a term which is legally defined.

Grape All grapes belong to the genus *vitis*. The most important species for wine production is VITIS VINIFERA of which there are almost unlimited varieties. Only about 20 varieties, however, are capable of producing really great wine. Different varieties respond well to certain climatic conditions, soil types etc. In the French APPELLATION D'ORIGINE system and the Italian DOC system, the varieties of grape to be grown in a certain area, or even in a vineyard, are strictly controlled. In Germany, where climatic conditions are hard, there has been considerable experimentation with new grape varieties. These are crosses between different varieties of VITIS VINIFERA and may have desirable qualities such as frost resistance or early ripening.

Graves (*Fr*) The word means gravel but is also used to describe several districts of Bordeaux. Gravelly soil provides excellent drainage for vines.

Grêle (*Fr*) Hail. Hail, like frost, is a natural disaster feared by winegrowers. A heavy hailstorm can destroy a crop in minutes and, because it is so localized, it may destroy one vineyard whilst the neighbouring vineyard is unaffected. Hail breaks the skin of the fruit and the leaves rendering them susceptible to attack by disease. Even a light hail can bruise berries giving wine, especially red wine, a 'goût de grêle', a faint taste of rot in the wine. Growers sometimes attempt to encourage hail to fall as rain by dispersing clouds with light aircraft. Occasionally nets are put over vines as protection against hail.

Grey rot In French pourriture grise—a disease caused by the same fungus, BOTRYTIS CINEREA, which causes noble rot. Only when conditions of high humidity are followed by hot weather does pourriture grise become pourriture noble, and this is only desirable on certain varieties of white grapes such as Riesling and Sémillon.

Grosslage (*Ger*) A large vineyard. Under the 1971 German wine law, it denotes a vineyard name which extends over several wine producing areas.

Hybrid This is the result of a cross between American and European vines. This is done in order to try to combine the best qualities of the parent plant or to produce an offspring which can contend with problems such as frost or phylloxera. Most French hybrids bear the name of the hybridizer and a number—for example Baco No. 1, Seibel 13053, Seyve-Villard 5/247. In France these hybrids are banned from all AC areas. There is a fundamental difference between a 'crossing' of two *vitis vinifera* grape varieties and a hybrid or 'producteur direct'.

Inoculation A term used to describe the introduction of a special yeast culture into the must. This is usually done after all the wild yeasts have been destroyed by sulphur dioxide.

Isinglass Purified fish glue—the most common fining agent for white wines.

Kabinett (*Ger*) The lowest of the Qualitätswein mit Prädikat divisions under the 1971 German wine law. It ranks just under Spätlese in quality.

Kellermeister (*Ger*) Cellar master.

Late picked Some grapes, such as Riesling, can be left on the vine after the main vintage. They will obviously be at the mercy of the weather, but the extra time on the vine allows the grapes to accumulate sugar. In Germany these late picked grapes are called Spätlese and in France vendange tardive. In Germany selected bunches are chosen for late harvesting—a process known as Auslese. Single grapes from a bunch may also be selected for the Beerenauslese, and, last of all, overripe grapes which have been attacked by BOTRYTIS CINEREA or Edelfaüle may be selected to make the rarest and sweetest of German wines the Trockenbeerenauslese.

Lees The heavy coarse sediment containing dead yeasts which is left in the barrel when young wine is racked off into other barrels.

Loess A fine fertile soil based on wind-blown dust—a mixture of loam, lime, sand and mica.

Maceration The process of extracting flavour and colour from grapes by steeping them in their own liquid before fermentation.

Malic acid One of the fruit acids—it is particularly tart. It is converted by a secondary fermentation known as MALOLACTIC FERMENTATION into lactic acid and carbon dioxide. This reduces the acid taste of the wine.

Marc (*Fr*) The solid matter consisting of skins, stalks and seeds which is left after the wine (red wine) or juice (white wine) has been extracted by pressing. It is also called pomace.

Must Grape juice or crushed grapes before fermentation.

Négociant (*Fr*) A wine merchant or shipper who buys in wine from growers and prepares it for sale.

Noble rot See BOTRYTIS CINEREA

Nose Term for a wine's bouquet or aroma.

Oechsle (*Ger*) Hydrometer scale for measuring the sugar content of must or grape juice before fermentation. From this the future alcoholic content of the wine can be calculated.

Oenologist A trained graduate of oenology, which is the science of wine, its production, care and handling.

Off-taste A rather broad wine tasting term to describe a wine which has an abnormal taste, for whatever reason. It usually denotes unhealthy wine.

Oidium One of the many diseases to which vines are prone, it is caused by a fungus which attacks the leaves, shoots and tendrils. It is now controlled by spraying with a sulphur solution.

Phylloxera A disease of vines caused by the aphid phylloxera vastatrix. It was accidentally introduced to Europe from America in about 1860. By the end of the century it had destroyed most European vineyards. Unfortunately, VITIS VINIFERA, from which most of the world's best wines are made, is particularly prone to its depredations. American vines which have tougher roots are resistant to phylloxera and most European vines are now grafted onto American stocks.

Pièces See BARREL

Piges Wooden poles used to keep the CAP of skins submerged in the fermenting must.

Pomace Mass of skins, seeds and stems left in the press after the wine or juice has been drawn off. The French term is MARC.

Pourriture noble See BOTRYTIS CINEREA

Pourriture vulgaire See GREY ROT

Pressing In red winemaking the skins are fermented with the juice, the wine is drawn off and the skins are then pressed. In the vinification of white wine, however, the juice is fermented without the skins so pressing occurs beforehand. There are several kinds of press in commercial use. The vertical hydraulic press (now rarely found) incorporates a cylinder into which the skins are put. A piston then raises the cylinder, pressing the skins against a metal disc above. The wine runs through the sides of the cylinder into a vat below. The horizontal press consists of a long horizontal wooden cylinder which holds the grapes. This cylinder is rotated and the two ends move slowly towards the centre exerting a slow but even pressure on the grapes. It is important that the pressure is not great enough to break open the pips. There is another horizontal press, the pneumatic press, which gives a gentle pressing by means of a rubber 'bladder' in the centre which swells pressing the grapes against the sides.

Press wine Also called vin de presse, this is the wine extracted by pressing the MARC after the free-run wine is drawn off. It is inclined to be hard and is sometimes mixed with vin de goutte in varying proportions to add body.

Qualitätswein bestimmter Anbaugebiete (QbA) (*Ger*) Wine which must come from a certain area and from particular grape varieties. It must have an Oechsle reading of 65° and have an official registration number on the bottle. Unlike the QUALITÄTSWEIN MIT PRÄDIKAT wines, it may be chaptalized.

Qualitätswein mit Prädikat (QmP) (*Ger*) The highest category of German wine under the 1971 wine law. The five predicates are: Kabinett, Spätlese, Auslese, Beerenauslese, and Trockenbeerenauslese. All the wines in these five categories must be made from unsugared musts. All QmP wine will have a registration number printed on the label.

Racking The drawing off of wine from one vat or barrel to another leaving behind the sediment or lees. This process is called soutirage in French, Abstich in German and travaso in Italian. Good wines are normally racked between two and six times before bottling. Racking usually follows fining. White wines are more delicate than red and are usually racked fewer times to avoid oxidation.

Remontage Removing a portion of the must from the bottom of the fermentation vat and pumping it over the top. This keeps the cap submerged and in contact with the fermenting must.

Riserva (*It*) Reserve—a word often found on wine labels which has legal standing in Italian DOC law.

Rootstock See GRAFTING.

Rotation The growing of different crops in a regular order to avoid soil exhaustion—a practice not often used in vineyards nowadays. Wine can also be rotated between new and older oak casks.

Sediment A deposit of solids and possibly crystals which accumulate in some wines as they age in bottle. Red wines deposit more sediment than white—it is composed of TANNIN, pigment and small quantities of mineral salts. The sediment in white wine usually consists of colourless crystals of cream of tartar—these are tasteless, harmless and will often disappear if the wine is left for some time at room temperature. Sediment is not a defect, and, indeed, indicates that a wine has not been over treated. It should be left in the bottle when the wine is poured.

Spätlese (*Ger*) The second of the QUALITÄTSWEIN MIT PRÄDIKAT categories above Kabinett in quality and below Auslese. It is made from late picked grapes and must have an Oechsle reading of 85°.

Stalks Stalks are sometimes added to the fermenting wine to give it extra 'backbone'. They are high in acid and may make the wine disagreeably astringent. They also tend to remove colour and alcohol. Whether they are added depends on the vintage and the traditional practice of the winemaker. The French term for destemming is egrappage.

Stemmer-crusher A rotating cylindrical machine which removes the stalks and lightly crushes the grapes.

Sulphiting Sulphur in the form of sulphuric acid, sulphur dioxide or sulphites is sometimes added to grape juice or must in order to delay or prevent fermentation. This is common practice in modern wineries in which a culture of a selected yeast is added later.

Sulphuring The sterilizing of barrels and casks and also the treatment of vines in the field to safeguard against OIDIUM. Sulphur is a disinfectant.

Süssreserve (*Ger*) Concentrated unfermented grape must which is kept back in order to be added to the wine just before bottling. This sweetens and softens the wine. The process which is allowed up to Spätlese is called Süssung. However, the Süssreserve must come from the same category of wine, for example, the Süssreserve for Spätlese must come from Spätlese juice, and from the same vineyard or district.

Tafelwein (*Ger*) Table wine—under the 1971 wine laws it ranks below QUALITÄTSWEIN in quality. It will always be sugared and will never carry the name of a vineyard.

Tannin A group of organic compounds found in wine—they are also found in the wood, bark, roots and stems of many plants. It is astringent to taste, making the mouth pucker. It is particularly pronounced in wines which have not had their stems removed before fermentation. Additional tannin may be picked up from the oak barrels in which wine is stored, especially from new barrels.

Tartaric acid The principal acid of wine made from ripe grapes. It is usually precipitated in the form of crystals of potassium bitartrate in the cask or in the bottle. This can be avoided by cold treatment before bottling. However, the crystals are harmless and do not detract from the wine's quality.

Tastevin This is a flat, shallow, silver winetaster's cup, used traditionally in Burgundy for judging new wine. When a wine is 'tasteviné', it is approved by the Confrérie des Chevaliers du Tastevin and carries a special label.

This applies only in Burgundy.

Tirage (*Fr*) A bottling. Several casks are usually assembled in a vat for a 'tirage' or bottling.

Topping up The filling up of casks or barrels of young wine with similar quality wine. It is usually carried out weekly to ensure that there is no air space or ullage between the wine and the bung. This prevents oxidation of the wine.

Trockenbeerenauslese (*Ger*) The top of the five QUALI- TÄTSWEIN predicated wines. The grapes are harvested very late and the berries are individually selected from the vine. Only those which have been attacked by BOTRYTIS CINEREA or Edelfäule will be used, ensuring that the wine will be very sweet.

Varietals Wines, especially American wines, which take their name from the name of the grape rather than from their place of origin. Under American law only a wine which has at least 51 per cent of a particular grape in it can take its name from that grape. However, fine wines from top wineries contain 100 per cent of the grape variety on the label.

Vat A large vessel in which wine is fermented or stored. Traditionally they were made of wood, but these days they may be of glass, enamel-lined concrete, or stainless steel.

Vin de goutte See FREE-RUN WINE

Vin de presse See PRESS WINE

Vintage The annual grape harvest and the wine made from those grapes. It has come to mean a wine year, thus, 1978 is a good vintage in Bordeaux and Burgundy.

Vitis vinifera The family of grape varieties from which almost all good European wines are made. Originally it came from the Middle East. Its great disadvantage is that it is susceptible to PHYLLOXERA, a problem overcome by grafting onto phylloxera-resistant rootstocks from America.

Yeasts Unicellular organisms some of which—saccharo- myces ellipsoideus—bring about fermentation in grape juice. It is the zymase in the yeasts which actually catalyzes the fermentation process and turns the grape juice into wine. There are many different yeast cultures suitable for particular conditions.

CONVERSION TABLES

Distance
1 kilometre = 0.621 miles
1 mile = 1.609 kms

Liquid measure
1 gallon (UK) = 1.2 gallons (US) = 4.5 litres
1 gallon (US) = 0.833 gallons (UK) = 3.785 litres
5 litres = 1.1 gallons (UK) = 1 gallon (US)
1 hectolitre = 100 litres

Area
1 sq kilometre = 0.386 sq miles
1 sq mile = 2.589 sq kms
1 hectare = 2.471 acres
1 acre = 4840 sq yards
1 are = 100 square metres

Measuring sugar and alcohol content
Most countries have their own ways of measuring the sugar content of grapes and the alcohol level in their wines. In France and Italy, measurement by degree has now been superseded by measurement by percentage potential alcohol by volume, which, however, gives an almost iden- tical reading. In Germany, the traditional Oechsle method is still employed.

Temperature
Degrees Centigrade
converted to
Degrees Fahrenheit
Multiply °C by 9/5 and add 32

Degrees C		Degrees F
37	=	98.6
50	=	122
100	=	212

Degrees Fahrenheit
converted to
Degrees Centigrade
Multiply °F by 5/9 after subtracting 32

Degrees F		Degrees C
−40	=	−40
32	=	0
59	=	15

specific gravity	1.060	1.065	1.070	1.075	1.080	1.085	1.090
° Oechsle	60	65	70	75	80	85	90
Baumé	8.2	8.8	9.4	10.1	10.7	11.3	11.9
Brix/Balling	14.7	15.8	17.0	18.1	19.3	20.4	21.5
% potential alcohol by volume	7.5	8.1	8.8	9.4	10.0	10.6	11.3
specific gravity	1.095	1.100	1.105	1.110	1.115	1.120	1.125
° Oechsle	95	100	105	110	115	120	125
Baumé	12.5	13.1	13.7	14.3	14.9	15.5	16.0
Brix/Balling	22.5	23.7	24.8	25.8	26.9	28.0	29.0
% potential alcohol by volume	11.9	12.5	13.1	13.8	14.4	15.0	15.6